ACS SYMPOSIUM SERIES 308

Polymeric Reagents and Catalysts

Warren T. Ford, EDITOR
Oklahoma State University

Developed from a symposium sponsored by
the Divisions of Organic and Polymer Chemistry
at the 189th Meeting
of the American Chemical Society,
Miami Beach, Florida,
April 28–May 3, 1985

American Chemical Society, Washington, DC 1986

sep/ae
CHEM

53177034

Library of Congress Cataloging-in-Publication Data

Polymeric reagents and catalysts.
(ACS symposium series; 308)

"Developed from a symposium sponsored by the Divisions of Organic and Polymer Chemistry at the 189th meeting of the American Chemical Society, Miami Beach, Florida, April 28–May 3, 1985."

Expanded version of the Symposium on Polymeric Reagents and Catalysts, held Apr. 30, 1985 in Miami Beach, Fla.

Includes bibliographies and index.

Contents: Overview / W. Ford—Soluble polymer-bound reagents and catalysts / D. Bergbreiter—Nafion-supported catalysts / F. Waller—[etc.]

1. Chemical tests and reagents—Congresses.
2. Catalysts—Congresses. 3. Polymers and polymerization—Congresses.

I. Ford, Warren T., 1942–   . II. American Chemical Society. Division of Organic Chemistry. III. American Chemical Society. Division of Polymer Chemistry. IV. Symposium on Polymeric Reagents and Catalysts (1985: Miami Beach, Fla.) V. Series.

QD77.P624   1986   543'.01   86–3521
ISBN 0–8412–0972–3

ACS Symposium Series

M. Joan Comstock, *Series Editor*

Advisory Board

FOREWORD

The ACS SYMPOSIUM SERIES was founded in 1974 to provide a medium for publishing symposia quickly in book form. The format of the Series parallels that of the continuing ADVANCES IN CHEMISTRY SERIES except that, in order to save time, the papers are not typeset but are reproduced as they are submitted by the authors in camera-ready form. Papers are reviewed under the supervision of the Editors with the assistance of the Series Advisory Board and are selected to maintain the integrity of the symposia; however, verbatim reproductions of previously published papers are not accepted. Both reviews and reports of research are acceptable, because symposia may embrace both types of presentation.

CONTENTS

PREFACE

Polymeric reagents and catalysts are of interest to organic chemists whose research could profit from their use. The purpose of this book is to bring this subject to their attention. Unlike most volumes in the ACS Symposium Series, this one contains no research papers, but only invited authoritative reviews.

Many synthetic chemists are reluctant to use polymeric reagents and catalysts because they do not understand polymers, or because they have been educated to believe that polymer chemistry is not pure. The development of polymer chemistry does lag behind that of organic chemistry. But after all, the concept of the macromolecule was generally accepted only around 1930, whereas urea was synthesized first in 1828. Its relative youth helps make polymer chemistry an exciting field of research, wide open for further exploration. For the benefit of readers who lack a fundamental background in polymer chemistry, the overview chapter includes a short section of basic terminology and concepts that should help in understanding the up-to-date reviews of research that follow.

A comprehensive work on polymeric reagents and catalysts would include coverage of immobilized enzymes, solid-phase peptide and nucleotide synthesis, and reagents and catalysts on inorganic supports. That work would also require several volumes. This single volume deliberately concentrates on ideas for the synthetic chemist working on new compounds, synthetic methods, or industrial chemical processes. At present, the organic chemical applications of polymeric reagents and catalysts are less well developed than immobilized enzymes and solid-phase peptide synthesis. The recent shift of research emphasis in industry from commodity chemicals to higher markup fine chemicals could change that situation. This recent shift presents opportunities for invention and use of many new polymeric reagents and catalysts and should lead to a rapid growth of the field.

I warmly thank the authors for their fine contributions that make this volume possible. The ACS Divisions of Organic and Polymer Chemistry, The Petroleum Research Fund, and the Rohm & Haas Company provided generous support for the original symposium. The reviewers and my colleague Erich Blossey contributed constructive suggestions that improved many of the chapters. Kris Shabestari did an excellent job of typing my own chapters. Robin Giroux of the ACS Books Department provided extensive organizational help.

WARREN T. FORD
Department of Chemistry
Oklahoma State University
Stillwater, OK 74078

November 11, 1985

Polymeric Reagents and Catalysts

An Overview

Warren T. Ford

Department of Chemistry, Oklahoma State University, Stillwater, OK 74078

R. Bruce Merrifield was honored with the Nobel Prize in Chemistry in 1984 for invention of peptide synthesis in polymer supports. Merrifield peptide synthesis is now used in scores of laboratories to produce natural peptides and, perhaps more importantly, analogs of natural peptides that are vital for the correlation of structures of peptides with their biological activity.

The basic concept of peptide synthesis in a support seems simple in retrospect, causing many of us to say "Why didn't I think of that?" Merrifield did, and the concept was clearly described in his research notebook (1) and dated 5/26/59, four years before his first paper on the subject in the Journal of the American Chemical Society (2).

> There is a need for a rapid, quantitative, automatic method for synthesis of long chain peptides. A possible approach may be the use of chromatographic columns where the peptide is attached to the polymeric packing and added to by an activated amino acid, followed by removal of the protecting group & with repetition of the process until the desired peptide is built up. Finally the peptide must be removed from the supporting medium.

Merrifield succeeded in doing exactly what he described. The basic steps are in Scheme 1 (3): 1) An N-protected amino acid is attached as an ester to a cross-linked polystyrene support. 2) The protecting group is removed. 3) An N-protected, activated amino acid is coupled to the amino group of the polymer-bound amino acid. Steps 2 and 3 are repeated with different amino acids to produce the desired peptide sequence. 4) The completed peptide is cleaved from the polymer, deprotected, and purified.

Within three years of the first publication Merrifield built a machine to automate the synthesis, and in six years he had synthesized ribonuclease A, an enzyme with 124 amino acid residues, having partial activity of the naturally isolated enzyme. The key features of the Merrifield method that have led to its widespread use are: 1) At each stage of the synthesis the polymer-bound peptide can be separated from all other components of the reaction mixture by filtration. This makes possible the use of a large excess of the soluble N-protected amino acid to drive each coupling step to high conversion. 2) The method can be automated.

Merrifield was not the only one to conceive independently syntheses with polymer supports. Also in 1963 Letsinger and Kornet (4) reported a peptide synthesis

0097–6156/86/0308–0001$06.00/0

Scheme 1

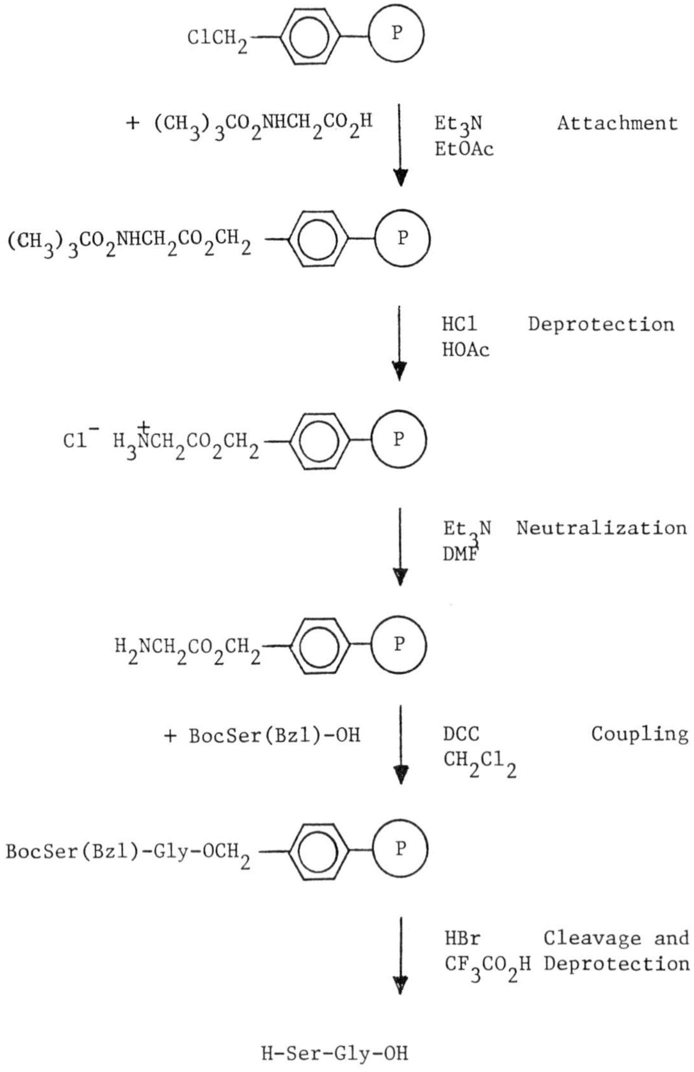

$ClCH_2$ —⬡— (P)

$+ (CH_3)_3CO_2NHCH_2CO_2H$ Et_3N Attachment
                                $EtOAc$

$(CH_3)_3CO_2NHCH_2CO_2CH_2$ —⬡— (P)

                              HCl Deprotection
                              $HOAc$

Cl^- $H_3\overset{+}{N}CH_2CO_2CH_2$ —⬡— (P)

                              Et_3N Neutralization
                              DMF

$H_2NCH_2CO_2CH_2$ —⬡— (P)

$+$ BocSer(Bzl)-OH DCC Coupling
                      CH_2Cl_2

BocSer(Bzl)-Gly-OCH_2 —⬡— (P)

                      HBr Cleavage and
                      CF_3CO_2H Deprotection

H-Ser-Gly-OH

Scheme 2

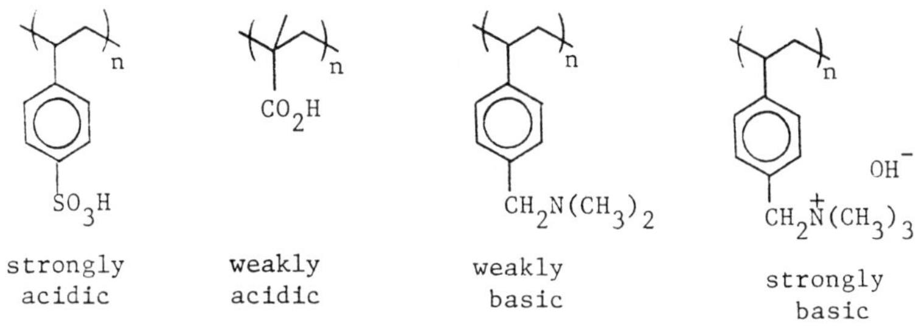

SO_3H	CO_2H	$CH_2N(CH_3)_2$	$CH_2\overset{+}{N}(CH_3)_3$ OH^-
strongly acidic	weakly acidic	weakly basic	strongly basic

with the amino terminus bound to the polymer. Fridkin, Patchornik, and Katchalski (5) used insoluble polymeric active esters of amino acids to acylate the N-termini of peptides in solution. Soon the concept was applied also to the syntheses of polynucleotides (6) and of polysaccharides (7). Many of the key papers in the development of polymer-supported syntheses of peptides, poly-nucleotides, and polysaccharides have been collected in a volume of reprints (8). Now in 1985, peptide synthesis in polymer supports is widely used (3), and polynucleotides are synthesized with silica gel as the support (9,10), but there has been much less development in synthesis of polysaccharides.

Merrifield peptide synthesis is the most highly developed method of synthesis with solid supports. Every step of the synthesis is carried out in the same polymer. Polymer-supported species also have been used as reagents and catalysts for single step synthetic transformations. Useful single step syntheses should be much easier to achieve than multi-step syntheses in polymer supports because they do not require >99% yield to be valuable. This book emphasizes one step processes in the synthesis of low molecular weight organic compounds carried out with polymer-supported reagents and catalysts.

The use of ion exchange resins as catalysts preceeds Merrifield peptide synthesis by more than a decade (11,12). These resins are most often functionalized, cross-linked polystyrenes. The primary commercial uses are water softening and deionization. For fundamentals of their chemistry see ref. (11). The common functional groups in ion exchange resins are strong and weak acids, tertiary amines, and quaternary ammonium salts (Scheme 2). (The latter are called "strongly basic", because they are used frequently with hydroxide as the counter ion. The quaternary ammonium ion is not a base.) The sulfonic acid resins are used as heterogeneous alternatives to soluble acid catalysts such as sulfuric acid or *p*-toluenesulfonic acid. The polymeric catalyst can be filtered out of a reaction mixture or used in a continuous flow process. The largest volume application at present is in the manufacture of the gasoline additive methyl *tert*-butyl ether by the addition of methanol to isobutylene catalyzed by a macroporous sulfonic acid resin (13). Chapter 3 by Waller describes catalytic uses of the much stronger acid Nafion, an insoluble polymeric perfluoroalkanesulfonic acid. The tertiary amine resins are heterogeneous alternatives to the common soluble tertiary amines used catalytically as weak bases and stoichiometrically as traps for strong acids liberated in reactions. The quaternary ammonium ion resins can be used with almost any desired counter ion as a reagent or catalyst, serving as sites for reactions of not only hydroxide ion, but also cyanide, halides, carboxylic acid anions, and carbanions.

The common ion exchange resins were developed in the 1950's, and Merrifield peptide synthesis was developed in the 1960's. Following a surge of research in transition metal-organic chemistry, in the 1970's heterogeneous analogs of a large number of homogeneous transition metal complexes were prepared by binding them to polymers and to silica gel. The aim was to create easily recovered catalysts with the high activity and selectivity of homogeneous catalysts for certain hydrogenation, hydroformylation, coupling, and oxidation reactions (Scheme 3) (14,15). Recovery is especially important because precious metals such as platinum, palladium, and rhodium often provide the best catalysts. The polymers used to bind transition metal species most often contain ligands such as phosphines or pyridines, or ion exchange sites to bind ionic metal complexes. The great initial hopes for such polymer-bound transition metal catalysts have not yet led to any commercial processes of which this author is aware. One problem is that the transition metals are not bound irreversibly, and their complexes are not as stable as they were once thought to be. Chapter 5 by Garrou analyzes the "leaching" of metals from and the chemical degradation of polymer-bound transition metal complexes.

Another development of the 1970's was the use of polymeric reagents in general organic synthesis (16,17,18). In principle, a polymeric analog can be devised for any useful soluble reagent. In practice, there is no reason to develop the polymeric reagent unless there are distinct advantages to be gained with the polymeric species, such as ease of separation of a by-product from a reaction mixture, use of the reagent in a continuous flow process instead of a batch process, the substitution of an odorous or toxic reagent with an analog that has no vapor pressure, or achievement of a chemical specificity not possible in solution. Chapter 8 on Wittig reagents by Ford and chapter 7 by Taylor on oxidizing agents concern some of the most common reagents in organic synthesis whose by-products are easier to separate from reaction mixtures when they are in insoluble, polymer-bound form. Chapter 6 by Neckers details the advantages of immobilization of the photosensitizer Rose Bengal. Polymer-supported reagents are potentially adaptable to continuous flow processes, although they have seldom been used that way in academic laboratories. Chapter 10 by Patchornik and co-workers describes a laboratory scale flow reactor for acyl transfer reactions. The problem of altering chemical reactivity by "isolation" of polymeric species from one another in a polymer support or by forcing them into "cooperation" in a polymer support are discussed in chapter 11 by Ford on site isolation syntheses. The concepts of isolation and cooperation of functional groups in polymer supports have not been easy to reduce to practice. The state of the art of specific binding of organic compounds by templates in polymer networks is reported in chapter 9 by Wulff.

A new version of catalysis by ion exchange resins appeared in the late 1970's. Insoluble polymers with much lower degrees of quaternary ammonium and phosphonium ion functionalization than the conventional ion exchange resins are active catalysts for reactions of anions with nonpolar organic compounds in triphase solid/liquid/liquid and solid/solid/liquid mixtures (Scheme 4). The organic reactant (and solvent if used) is one liquid phase, the polymeric catalyst is a solid (actually a gel) phase, and the inorganic reagent is used as either an insoluble solid or an aqueous solution. Polymer-bound chelating agents of metal ions such as crown ethers, cryptands, and polyethers show similar activity. The processes have been called "triphase catalysis" (19) and polymer-supported phase transfer catalysis (20). The latter term stresses the analogy to catalysis of reactions between hydrophilic anions and nonpolar organic molecules by lipophilic quaternary onium ions and crown ethers in liquid/liquid and solid/liquid mixtures. Subsequent kinetic analyses of polymer-supported phase transfer catalysis have shown that the reactions behave as normal heterogeneous catalysis in the sense that often there are diffusional limitations to catalytic activity.

Scheme 3

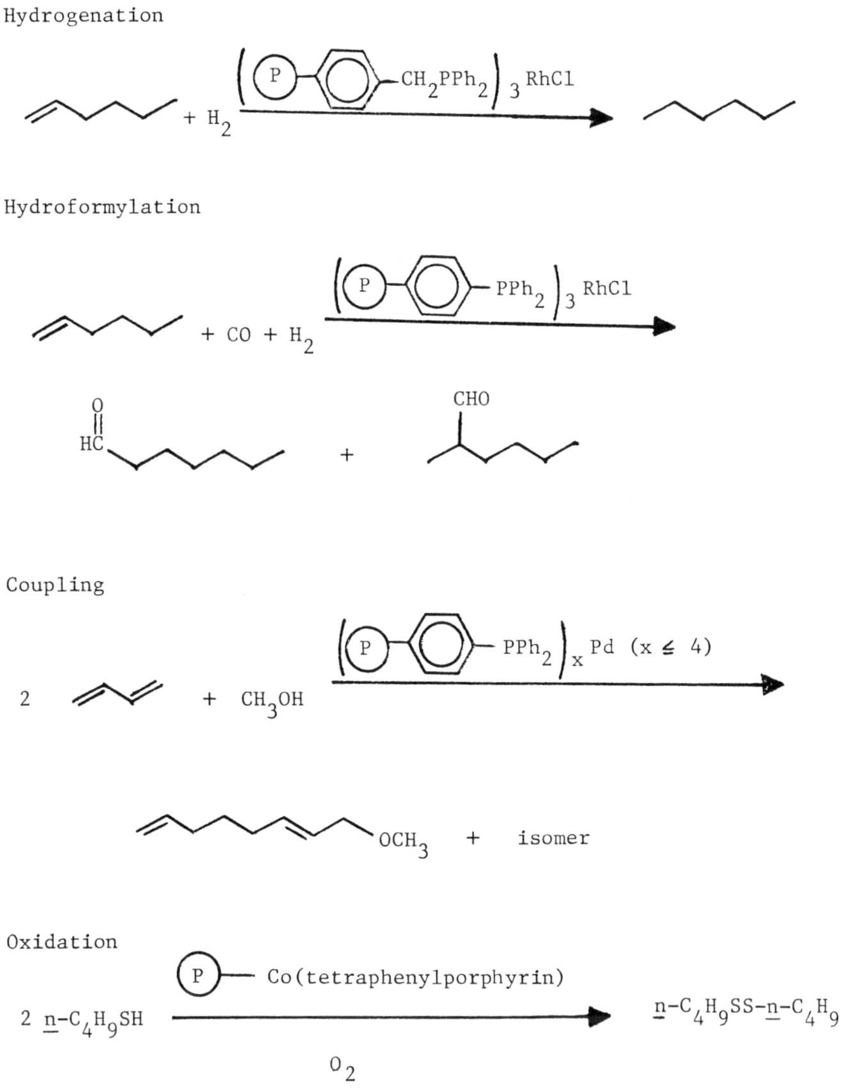

Hydrogenation

$$\text{(P)}-\langle\text{Ar}\rangle-CH_2PPh_2\Big)_3 RhCl$$

Alkene $+ H_2 \longrightarrow$ alkane

Hydroformylation

$$\text{(P)}-\langle\text{Ar}\rangle-PPh_2\Big)_3 RhCl$$

$+ CO + H_2 \longrightarrow$

$$HC(=O)\text{—}\text{chain} \quad + \quad \text{CHO branched isomer}$$

Coupling

$$\text{(P)}-\langle\text{Ar}\rangle-PPh_2\Big)_x Pd \ (x \leq 4)$$

$2 \ \text{CH}_2=\text{CHCH}_3 + CH_3OH \longrightarrow$

$\text{—OCH}_3 \quad + \quad \text{isomer}$

Oxidation

$$\text{(P)}-Co(\text{tetraphenylporphyrin})$$

$2 \ \underline{n}\text{-}C_4H_9SH \xrightarrow{\hspace{3cm}} \underline{n}\text{-}C_4H_9SS\text{-}\underline{n}\text{-}C_4H_9$

O_2

Scheme 4

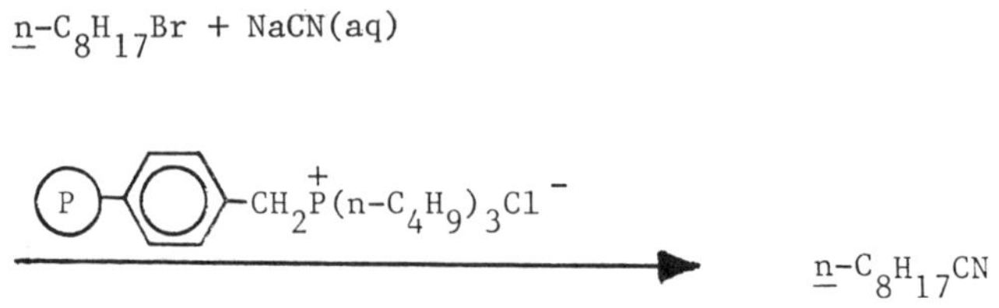

$$\underline{n}\text{-}C_8H_{17}Br + NaCN(aq)$$

$$\overset{(P)}{}\text{—}\overset{}{\bigcirc}\text{—}CH_2\overset{+}{P}(\underline{n}\text{-}C_4H_9)_3Cl^-$$

$$\xrightarrow{\hspace{5cm}} \underline{n}\text{-}C_8H_{17}CN$$

The state of polymer-supported reactions in organic chemistry at the end of the 1970's is described well in two books which the reader may wish to consult as further background for this volume (21,22).

<u>Basic Polymer Chemistry</u>

Far too often the organic chemist's concept of a polymer is a residue from a distillation. A polymer is considered by many non-polymer chemists to be an intractible mixture of compounds, difficultly soluble at best, and completely uncharacterizable. Although there are elements of truth in this concept, it is largely based on ignorance due to the failure of chemistry professors to teach students and experienced chemists to teach themselves a few basic principles about the nature of synthetic macromolecular materials. Readers who have not studied basic polymer chemistry are encouraged to consult and study one of the basic textbooks in the references (23,24,25,26). For principles of polymerization reactions the book by Odian (27) is especially recommended, but it lacks coverage of basic physical principles of macromolecules found in the other textbooks. A few fundamental definitions are covered in the following paragraphs to refresh some readers' memories. Anyone familiar with polymers should skip this section.

<u>Polymer Structures and Properties</u>. Synthetic polymers are mixtures of compounds composed of the same repeating structural units but differing in molecular weight. Thus polystyrene has the *repeat unit* structure

$$(CH_2\ CH)_n$$
$$|$$
$$C_6H_5$$

where n is the number average *degree of polymerization*, the average number of repeat units per molecule. A polymer is made by polymerization of *monomer*, a small molecule that reacts in regular, repeating fashion. The *number average molecular weight*, $\overline{M}_n$, is n times the formula weight of the repeat unit. When n exceeds some lower limiting value (which depends on the structure of the polymer and usually corresponds to $\overline{M}_n$ of 20,000 or more), the polymeric material has the desirable physical characteristics needed to produce a plastic, fiber, or elastomer. A material with regular repeating structure but too low an $\overline{M}_n$ to achieve these properties is usually called an *oligomer*.

A polymer having pseudochiral carbon atoms in its main chain is *isotactic* if adjacent centers are related as in a meso compound, *syndiotactic* if adjacent centers are related as in a (d,l) compound, and *atactic* if stereochemically random. Seldom is a polymer completely isotactic or completely syndiotactic. Most polymers made by free radical polymerization of vinyl monomers are atactic.

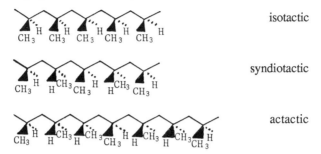

isotactic

syndiotactic

actactic

The structures of the *end groups* of the polymer chains are usually not specified because they have no effect on the bulk properties of the polymer, and often they are not known.

A *copolymer* is comprised of more than one kind of repeat unit and is synthesized from two or more monomers. Those repeat units, A and B, may be arranged in *alternating*, *block*, or *random* fashion.

ABABABABABABABAB	alternating
AAAAAAAABBBBBBBB	block
ABAAABBABABBBABA	random

Polymers are less soluble than low molecular weight materials because of the smaller entropy of mixing. They are totally insoluble when they are *cross-linked* by bonding the primary macromolecular chains into a *network*. Copolymerization of styrene with divinylbenzene provides the cross-linked polymers most often used for polymer-supported reagents and catalysts (Scheme 5). The insolubility makes them easy to separate from a reaction mixture but complicates their analysis. Their molecular weights are effectively infinite. A solvent that dissolves a homopolymer only swells a cross-linked polymer. Some swelling of a polymeric reagent is usually necessary to permit transport of the reagents to the reactive sites within the polymer network. A polymer swollen but not dissolved by solvent is a *gel*, a state of matter that as a solid does not exhibit macroscopic flow, but has properties of a liquid at a molecular level.

A *thermoplastic* may have either a partly crystalline or a completely amorphous structure in the solid state. An amorphous glassy solid is converted to a rubbery liquid upon heating above its *glass transition temperature*, T_g. A semicrystalline polymer has thermoplastic properties when it is above the T_g of its amorphous regions but still below its crystalline *melting temperature*, T_m. Upon heating above T_m a semicrystalline polymer becomes a viscous liquid. Network polymers are usually completely amorphous. The heating of a polymer network to above its T_g leads to an *elastomer* that can be deformed but does not flow. The presence of a low molecular weight solvent in a network markedly reduces its T_g, and under most conditions of use of polymeric reagents and catalysts, the solvent-swollen polymer gel is in an elastomeric state.

Scheme 5

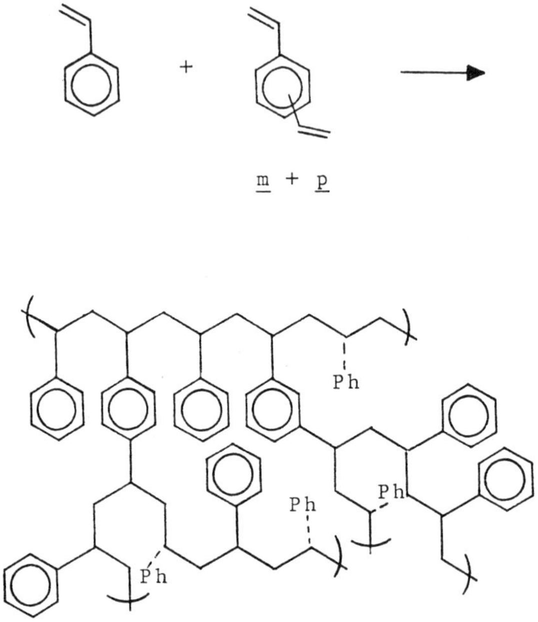

m + p

The term *resin* may refer to particles of any plastic or elastomeric polymer, as an ion exchange resin.

Polymer Synthesis. Polymer synthesis could be as varied as all of synthetic organic chemistry, but it is not, because the conditions required for successful production of macromolecules are far more exacting than for most micromolecules. In a polymerization the chain-forming step must occur 100 times for every termination step to achieve a degree of polymerization of 100. Not many organic reactions proceed in >99% yield.

Most polymerization reactions fall into two large classes, *chain growth* (Scheme 6) and *step growth* (Scheme 7) polymerizations. In a chain growth process a macromolecule is formed rapidly from monomer via a highly reactive intermediate such as a free radical, a carbanion, a carbenium ion, or a transition metal alkyl complex. After partial conversion of monomer in a free radical polymerization, the reacting mixture contains monomer and high molecular weight polymer. A step growth polymerization involves slow reactions of two monomers to form dimer, dimer and monomer to form trimer, two dimers to form tetramer, and so on in all possible combinations until macromolecules are formed. High molecular weight is achieved only after high conversion of monomer. After partial conversion the reacting mixture contains a broad distribution of monomers and oligomers.

Scheme 6

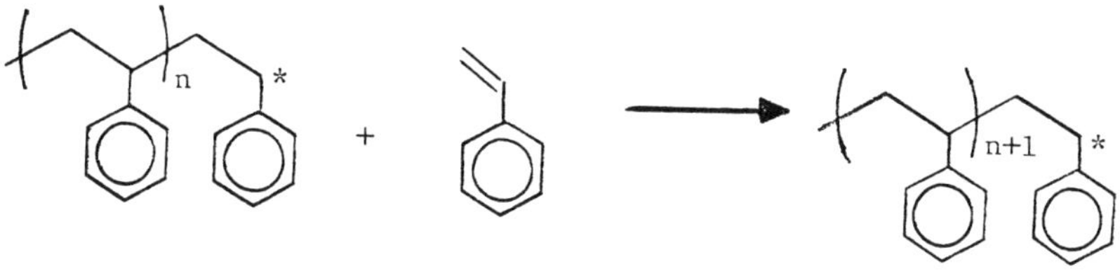

$$* = \cdot \; , \; + \; , \; -$$

Scheme 7

$n \; H_3\overset{+}{N}(CH_2)_6\overset{+}{N}H_3 \qquad \longrightarrow \qquad H_3\overset{+}{N}(CH_2)_6NHCO(CH_2)_4CO_2^- + H_2O$

$\overset{-}{O}_2C(CH_2)_4CO_2^-$

$\longrightarrow \qquad +NH(CH_2)_6NHCO(CH_2)_4OC +_n$

$n \; HOCH_2CH_2OH + n \; CH_3O_2C\text{—}\langle\bigcirc\rangle\text{—}CO_2CH_3 \qquad \longrightarrow$

$H \text{—}+OCH_2CH_2O_2C\text{—}\langle\bigcirc\rangle\text{—}CO+_n OH + \;(2n-1)\; CH_3OH$

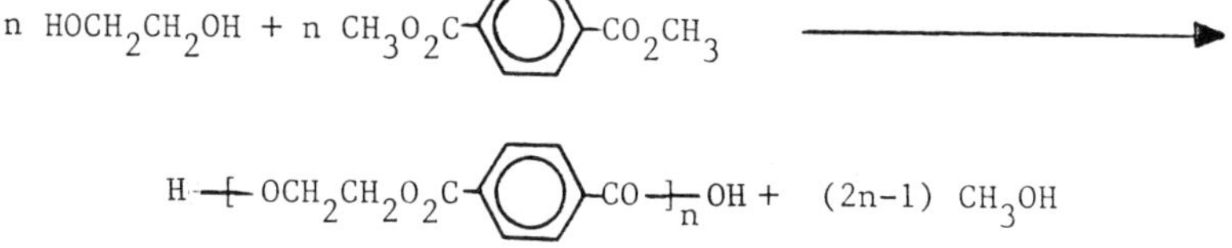

The experimental conditions for chain growth polymerization normally require exclusion of all contaminants that may react with the free radical, carbanion, carbenium ion, and transition metal alkyl imtermediates. Oxygen, water, carbon dioxide, and many other compounds can stop the chain reactions. Monomers should be highly purified to exclude chain-stopping contaminants. Free radical processes are generally the easiest to execute because only oxygen needs to be excluded. Polymerizations usually are carried out under rigorous inert atmosphere conditions.

Step growth polymerizations require high purity monomers and chemical reactions that proceed in >99% yield. These requirements have limited the large volume production of polymers by step growth reactions to esterifications of alcohols with carboxylic acids and carboxylic acid chlorides, amidations of amines with carboxylic acids and their acid chlorides, and reactions of amines and alcohols with isocyanates to form ureas and urethanes. Many other step growth processes have been used to produce polymers in smaller volume. When two monomers are used to form a copolymer such as nylon-6,6, the monomers must be used in nearly perfect stoichiometric balance to achieve high molecular weight, or else a process must be devised to provide that stoichiometric balance as the reaction proceeds to high conversion.

The most common supports for polymeric reagents and catalysts are produced by free radical copolymerization of a vinyl monomer such as styrene, 4-vinylpyridine, or 3- and 4-chloromethylstyrene with a cross-linking monomer such as divinyl-benzene. Uniform distribution of the functional sites on a microscopic level occurs only if the copolymerization is nearly random rather than alternating. *Bulk* and *solution* polymerizations produce solid and precipitated plastics that are ground into irregularly-shaped particles of the size desired for further use. *Suspension* polymerization of droplets of monomers dispersed in water with an oil-soluble initiator produces spherical polymer beads with a distribution of particle sizes somewhere in the range of 20-2000 μm that depends on the conditions of the synthesis. Spherical particles of narrow size range are preferred for use in columns, ease of filtration, and reproducibility of experiments with bound reagents and catalysts. *Emulsion* polymerization of monomer present initially in surfactant micelles in water with a water-soluble initiator produces a polymer latex, a colloid with a narrow distribution of particle sizes somewhere in the range of 0.02-1.0 μm.

Functional sites are often introduced by further reactions in the pre-formed copolymers. Those reactions must proceed in high yield to avoid unreacted and by-product functional groups in a polymeric reagent or catalyst. New types of polymer chains can be attached to or polymerized onto preformed polymers to produce graft copolymers. Microscopic heterogeneity of cross-linked polymers leads to complex kinetics of functional group conversions, usually with the rate of reaction slowing drastically at some time before complete conversion, because some of the sites are located in relatively inaccessible microscopic regions of the polymer network. Successful functionalization reactions usually require solvent-swollen polymers. Without swelling, reagents from solution fail to penetrate the polymer network to the functional sites.

Analysis of Polymers. The repeat unit structure of a synthetic polymer usually is known from the method of synthesis. Compositions of copolymers often are determined by elemental analysis when one monomer contains an element not present in the other monomer. Soluble polymers are often characterized by their molecular weights and *molecular weight distribution*, but insolubility prevents such characterization of cross-linked polymers.

Polymers can be characterized by infrared and UV/visible spectroscopy just as low molecular weight organic compounds are, although polymers are usually sampled as films or KBr pellets.

^{1}H and ^{13}C NMR spectroscopy are highly useful for soluble polymers (28). NMR spectra can often be used to identify the distribution of microstructural sequences in copolymers and stereochemical microstructures of polymers with pseudochiral carbon atoms in the main chain. Low levels of structural defects and end groups can sometimes be identified. NMR is more difficult but still can be used with cross-linked polymers (29). Lightly cross-linked (<4% divinylbenzene in polystyrene), highly swollen gels by liquid state techniques give definitive ^{13}C spectra whose lines are typically ten times wider than those of low molecular weight organic compounds. Similar line broadening causes so much overlap of the lines in ^{1}H spectra that the spectra are not useful. The solid state technique of cross-polarization with high speed magic angle spinning and dipolar decoupling (CP/MAS NMR) can give ^{13}C spectra of dry solids with line widths on the order of 1 ppm. Other spin 1/2 nuclei, such as ^{19}F and ^{31}P, also are easily detected.

Electron impact mass spectra can be obtained, although of course only small fragments of the polymer are detected. The technique of pyrolysis/gas chromatography has been used to analyze the sequences of repeat units in cross-linked polystyrenes (30).

Single crystal X-ray analysis is not useful because polymers rarely form single crystals. The microcrystallites in polymers detected by X-ray powder diffraction typically have dimensions of 100-1000 Å.

Industrial Use of Polymeric Reagents and Catalysts

The key to wider use of polymeric reagents and catalysts is their adoption in industry for fine chemical and pharmaceutical manufacturing. Without industrial use they will remain items of academic curiosity but little practical importance. A look at some related materials may give us some idea of the prospects for industrial application of the materials and processes described in this book.

The term phase transfer catalysis was coined by Starks to describe the mechanism of catalysis of reactions between water-soluble inorganic salts and water-insoluble organic substrates by lipophilic quaternary ammonium and phosphonium ions (31). His investigations of nucleophilic displacement reactions, such as that of aqueous sodium cyanide with 1-chlorooctane, and the investigations of Makosza on reactions of aqueous sodium hydroxide with chloroform to generate dichlorocarbene, and with active ketones and nitriles to generate carbanions, pioneered the field in the mid-1960's. It was nearly fifteen years before many such processes were adopted in industry. Starks now estimates there are about sixty phase transfer catalytic processes in use worldwide, mostly in pharmaceutical and fine chemical manufacturing (32).

Merrifield peptide synthesis was first disclosed in late 1962. It has been used widely in research laboratories to prepare both natural and synthetic peptides. The method is automated. But only two peptides are synthesized commercially by the Merrifield method (33). Lower yield solution coupling methods are used more widely.

Supported enzymes are a major class of catalysts not covered in this book because they have become a large field in their own right (34). Polymers and silica have been used as supports. There are now four major and many more minor industrial processes using supported enzymes (35). The largest scale process is the partial conversion of glucose to fructose in corn syrup with glucose isomerase as catalyst.

The histories of phase transfer catalysis, Merrifield peptide synthesis, and supported enzymes suggest that synthetic processes using polymeric catalysts may be adopted in commercial production, but that methods for multistep syntheses are less likely. The reasons for this prediction follow.

Advantages and Disadvantages. The advantages of supported reagents and catalysts are cited repeatedly throughout this book. They are 1) ease of separation of the supported species from a reaction mixture, 2) reuse of a catalyst or of a supported reagent after regeneration, 3) adaptability to continuous flow processes, 4) reduced toxicity and odor of supported species compared with low molecular weight species, and 5) chemical differences, such as prolonged activity or altered selectivity of a catalyst, in supported form compared with its soluble analog. The main disadvantages of supported reagents and catalysts are 1) higher cost, 2) lower reactivities due to diffusional limitations, 3) greater difficulty of analysis of the structure of the supported species and of impurities, 4) inability to separate polymer-bound impurities, and 5) lesser stability of organic supports than of inorganic supports.

The polymer-bound species, a catalyst, a product, or a by-product, can be separated from the reaction mixture. Macroscopic solids and gels can usually be separated from liquids by filtration. Soluble polymers and colloids can be separated from low molecular weight compounds by ultrafiltration. Colloids can be coagulated and filtered. The functionalized polyethylenes described by Bergbreiter in chapter 2 are soluble hot and insoluble cold so that they can be filtered.

Filtration is not always easy. On a laboratory scale fine particles (fines) that slough off from larger polymer beads or ground resin often clog the pores of glass frits and filter papers. In a fluidized bed fines would flow out in the product stream. Polymers can be ground to fines between magnetic stirring bars and the walls of glass flasks. Mechanical stirring and shaking are recommended to minimize physical breakdown of the particles. Osmotic forces also can lead to physical attrition of polymer particles. Gradual swelling and deswelling by gradient exposure to a new solvent lessens the osmotic shock. Precipitated polymers generally pose more difficult filtration problems than polymer beads. Large scale filtration requires that the polymer not be so gelatinous that it deforms markedly under pressure. Generally polystyrenes cross-linked with $\leq 2\%$ divinylbenzene and swollen in solvents will not meet this requirement. More highly cross-linked, less swellable polymers will be needed.

Possible alternatives to cross-linked polymer supports are soluble and colloidal polymers. They would require large scale ultrafiltration for industrial use. Although ultrafiltration is not yet economical for desalination of seawater, it might be for a separation of a more expensive product. One example is the catalytic partial hydrogenation of soybean oil (36) with soluble polymer-bound transition metal complexes. Solid inorganic supports such as silica gel and alumina are usually not subject to these physical attrition and filtration problems.

Considerable effort has been devoted to the modification of polymeric catalysts and reagents to give chemoselectivities different from those of soluble catalysts and reagents. These modifications often depend upon the binding of more (or fewer) ligands to a metal or upon the reduced mobility of polymer-bound species compared with micromolecular species. A review of these efforts is in chapter 11 by Ford on site isolation syntheses. Altered chemoselectivity has appeared unexpectedly in attempts to construct polymeric reagents and catalysts with structures analogous to those of successful soluble species.

Reuse of a polymer-bound catalyst or reagent is a key economic factor. The preparation of the polymeric species generally requires at least one more synthetic step than the preparation of a soluble analog, adding to the initial cost. Recycling of the catalyst or recovery of the metal may be necessary for the polymeric species to be cost-competitive. If the polymer-bound catalyst retains its activity during long use, and the analogous soluble catalysts are hard to recover, the polymeric catalyst may be cost-effective. However, many of the common polymer-bound transition metal catalysts show metal leaching or degradation, as described in chapter 5 by Garrou. The common polymer-bound phase transfer catalysts, quaternary ammonium and phosphonium ions, also undergo degradation under basic conditions and even under neutral conditions at elevated temperature (20). The most important factor in the cost of a polymer-bound catalyst is its lifetime.

The regenerability and efficiency of polymer usage are most important with polymer-supported reagents. Among the easily regenerated reagents are those based on ion exchange resins, which require only a treatment with an excess of the ionic reactant for regeneration, the phosphine oxide by-products of Wittig reactions, which are easily reduced to phosphines (Chapter 8), and the arsenic peracids, which can be regenerated with hydrogen peroxide (Chapter 7). The buildup of polymer-bound by-products as a catalyst or reagent is recycled may lead to reduced activity of a catalyst or capacity of a reagent, but analysis of the by-products is attempted rarely.

Another factor, often overrated, in the cost of using polymer-supported reagents is the degree of functionalization (DF) of the polymer. The higher the DF, the higher the effective concentration that can be used in a reaction. However, lower DF reagents are not necessarily much more expensive to prepare, because the added weight in most cases is only inexpensive polystyrene. A more detailed analysis of the decreased capacity of low DF Wittig reagents is in Chapter 8. Sometimes very low DF reagents (DF <0.02) and catalysts have been explored for site-isolation syntheses. If they succeed when all other methods fail, even such low DF reagents may be cost-effective.

Lesser reactivity of insoluble polymer-bound reagents and catalysts is normal, because diffusion of reactants through the polymer network or through the pores of a macroporous catalyst becomes partially rate-limiting for fast reactions. This diffusional limitation of catalyst activity has been analyzed in detail for phase transfer catalysis (20) and for transition metal catalysis in chapter 4 by Ekerdt.

All of these advantages and disadvantages must be considered in the development of a polymer-supported catalyst or reagent. An industrial process must meet the requirements of 1) technical performance and 2) lower cost than alternative processes. The overwhelming majority of organic reactions that are successful in the laboratory fail the test of low cost. Polymeric reagents and catalysts offer real opportunities for cost savings if they can be recycled. Particular attention must be paid to continuous flow experiments, which academic chemists have neglected.

Predictions. The prospects for development of processes using polymeric catalysts and reagents depend on today's industrial chemists. It is unlikely that current manufacturing processes will be modified to use polymeric species. A plant process than makes money may be fine-tuned, but it will not be abandoned for a new process. If polymer-bound species are to be used, they will be in new processes, usually for new products.

The first commercial uses of polymeric reagents and catalysts are likely to be for production of pharmaceuticals, agricultural chemicals, and other high priced fine chemicals. The polymeric species most likely to be used are those already available. Catalytic applications are likely to be more common than reagent applications. The preparation of the catalyst or reagent is inherently expensive because of extra synthetic steps. If it can be used in only catalytic amounts and has a long lifetime, its initial cost will not inhibit use. Reagents, on the other hand, must be used in stoichiometric amounts and must be regenerated.

A variety of ion exchange resins with strong and weak acid, weak base, and quaternary ammonium ion functionality are available in bead form well suited for filtration from reaction mixtures and for use in continuous flow processes. They have been used for >30 years in flow systems for water deionization. Sulfonic acid resins are already used on a large scale as catalysts for the addition of methanol to isobutylene to form methyl *tert*-butyl ether, for the hydration of propene to isopropyl alcohol, and for a variety of smaller scale processes. Tertiary amine resins have been used as catalysts for the addition of alcohols to isocyanates to form urethanes. The quaternary ammonium ion resins could be used as reagents with any of a large number of counter ions, and as catalysts in two and three phase reaction mixtures, although the author is not aware of any commercial process of this sort at present.

Next to ion exchange resins, the polymeric supports most likely to be used for catalysts are other cross-linked polystyrenes or silica gels. Both are inexpensive, easy to functionalize, and void of other reactive functional groups. Their limitations are the thermal and physical stability of polystyrene and the solubility of silica in alkali. Polystyrene can be derivatized by almost every known reaction of mononuclear aromatic hydrocarbons, and the conditions for those reactions on polymers have been published and reviewed (37). The surface of silica gel can be covered with a wide range of organic materials by reaction of its hydroxyl groups with silyl esters and chlorosilanes (38).

A modest number of polymeric reagents and catalysts are currently available from suppliers of research chemicals such as Aldrich Chemical Co., Alfa Products, Bio-Rad Laboratories, Chemical Dynamics Corp., Fluka Chemical Corp., Petrarch Systems, Inc., Pierce Chemical Co., Polysciences, Inc., and Strem Chemicals. The available materials include 1) many standard ion exchange resins, 2) quaternary ammonium ion resins with many counterions, 3) linear and cross-linked microporous and macroporous polystyrenes, 4) cross-linked polystyrenes functionalized with halogens, chloromethyl groups, diphenylphosphino groups, Rose Bengal, aluminum chloride, crown ethers, quaternary ammonium and phosphonium ions for phase transfer catalysis, and a variety of groups for peptide attachment, 5) several polystyryldiphenylphosphine-complexed transition metal complexes, 6) linear and cross-linked poly(4-vinylpyridine) and its 2- isomer, 7) ion exchange resins and poly(vinylpyridine)s bearing $Cr(VI)$ oxidizing agents, and 8) poly(ethylene glycol)s.

A wide variety of functionalized silica gels is available, as are many other supports such as alumina, clays and graphite. Many of these other supports are not covered in this volume, but will be in a broader volume on supported reagents (39).

Soluble polymers and colloids might also be used as supports for catalysts if ultrafiltration can be used to reclaim the catalysts efficiently.

New Directions in Academic Research. The future of academic research on polymer-supported catalysts and reagents may depend on the success of currently available species in industrial processes. Large scale successes would encourage many more chemists to employ polymer-supported species routinely in the laboratory and to invent new polymeric catalysts and reagents.

The directions new academic research are likely to follow will depend upon new reactions in synthetic organic, inorganic, and polymer chemistry, on industrial successes, and on new inventions of creative chemists. Many of these developments are impossible to predict. The most predictable lines of current research likely to yield new polymer-supported synthetic methods are 1) catalysts that overcome the current stability problems with transition metal phosphine complexes and with quaternary onium ions, 2) reagents and catalysts for asymmetric organic synthesis, 3) automated synthetic methods, and 4) tailored polymers for selective separations of organic and inorganic compounds on the basis of structure and stereochemistry. Readers are encouraged to experiment with new supported catalysts and reagents, new monomers, and new polymers.

Literature Cited

1. Merrifield, R. B., cited in Chem. Eng. News, 1971(Aug. 2), 22.
2. Merrifield, R. B. J. Am. Chem. Soc. 1963, 85, 2149.
3. Barany, G.; Merrifield, R. B. In: "The Peptides", Vol. 2, Part A, Gross, E.; Meienhofer, J., Eds.; Academic Press, New York, 1979, Chap. 1.
4. Letsinger, R. L.; Kornet, M. J. J. Am. Chem. Soc. 1963, 85, 3045.
5. Fridkin, M.; Patchornik, A.; Katchalski, E. J. Am. Chem. Soc. 1965, 87, 4646.
6. Gait, M. J. In: "Polymer-supported Reactions in Organic Synthesis"; Hodge, P.; Sherrington, D. C., Eds.; Wiley-Interscience, New York, 1980; Chap. 9.
7. Frechet, J. M. J. In: ref. 6, Chap. 8.
8. Blossey, E. C.; Neckers, D. C., Eds.; "Solid Phase Synthesis", Dowden, Hutchinson & Ross, Stroudsburg, PA, 1975.
9. Caruthers, M. H. Science 1985, 230, 281.
10. Koster, H.; Biernat, J.; McManus, J.; Wolter, A.; Stumpe, A.; Narang, C. K.; Sinha, N. D. Tetrahedron 1984, 40, 103.
11. Helfferich, F. "Ion Exchange", McGraw-Hill, New York, 1962, Chap. 11.
12. Sherrington, D. C. In: ref. 6, Chap. 3.
13. Ancillotti, F.; Mauri, M. M.; Pescarollo, E.; J. Catal. 1977, 46, 49.
14. Pittman, C. U., Jr. In: ref. 6, Chap 5.
15. Chauvin, Y.; Commereuc, D.; Dawans, F. Prog. Polym. Sci. 1977, 5, 95.
16. Hodge, P. In: ref. 6, Chap 2.
17. Akaleh, A.; Sherrington, D. C. Chem. Rev. 1981, 81, 557.
18. Akaleh, A.; Sherrington, D. C. Polymer 1983, 24, 1369.
19. Regen, S. L. Angew. Chem., Int. Ed. Engl. 1979, 18, 421.
20. Ford, W. T.; Tomoi, M. Adv. Polym. Sci. 1984, 55, 49.
21. Hodge, P.; Sherrington, D. C., Eds., "Polymer-supported Reactions in Organic Synthesis", Wiley-Interscience, New York, 1980.
22. Mathur, N. K.; Narang, C. K.; Williams, R. E. "Polymers as Aids in Organic Chemistry", Academic Press, New York, 1980.
23. Allcock, H. R.; Lampe, F. W. "Contemporary Polymer Chemistry", Prentice-Hall, Englewood Cliffs, NJ, 1981.

24. Billmeyer, F. W., Jr. "Textbook of Polymer Science", 3rd ed., Wiley-Interscience, New York, 1984.
25. Bovey, F. A.; Winslow, F. H., Eds., "Macromolecules: An Introduction to Polymer Science", Academic Press, New York, 1979.
26. Rudin, A. "The Elements of Polymer Science and Engineering", Academic Press, New York, 1982.
27. Odian, G. "Principles of Polymerization", 2nd ed., Wiley-Interscience, New York, 1981.
28. Bovey, F. A. "Chain Structure and Conformation of Macromolecules", Academic Press, New York, 1982.
29. Ford, W. T.; Mohanraj, S.; Periyasamy, M. Brit. Polym. J. 1984, 16, 179.
30. Nakagawa, H.; Tsuge, S. Macromolecules 1985, 18, 2068.
31. Starks, C. M.; Liotta, C. "Phase Transfer Catalysis", Academic Press, New York, 1978.
32. Starks, C. M., Vista Chemical Co., personal communication.
33. Hughes, J., Beckman Instruments Co., personal communication.
34. Chibata, I. "Immobilized Enzymes: Research and Development", Wiley, New York, 1978.
35. Maugh, T. H., II Science 1984, 223, 474.
36. Bayer, E.; Schurig, V. CHEMTECH 1976, 212.
37. Frechet, J. M. J. Tetrahedron 1981, 37, 663.
38. Unger, K. K. "The Journal of Chromatography Library", Vol. 16, Elsevier, Amsterdam, 1979, pp. 57-146.
39. Laszlo, P., Ed., book in preparation.

RECEIVED September 26, 1985

Soluble Polymer-Bound Reagents and Catalysts

David E. Bergbreiter

Department of Chemistry, Texas A&M University, College Station, TX 77843

Synthetic applications of soluble polymer-bound reagents and catalysts are reviewed. Examples show
these soluble macromolecular reagents have many of
the same advantages as insoluble polymeric reagents
as replacements for conventional low molecular
weight reagents or catalyst ligands. The homogeneity of reaction solutions employing such reagents
or catalysts is their principal advantage over comparable chemistry using an insoluble reagent or
catalyst derived from a cross-linked polymer.

The use of soluble polymer-bound reagents or catalysts is an
attractive alternative to the use of insoluble polymer-bound
reagents or catalysts when a substitute for a conventional homogeneous reagent or catalyst is needed or is appropriate for a
given application (1-5). Insoluble cross-linked polymer-bound
reagents are most useful when an expensive or toxic reagent is
used and it is important to recover the reagent quantitatively at
the end of a reaction (1-3). Similarly, when a reaction produces
a by-product which is separated only with difficulty from the
desired product, the facile separation of an insoluble cross-
linked polymeric reagent can have advantages. Non-cross-linked
polymers can be used in much the same way as their heterogeneous
counterparts. While a linear polymer can often be dissolved under
certain conditions, addition of a second poorer solvent or cooling
can in many cases quantitatively precipitate such linear polymers.
In other cases, the modification of a polymer-bound functional
group which occurs during consumption of such a reagent during a
stoichiometric reaction sufficiently changes the properties of the

0097-6156/86/0308-0017$07.25/0

polymer that the polymer precipitates from solution of its own
accord as the reaction proceeds. Thus, a reagent attached to a
linear polymer can be used in homogeneous solutions and can still
be recovered at the end of a reaction by precipitation and filtration. Catalysts attached to insoluble cross-linked polymers such
as divinylbenzene (DVB) cross-linked polystyrene have the
advantages of easy recovery of the catalyst and easy separation of
the catalyst from the reaction products. In addition, insoluble
polymer-bound catalysts can, at least in principle, be used in
continuous reactors much like conventional heterogeneous
catalysts. Catalysts bound to soluble polymers have the advantages of easy recovery and separation from reaction products.
However, like other homogeneous catalysts, they are not as easily
used in a continuous reactor.

Soluble polymer-bound reagents and catalysts have received
less attention than their heterogeneous counterparts. It is not
necessary or reasonable that this should be the case considering
some of the advantages of using soluble polymers. One reason for
this lack of attention is the perceived difficulty of product
isolation and separation from a soluble polymer. However, there
are in practice several reliable and simple ways in which soluble
macromolecules can be separated from the products of stoichiometric or catalytic reactions. Techniques available for recovery
and separation of a linear soluble polymer-bound reagent or
catalyst from low molecular weight reaction products include
selective precipitation of the polymeric reagent by addition of a
non-solvent followed by filtration, thermal precipitation of the
polymeric reagent and its removal by filtration, the use of membrane filtration with membranes whose porosity is such that only
low molecular weight species can readily diffuse through the
membrane, and simple filtration when a polymeric by-product of a
stoichiometric reaction is insoluble. Membrane filtration and
solvent precipitation are the most generally applicable of these
four separation methods. Centrifugation can be used as an alternative to conventional filtration in these procedures.

Soluble macromolecular reagents have many of the advantages
and disadvantages of their insoluble macromolecular counterparts.
The most general advantage of each of these classes of reagents is
their facile separation from low molecular weight reaction
products as discussed above. Soluble polymeric reagents may have
other unique advantages in individual cases and some examples are
described below. A potential disadvantage of both soluble or
insoluble macromolecular reagents is the higher molecular weight
of a macromolecular reagent versus a conventional reagent. While
lower reagent equivalent weights (higher loadings of the reactive
functional groups on a polymer) are possible and have been used,
typical equivalent weights for a macromolecular reagent are 1000
or more (1 mequiv of reactive functionality/g of polymer). Such
loadings are practical in both small scale and medium scale reactions. Large scale syntheses would either require polymeric rea-

gents of higher loading or the use of polymer-bound catalysts in place of stoichiometric reagents. The degree of functionalization of a polymeric reagent is less of a problem in the case of polymer-bound catalysts. It is usually possible to use sufficient amounts of catalysts to achieve reaction rates comparable to conventional catalysts with most polymer-bound catalysts. Perhaps more importantly, in the case of soluble polymer-bound catalysts, the concentration of soluble polymer-bound catalysts can readily be adjusted to compare with concentrations attainable with conventional catalysts.

Soluble macromolecular reagents and catalysts have some general advantages over their heterogeneous counterparts. First, because they are soluble, many of the diffusional constraints which affect the utility of reagents or catalysts bound to insoluble polymer resins are minimized. However, the behavior of soluble polymeric reagents is complicated by the possibility that polymer chains can aggregate in solution and by the demonstrated lower diffusion rates of soluble macromolecules versus a small organic molecule in solution (vide infra). Nonetheless, reaction of a small molecule with a soluble macromolecule has been shown to be more facile than similar reactions with an analogous insoluble polymeric reagent. For example, the reaction of primary alkyl bromides with nucleophiles catalyzed by gel-type DVB cross-linked polystyrenes is affected by the size of the alkyl halide because of the kinetic significance of diffusion of the reactant molecule through the gel polymer to an active site (6). Our work has shown that such effects are decreased in reaction of sodium iodide with primary alkyl bromides catalyzed by soluble alkene oligomer-bound crown ethers (7). The use of macroporous DVB cross-linked polystyrene should also diminish effects of intraparticle diffusion. The reactivity of the reactive sites in a soluble polymeric reagent has also been shown to be comparable for all or nearly all of their reactive sites while similar studies of insoluble polymeric reagents have shown that their reactive sites have a wider range of reactivity (8). These soluble reagents or catalysts should also be more practical in exothermic reactions because dissipation of reaction heat into the surrounding solvent is more efficient. Local heating effects resulting from an exothermic reaction have been discussed as a disadvantage for insoluble catalysts (9) and could lead to degradation of the organic polymer supports or more likely to degradation of the catalyst complexes or reagents attached to such polymers. Third, soluble macromolecular reagents or catalysts can be more readily characterized than their heterogeneous counterparts. While solid state NMR spectroscopy is developing into a useful tool for characterizing insoluble polymer-bound reagents and catalysts, (10). soluble macromolecular reagents and catalysts can be characterized in a much more routine manner using solution state NMR spectroscopy. Lightly cross-linked polystyrenes have been analyzed by both ^{13}C and ^{31}P NMR spectroscopy (11,12). However,

soluble polymers can be analyzed routinely by ^{1}H NMR spectroscopy
as well. The application of the other spectroscopic techniques
commonly used in organic chemistry is similarly facilitated when a
polymeric reagent or catalyst can be studied either as a solid or
as a solution. The use of a cross-linked polymeric reagent or
catalyst can also face some physical problems relating to
mechanical breakdown of very rigid polymers. Less rigid polymers,
which often have less intrinsic porosity, require good swelling
solvents to insure that the polymer-bound species have
good access to species in solution. Linear polymers face a
similar problem in that a solvent and temperature must be chosen
such that both the polymer and the reagents are in solution.
Finally, chemists are accustomed to dealing both empirically and
quantitatively with kinetics and thermodynamics of homogeneous
reactions and, as a result, can often make minor modifications in
reagent or catalyst structure or reaction conditions to maximize
the yield of a desired process. Effecting similar improvements in
reactions which use heterogeneous reagents often proves to be more
difficult. For example, we have found that it is feasible to
readily prepare a range of structurally diverse phosphite ligands
bound to alkene oligomers, and we have examined their use in
nickel(0) catalyzed diene cyclooligomerization (13). At least in
our hands, attempts to prepare a series of structurally different
well characterized insoluble polymeric ligands were more
difficult.

Preparation of Non-cross-linked Polymeric Reagents

Derivatization of linear polymers to form a soluble polymeric
reagent can be accomplished by the use of conventional chemical
reactions. Several strategies have been used. As is the case
with preparation of insoluble polymer-bound reagents and
catalysts, the most common strategy illustrated by reactions 1-4
below is to use commercially available polymers or to prepare an
unsubstituted addition or condensation polymer and to then introduce the desired functional groups. In the case of linear polystyrene, bromination can be used to prepare a functionalized
polystyrene which contains electrophilic sites. These electrophilic sites can then be further transformed by reaction with
nucleophiles into a polymeric reagent. In this example and in
Equations 3, 4 and 5, relatively high loadings of functionality
are possible since functional groups can be introduced at nearly
every monomer unit. Alternatively functional groups can be introduced at a chain terminus. It has been reported that this latter
procedure illustrated by Equations 2 and 6 produces a soluble
polymeric reagent whose reactivity more closely resembles that of
a low molecular weight reagent (8,14).
    There are also examples where the non-cross-linked polymeric
reagent is most readily available by direct polymerization of
suitably functionalized monomers. For example, polymerization of
a crown ether which contains styrene units is possible by free
radical or anionic methods (15). In the case of anionic oligomerization (Equation 6) or in the case of the metalation shown in
Equation 4, the reactive intermediate macromolecule is a nucleophilic polymer which can be derivatized with a variety of electro-

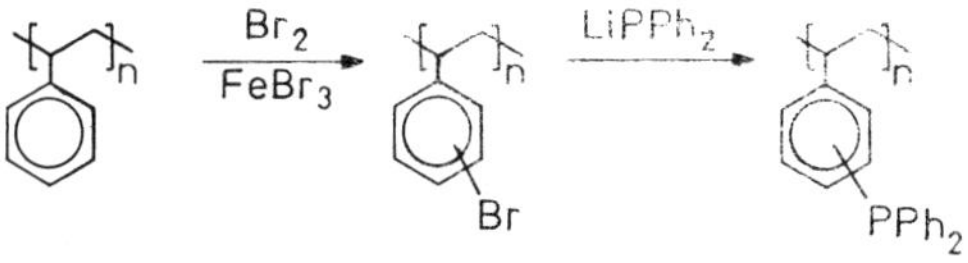

$$\text{(1)}$$

$$\text{(2)}$$

$$\text{(3)}$$

$$\text{(4)}$$

$$\text{(5)}$$

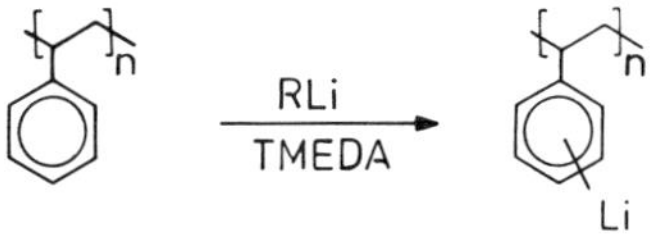

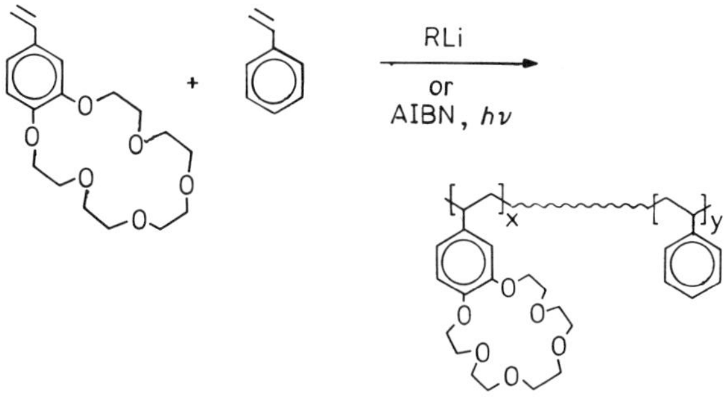

$$\text{(6)}$$

philes (16). The molecular weight of the linear polymers
prepared in Equations 1-6 varies widely. For example, "n" can
vary from 60 in Equation 6 to 1000 or more in Equation 4.

Stoichiometric Halogenating Reagents

Soluble polystyrene-bound diphenylphosphine reagents have been
used in several applications in place of triphenylphosphine in
substitution of halogen for hydroxyl groups. In these cases, the
use of a polymeric reagent permits ready separation of the by-
product phosphine oxide from the halogenated organic derivative.
For example, Hodge has reported that the use of non-cross-linked
polystyrenediphenylphosphine is nearly as effective as the use of
DVB cross-linked polystyrenediphenylphosphine in formation of
alkyl chlorides from an alcohol and carbon tetrachloride, dichloromethane or hexachloroethane (21). Phosphine containing
linear polystyrene, prepared such that there is 2.7-3.0 mmol of
phosphine/g of polymer (which corresponds to having a phosphine
group bound to every other styrene group in the polymeric reagent), can be used as shown in Equation 7. Typically these reactions were carried out using 2 equiv of the phosphinated polymer
and 1 equiv of carbon tetrachloride (or one of the other chlorinated solvents) at 60-77 °C. These starting linear polystyrene
reagents were soluble initially but precipitated during the reaction. The polymeric reagent was thus easily separated from the
reaction product by filtration. The spent linear polymeric reagent was found to contain 0.8 Cl/P atom and was soluble in
methanol. Hodge postulated that the spent polymer contained both
phosphine oxide and chloro- or dichloromethylphosphonium salts.
In qualitative kinetic studies, Hodge was also able to show that
the linear soluble phosphinated polystyrene reagent was only
slightly less reactive than a similar DVB cross-linked polystyrene reagent with the same alcohol substrate. Recycling of the
soluble polystyrene reagent was not explicitly described.
    Recently our group has found that ethylene oligomers are an
alternative to the use of non-cross-linked polystyrene as a polymer to which to bind a reagent (20). For example, we have found
that diphenylphosphinated ethylene oligomers prepared by reaction
6 above can be used in the same way as 1% DVB cross-linked (polystyryl)methyldiphenylphosphine to prepare alkyl chlorides from alcohols and carbon tetrachloride (Equation 8). These polyethylenediphenylphosphine reagents are comparable in activity to these
insoluble polymeric phosphines and could be recycled after
reduction of the by-product polyethylenediphenylphosphine oxide
with trichlorosilane, although the recycled polymeric reagent only
had 65% of the activity of the fresh polyethylenediphenylphosphine
reagent. Hodge earlier reported that similar reduction allowed
recycling of insoluble DVB cross-linked polystyrenediphenylphosphine with only 40% of the original activity. Presumably the
difficulties in recycling these soluble polyethylene-bound phosphine reagents are due to reactions in which halogenated phosphonium salts form as unwanted and unreducible by-products as
suggested by Hodge (21). Representative examples of alkyl
chloride syntheses using these polystyrene- and polyethylenediphenylphosphine reagents are listed in Table I. It also seems

probable that these reagents could be used in other reactions where cross-linked diphenylphosphinated polystyrene has been shown to be useful, although we have not yet specifically examined this question.

Table I. Synthesis of Alkyl Chlorides form Carbon Tetrachloride, a Phosphinated Soluble Polymer and an Alcohol or Thiol

Alcohol or Thiol	Polymeric Reagent[a]	Alkyl Chloride Yield (%)	Reference
1-octanol	"PE"-PPh$_2$	96	17
	"PE"-PPh$_2$[c]	65	17
	"PE"-PPh$_2$[d]	41	17
	"DVB-PS"-CH$_2$PPh$_2$	93	17
	"PS"-PPh$_2$	93	18
benzyl alcohol	"PE"-PPh$_2$	91	17
	"PS"-PPh$_2$	91	18
octadecanol	"PE"-PPh$_2$	69	17
	"PS"-PPh$_2$	76	18
cyclohexanol	"PE"-PPh$_2$	42	17
cyclododecanol	"PE"-PPh$_2$	61	17
	"DVB-PS"-CH$_2$PPh$_2$	57	17
phenylmethanethiol	"PS"-PPh$_2$	91	18

[a]"PE"-PPh$_2$ stands for phosphinated ethylene oligomers containing 0.8 mmol of -PPh$_2$/g of polymer; "PS"-PPh$_2$ stands for linear polystyrene containing 2.69 mmol of -PPh$_2$/g of polymer; and "DVB-PS"-CH$_2$PPh$_2$ stands for diphenylphosphinated polystyrene derived from chloromethylated 2% DVB cross-linked polystyrene by reaction of this commercially available resin with lithium diphenylphosphide.

Wittig Reagents

Many groups have described examples of polymer-bound Wittig reagents useful in synthesis (this volume includes a comprehensive review by W. T. of this subject) (24-28). The principal advantage cited for the use of a polymeric phosphine for formation of an phosphonium ion in these cases is the facile separation of the alkene product from the phosphine oxide by-product and the recyclability of the polymeric phosphine oxide by trichlorosilane reduction (28). While DVB cross-linked polystyrene is most commonly used as the support for polymer-bound Wittig reagents, several reports describe the use of linear polystyrene. One particularly interesting example described by Hodge and co-workers is the use of diphenylphosphinated linear polystyrene (M_w = 100,000) containing 1.0 mequiv of -PPh$_2$/g of polymer (27) Using this soluble macromolecular phosphine and benzylchloride or 2-bromomethylnaphthalene, a phosphonium salt was readily prepared (Equation 9). In this example, the comparatively high acidity of the benzylic C-H's enabled weaker bases such as NaOH to be used to generate the ylid intermediate. Similar reactions were also

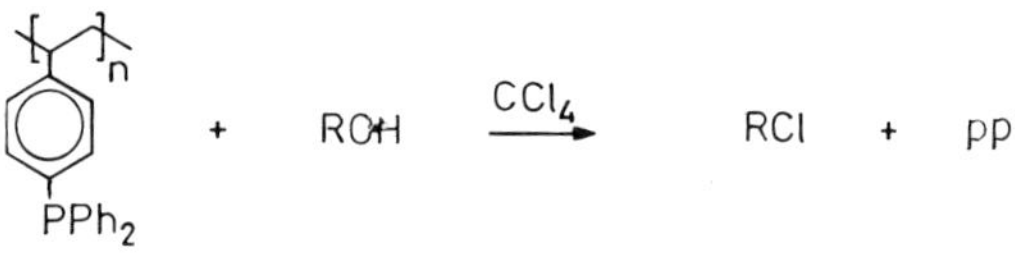

$$\text{[polymer]—PPh}_2 \;+\; ROH \;\xrightarrow{\; CCl_4 \;}\; RCl \;+\; ppt \qquad (7)$$

$$\begin{array}{c} \{CH_2CH_2\}_n{-}CH_2CH_2PPh_2 \;+\; ROH \xrightarrow{\; CCl_4 \;} \\[4pt] \overset{\displaystyle O}{\{CH_2CH_2\}_n{-}CH_2CH_2\overset{..}{P}Ph_2} \;+\; RCl \end{array} \qquad (8)$$

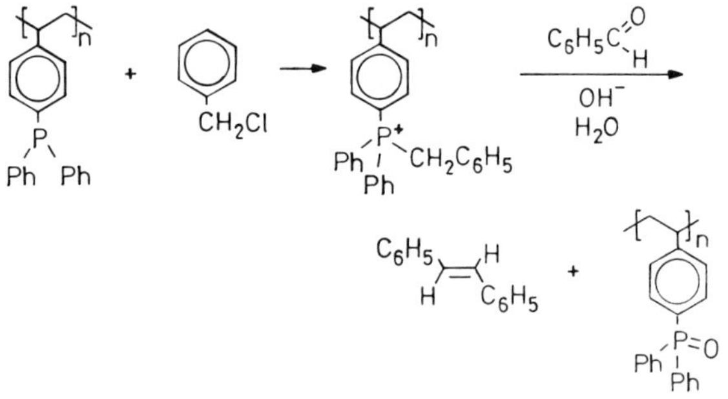

$$(9)$$

effected using DVB cross-linked polystyrene. Ketone substrates examined included 9-formylanthracene and cinnamaldehyde (Table II). Since the the mild bases used in these Wittig reactions were only soluble in the aqueous phase, Wittig reactions using the insoluble cross-linked macromolecular phosphonium salt required the presence of a phase-transfer catalyst. However, in contrast, the soluble macromolecular phosphonium salt did not require this added reagent, possibly because the soluble macromolecular reagent acted as its own phase transfer catalyst. Isolation of alkene products in the examples using soluble macromolecular phosphonium ions required precipitation of the by-product polystyrenediphenylphosphine oxide by addition of methanol. A 50% excess of the polymeric phosphonium salts was used in these reactions and gave high yields of alkene product in 2-3 h at 20 °C.

Hodge's group has also used linear phosphinated polystyrene to form haloolefins from carbon tetrabromide and aromatic aldehydes (27). Using 2 mol equiv of phosphine, 1 mol equiv of CBr_4 and 1 mol equiv of p-tolualdehyde at 50 °C for 16 h formed a 54% yield of the dibrominated alkene. Substitution of 1% or 8% DVB cross-linked polystyrene for linear polystyrene yielded 67% and 12% of dibrominated alkene under the same conditions (Table II).

Soluble Polymer-Bound Oxidants

A variety of groups have reported examples of the use of oxidizing agents in which an organic polymer matrix is used to ionically or to covalently bind an oxidizing reagent. Soluble polymers have also been used. For example, Schuttenberg has described the preparation and application of N-chlorinated nylon polymers which contained a high loading of N-chloro groups and which could be used to oxidize primary or secondary alcohols to aldehydes and ketones and which oxidized sulfides to sulfoxides (18). The N-chlorinated nylons were prepared by chlorination of linear polyamides using tert-butyl hypochlorite or chlorine monooxide in CCl_4. These halogenation reactions required 3 h at 15 °C using Nylon 66 and converted 94% of the N-H bonds in the original polyamide into N-Cl bonds. In addition, the polyamide which was originally insoluble became readily soluble in chloroform, perhaps because of diminished intramolecular hydrogen bonding once the N-H bonds were replaced by N-Cl bonds. In a typical oxidation such as is shown below the chlorinated polymer was converted back into the starting polymer. Since the starting polyamide had poor solubility in benzene, it was readily removed from the reaction products by filtration. Alcohol oxidations were performed using endo-1,7,7-trimethylbicyclo[2.2.1]heptan-2-ol, cyclohexanol, 1-phenyl-2-propanol, 1-phenyl-3-butanol and with other secondary alcohols and with benzyl alcohol. Yields of ketone or benzaldehyde were typically >90% as measured by GC in oxidation reactions in benzene with reaction times of 24 h at 35 °C. Unactivated primary alcohols did not react appreciably with this polymeric oxidant under these conditions. Sulfide oxidation in methanol was also successfully accomplished, although in this case the conversion of sulfide to sulfoxide was incomplete due to methanol oxidation and sulfoxide rearrangement. Unlike the alcohol oxidation reactions, the sulfide oxidations were carried out under conditions where the N-chloropolyamide was insoluble. In this work, Schuttenberg also described attempts to carry out asymmetric oxidations of sulfides using a chiral N-chloropolyamide derived from (-)-poly-(S)-(-)-4-methylazetidinone. However, the sulfoxides derived from this reaction were optically inactive.

Table II. Wittig Reactions of Soluble Macromolecular Ylids[a]

Substrate	Phosphonium Salt	Reaction Time (h)	Yield (%)
9-formylanthracene	"PS"-PPh$_2$CH$_2$C$_6$H$_5$$^+$ Cl$^-$	2	92
	"1% DVB-PS"-PPh$_2$CH$_2$C$_6$H$_5$$^+$ Cl$^-$ [b]	2	98
	"1% DVB-PS"-PPh$_2$CH$_2$C$_6$H$_5$$^+$ Cl$^-$ [c]	2	35
	"PS"-PPh$_2$CH$_2$C$_{10}$H$_7$$^+$ Br$^-$	2	100
cinnamaldehyde	"PS"-PPh$_2$CH$_2$C$_6$H$_5$$^+$ Cl$^-$	2	75
para-tolualdehyde	"PS"-PPh$_2$CBr$_2$$^+$ Br$^-$	16	54
	"1% DVB-PS"-PPh$_2$CBr$_2$$^+$ Br$^-$ [d]	16	67
	"8% DVB-PS"-PPh$_2$CBr$_2$$^+$ Br$^-$ [e]	16	12

[a] Reactions of benzylic phosphonium salts were carried out at 20 °C using 10 mL of methylene chloride, 1.5 mmol of the polymeric phosphonium salt, and 3 mL of 50% NaOH (aq). The linear polystyrene had a MW of 150,000 with 2.7 mmol of -PPh$_2$/g of polymer. The cross-linked polystyrene contained 3.0-3.5 mmol of -PPh$_2$/g of polymer. The halogenated phosphonium ion was prepared from phosphinated polystyrene having 0.4 mequiv of -PPh$_2$/g of polymer and was allowed to react with para-tolualdehyde at 50 °C for 16 h. [b] 2 mmol-% hexadecyltrimethylammonium bromide was added as a phase transfer catalyst. [c] No phase transfer catalyst was present. [d] 1% DVB cross-linked polystyrene containing 0.81 mequiv of phosphonium ion/g of polymer. [e] 8% DVB cross-linked polystyrene containing 0.30 mequiv of phosphonium ion/g of polymer.

Soluble polymer catalysts for oxidation have also been described. One such example would be the use of various polybasic polymers as polydentate ligands for copper(II) in oxidative polymerization of phenols (29,30). Polybasic polymers such as poly-(vinylpyridine) have been used. In this example, the neighboring group effect consisting of having adjacent pyridine groups on the poly(vinylpyridine) capable of complexing a copper(II) ion led to a significantly higher complex formation constant for complexation of copper(II) versus the complexation constant measured for pyridine. The complex formed contained four pendant pyridine units of this polybasic polymer.

Soluble Polymer-Bound Reducing Agents

The use of polymer-bound stoichiometric reducing agents has not received much attention, perhaps because removal of impurities or recycling a reagent is of lesser importance in most reduction reactions using Main Group metal hydride complexes. One example of a soluble polymeric metal hydride that has been reported is the use of linear poly(vinylpyridine) to bind BH_3 (31). Hallensleben has shown that such a borane complex behaves like pyridine-borane, reducing carbonyl groups to hydroxy groups. Examples of such reductions include (time in h in refluxing benzene, alcohol yield using pyridine-borane, alcohol yield using polystyrene-poly(vinyl-pyridine)-borane) benzaldehyde (0.5, 76, 74), para-chlorobenz-aldehyde (1.25, 75, 51), benzophenone (2.5, - , 40), and cyclo-pentanone (2.5, 25, 12).

The use of insoluble polystyrene-bound alkali metal aromatic radical anions, related alkali metal-graphite intercalation com-pounds and alkali metal derivatives of weakly acidic polystyrene derivatives in reactions like those described for alkali metal aromatic radical anions and alkali metal organometallics in ether solutions has been reported (19,32). Similar soluble alkali metal aromatic radical anions derived from poly(vinylnaphthalene) and polyacenaphthylene have been reported (33). Alkali metal derivatives of poly(vinylnaphthalene) prepared reportedly included the dilithium salt, the sodium salt and the potassium salt, all prepared by reaction of a solution of the polymeric naphthalene derivative with the alkali metal at 25 °C for 24 h. The lithium poly(vinylnaphthalene) was found to react quantitatively with some organic halides such as benzyl chloride, butyl bromide and allyl chloride but not at all with iodobenzene or other halogenated arenes or with cyclohexyl chloride. The lack of reaction of the latter halides was ascribed to their having a ring structure which the authors said resulted in steric hindrance with the naphthalene groups attached to the vinyl polymer backbone. In contrast, the alkali metal salts of polyacenaphthylene did react with these halides. The authors rationalized this difference in reactivity in terms of the greater flexibility and resultant diminished steric hindrance of the naphthalene groups in the latter polymer. Regardless of the correctness of this explanation, the lack of reaction of iodobenzene with a dilithionaphthalene derivative is remarkable. A thorough study of all the products of these reac-tions including the nature of the poly(vinylnaphthalene) after reaction was not performed so it is difficult to ascertain if this apparent differing reactivity of a soluble polymeric alkali metal aromatic radical anion and a simple aromatic radical anion is general or of some particular synthetic value.

An unusual synthetic application of a soluble macromolecular reducing reagent described by Smith is the use of aqueous solu-tions of hydrazonium polyacrylate or hydrazonium poly[2-(acryl-amido)-2-methylpropanesulfonate] to prepare stable colloidal dispersions of red, amorphous selenium. (34). In these reac-tions, the solutions were prepared sufficiently dilute so that each macromolecular hydrazonium polyacrylate could react individually. When a solution of H_2SeO_3 was added to this poly-meric reducing agent, the selenious acid molecules in the

vicinity of a given hydrazonium polyacrylate macromolecule were
reduced to selenium atoms. Presumably these selenium atoms then
aggregated to form a single hydrophobic selenium particle which
remains bound to the polyacrylate. In these reactions, only 75%
of the pendant acid groups of the polyacrylic acid were neutralized by added hydrazine to insure that the macromolecular
reagents would contain some residual acid. These residual acid
groups presumably catalyzed the reduction of H_2SeO_3 by the polymeric hydrazine.

Site Isolation Using Soluble Polymer-Bound Species

The practice of using an insoluble polymer to isolate and kinetically stabilize a reactive intermediate has been addressed in
several reports, most commonly using DVB cross-linked polystyrene
as a support. In these cases, the three dimensional structure of
the polymer and rigidity of the polymer backbone diminish intramolecular reactivity between two sites on the same polymer bead.
Physical constraints preclude any significant reaction between two
different polymer beads. Similar, less dramatic reduced intramolecular reactivity has also been noted for reactive intermediates bound to linear polystyrene. For example, o-benzyne
bound to linear polystyrene has been shown by Mazur to have enhanced stability relative to non-polymer-bound o-benzyne (35). In
this case, o-benzyne was generated by lead tetraacetate oxidation
of a 2-aminobenzotriazole precursor, 1. Analysis of the reaction
products after cleaving the benzyne derived products from the
polymer by hydrolysis showed a 60% yield of aryl acetates was
obtained (Equation 11). In contrast, the monomeric aryne forms
only coupled products under similar conditions. Further compari-
sons of the reactivity of o-benzyne bound to insoluble 2% or 20%
DVB cross-linked polystyrene showed the intermediate o-benzyne had
an even longer lifetime. Overall, the reactivity of o-benzyne
bound to the soluble polymer was found to be intermediate between
that of non-polymer bound benzyne and benzyne bound to 20% DVB
cross-linked polystyrene. The 10^{-2} sec lifetime reported for
benzyne in this case presumably reflects diffusional constraints
associated with the polymer chain whose M_w was 10^5.
     Isolation of a cobalt phthalocyanine catalyst known to be
active in autooxidation and to be deactivated by dimerization has
been reported by Schutten (36). In this case, a polyvinylamine polydentate ligand was added to a dilute aqueous solution of the
cobalt(II) phthalocyanine tetra(sodium sulfonate) in order to
prepare a thiol oxidation catalyst. By employing dilute solutions, the polydentate polyamine polymer in effect isolated the
cobalt(II) catalyst within an individual polyamine coil minimizing
dimerization and significantly increasing catalyst activity.
     Soluble polymer-bound substrates have also been used as part
of an experimental protocol to probe the homogeneity of a
catalyst (37). In these experiments, the reactivity of a substrate
bound to a soluble or insoluble polymer is compared to the reactivity of the same substrate not bound to a polymer. No reaction
of an insoluble polymer-bound substrate with a catalyst under
conditions where a soluble polymer-bound substrate or a non-
polymer-bound substrate did react with the same catalyst would

imply that the catalyst in question was heterogeneous. In these experiments, as in Mazur's work described above, the soluble polymer-bound substrate's reactivity was found to be intermediate between that of a free substrate and that of a substrate bound to a cross-linked polymer.

Influences of Polymer Size on Reactivity
While many of the reactions discussed in this review can be viewed as direct analogies of similar reactions of small molecules, it is important to remember that the reactivity of a soluble polymeric reagent can be affected by factors other than those affecting small molecules (38,39). Specifically, in dilute solutions soluble macromolecular reagents exist as isolated coils separated one from another by solvent. The size of the coil (and therefore the concentration of reagent within the coil volume) vary with solvent and temperature. At higher concentrations of polymer, aggregation of the polymer chains can occur. These sorts of effects have been most thoroughly studied and discussed for reactions involving polybasic catalysts and polyelectrolytes but are of equal importance and significance for reactions involving soluble polymeric reagents and catalysts in nonaqueous systems.

Peptide Synthesis Using Soluble Polymeric Reagents

Peptide syntheses using polymeric reagents have served as a stimulus to develop the general area of polymer supported reagents and catalysts useful in organic synthesis. While solid phase peptide syntheses pioneered by Merrifield have developed into a widely recognized and used peptide synthesis strategy (3), alternative procedures employing non-cross-linked polymers have been developed into a useful peptide synthesis procedures (3,14). Several polymers have been studied for this purpose, including both polystyrene and poly(ethylene glycol). However, for reasons enumerated below, poly(ethylene glycol) is the polymer support of choice in these liquid phase peptide syntheses.
The general strategy in all of these procedures is to carry out a reaction in a homogeneous solution and to thereby avoid the disadvantage of diffusional constraints and reactivity problems often encountered using the now classical solid phase peptide synthesis strategy. While these peptide syntheses employing soluble reagents thus have the advantages normally associated with reactions of a low molecular weight substrate in a homogeneous solution, they also confer desirable solubility properties on an attached growing peptide chain. Specifically, a poly(ethylene glycol)-bound peptide possesses the desirable features of solubility during peptide bond formation but insolubility after a reaction. Such a change in solubility can be induced by addition of a solvent in which the linear polymer support is insoluble. As in the Merrifield synthesis, advantage is taken of this insolubility to more conveniently separate excess starting materials or soluble reaction by-products from the growing peptide chain attached to the macromolecule. A disadvantage of polyfunctional supports such as functionalized linear polystyrene is the observation that the functional groups do not all have equivalent reactivity in spite of linear polystyrene's solubility.

(40). This problem has been successfully overcome through the use of terminally functionalized poly(ethylene glycol). Presumably, terminally functionalized polystyrene derived from anionic polymerization reactions would also avoid this problem. However, poly(ethylene glycol) is also a more polar polymer and is readily available in various relatively monodisperse fractions in various molecular weight ranges. Poly(ethylene glycol) also has a terminal hydroxyl group which can easily be further modified to incorporate an anchor for a growing peptide chain or a polymeric reagent.

The drawings below illustrate typical groups used to anchor a growing peptide onto poly(ethylene glycol) as a passive support to facilitate purification and isolation of the growing polypeptide. In syntheses using these groups, the poly(ethylene glycol) portion of the macromolecule being formed determines the solubility properties of the molecule as a whole (41). Thus it is possible to selectively precipitate the poly(ethylene glycol)-peptide product in the presence of other low molecular weight impurities. Moreover, while isolation of a peptide after an individual coupling or activation step has been facilitated, kinetic studies show that the actual chemical reactions proceed as efficaciously as their analogs which use low molecular weight reagents (42)

Peptide synthesis using soluble polymers commonly involves attachment of the growing peptide's carboxyl terminus to the poly-(ethylene glycol). While attachment can be accomplished via a simple ester linkage, the eventual requirement for cleavage of the final peptide product usually requires use of other anchoring groups which can be cleaved under mild conditions. Milder conditions minimize epimerization possible in a conventional alkaline ester hydrolysis. Various anchor groups have been used including both reactive benzyl esters and photolabile ester and amide groups (cf. 2-4). Many of the same groups are used in solid phase peptide syntheses. However, the homogeneity of the polymer-peptide adduct in the liquid phase method permits the use of other heterogeneous reagents such as Pd catalysts for hydrogenolysis of poly-(ethylene glycol)-bound peptides (43).

The methodology for forming peptide bonds in the liquid phase method is the same as that used in conventional peptide bond syntheses using low molecular weight reagents. Most commonly, the N-groups of the added amino acids are protected by tert-butoxy-carbonyl groups during the coupling of a new amino acid residue to the free amino group of the polymer attached peptide. One interesting difference between the liquid phase method and the solid phase synthesis is the ability to use reagents such as dicyclohexylcarbodiimide to effect the coupling reaction. The insoluble urea by-product is readily removed from solutions of the peptide-polymer adduct and other soluble reagents and the peptide polymer adduct can then in turn be separated from the remaining soluble species by selective precipitation with diethyl ether and filtration. Typical solvents used during the peptide synthesis include polar aprotic solvents such as dimethylformamide and dimethyl sulfoxide as well as methylene chloride.

Automation of liquid phase peptide synthesis is also possible. However, as the size of the peptide attached to the poly(ethylene glycol) support increases, the properties of the

polymer-peptide complex change. These changes in solubility and
permeability have served to limit developments in automation to
syntheses in which smaller peptides are prepared.

A major advantage of the liquid phase peptide synthesis over
a solid phase peptide synthesis is the facility with which the
progress of the reaction can be monitored. Quantitative fluorescence analysis of the products of reaction of the polymer-peptide
complex with fluorescamine or ninhydrin can be used as simple and
direct measures of the extent of reaction (14). Another
significant advantage of the liquid phase synthesis is its
potential for ready analysis by solution state NMR spectroscopy.
While long acquisition times for ^{13}C NMR spectra were required in
reported examples of application of ^{13}C NMR spectroscopy to
analysis of poly(ethylene glycol)-peptide complexes, the increased
availability of 400 and 500 MHz NMR instrumentation will
significantly increase the facility of these analyses. One can
also anticipate that ^{1}H NMR spectroscopy will become extremely
useful at these higher fields because of the dispersion afforded
by these fields and because of the sensitivity of ^{1}H NMR
spectroscopy.

Water Soluble Macromolecular Catalysts

Enzymes are the archetypal water-soluble macromolecular catalysts.
Indeed, the application of such soluble biochemical catalysts to
reactions both in aqueous and in organic media is a topic of great
current interest. While enzymatic catalysis is outside the scope
of this review (this area has been recently reviewed) (44,45), pro-
teins have been employed as macromolecular ligands to increase or
alter the selectivity of more traditional catalysts. For example,
homogeneous asymmetric hydrogenations of amino acid precursors
have been reported in which the cationic rhodium catalyst was
ligated by N,N-bis(2-diphenylphosphinoethyl)biotinamide which had
been irreversibly complexed to the protein avidin (46). The best
reported examples had turnover numbers in excess of 500 and enantiomeric excesses ranging from 33-44% in reduction of α-acetamidoacrylic acid to N-acetylalanine. In this case the products
of the reaction were separated from the protein-rhodium conjugate
by filtration through a 10,000 molecular weight cutoff ultrafiltration membrane.

Other protein-rhodium conjugates containing cationic rhodium
catalysts have also been prepared using bis(diphenylphosphino-
ethyl)amino derivatives 5-7 and solutions of these bis(phosphine)
ligands in the presence of carbonic anhydrase, α-chymotrypsin and
bovine serum albumin (47). However, the exact nature of the complexes formed has not been discerned in any of these cases, and
these latter enzyme-transition metal complexes evidently do not
exhibit enantioselectivity in hydrogenation of α-acetamidoacrylic
acid.

Water soluble ion exchangers have been used by Pittman's
group as supports for conventional homogeneous Reppe and hydroformylation catalysts (48). These procedures employ resin
particles of poly(vinyl alcohol) and poly(vinyl acetate) which
contained a mixture of a cross-linked poly(acrylic acid) or poly-
(methacrylic acid) and a cross-linked polymeric secondary or

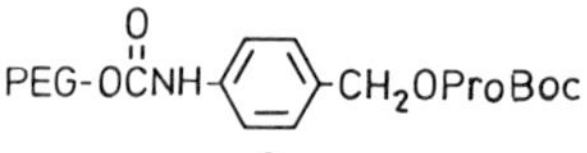

$$PEG\text{-}O\overset{O}{\overset{\|}{C}}NH\text{-}\langle\rangle\text{-}CH_2OProBoc$$

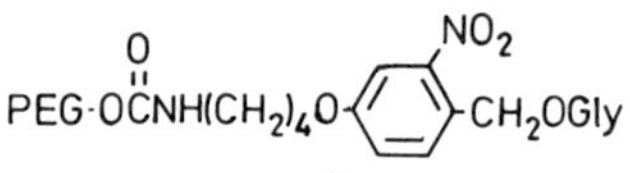

2

$$PEG\cdot O\overset{O}{\overset{\|}{C}}NH(CH_2)_4O\text{-}\langle\rangle\text{-}CH_2OGly \quad (NO_2)$$

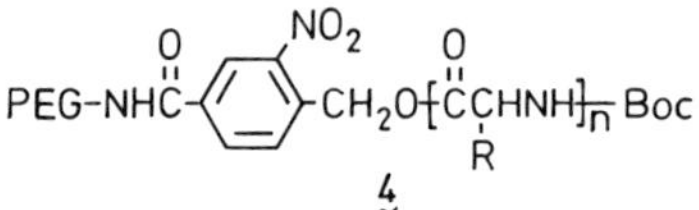

3

$$PEG\text{-}NH\overset{O}{\overset{\|}{C}}\text{-}\langle\rangle\text{-}CH_2O\overset{O}{\overset{\|}{[C}}\overset{}{\underset{R}{C}HNH]}_{n}\,Boc \quad (NO_2)$$

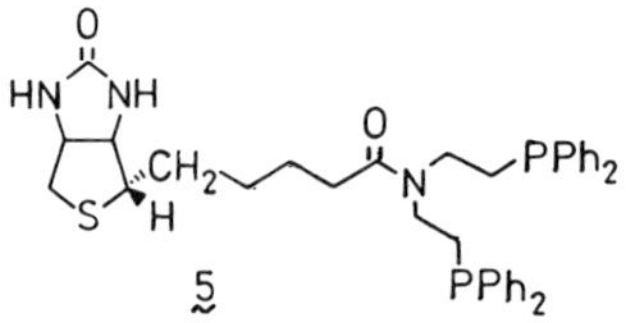

4

5

6

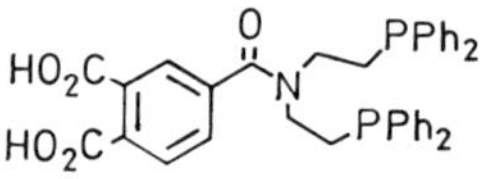

7

$$-[CO(CH_2)_4CONCl(CH_2)_6NCl]_n- \;+\; \langle\rangle\overset{OH}{\underset{}{CH}}CH_3 \xrightarrow{\text{pyridine}}$$

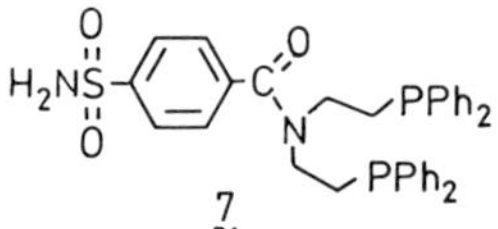

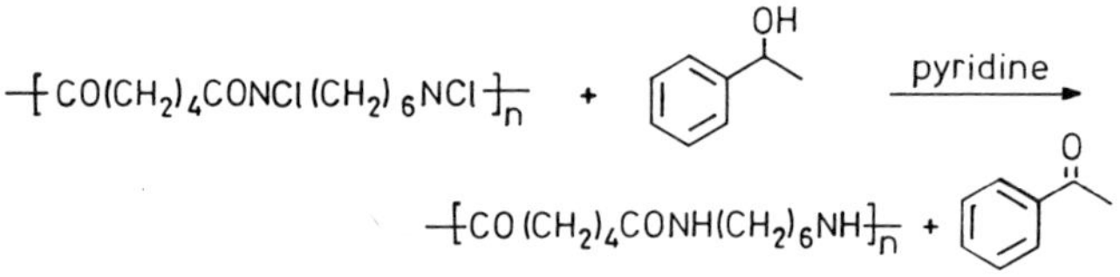

$$-[CO(CH_2)_4CONH(CH_2)_6NH]_n- \;+\; \langle\rangle\overset{O}{\overset{\|}{C}}CH_3 \qquad (10)$$

$$\xrightarrow[\begin{array}{l}1.\ Pb(OCOCH_3)_4,\ CH_3COOH\\ 2.\ CH_3MgBr\\ 3.\ H_3O^+\end{array}]{} \quad HOCH_2CH_2\text{-}\langle\rangle\text{-}OH \;+\; HOCH_2CH_2\text{-}\langle\rangle\text{-}OH \qquad (11)$$

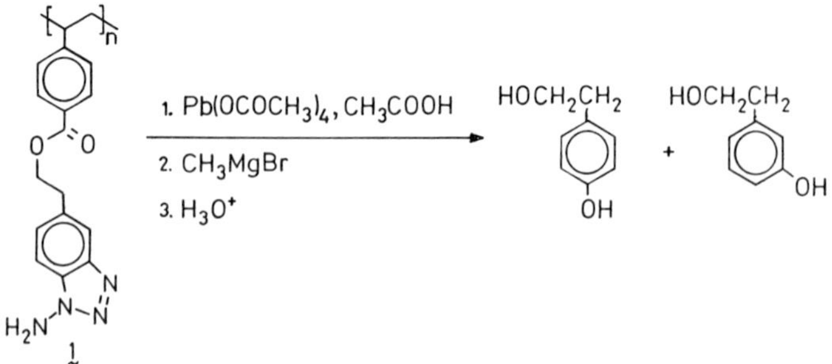

1

tertiary amine. By changing the temperature, the acid-base
equilibrium below can be shifted to entrap a cationic or an

$$\text{"Poly"-NR}_2 \quad + \quad \text{"Poly"-COOH} \quad + \quad Na^+ \quad + \quad Cl^-$$

$$\rightleftharpoons \quad \text{"Poly"-NR}_2HCl \quad + \quad \text{"Poly"-COONa} \tag{12}$$

anionic transition metal complex either as a carboxylate or an
ammonium salt, respectively. At higher temperatures, the
equilibrium 12 shifts to form more of the amine and carboxylic
acid, thus releasing a transition metal into solution. At the end
of a reaction, cooling the reaction mixture favors formation of
the insoluble resin containing ammonium and carboxylate salts.
Using this scheme, Pittman was able to use $RhCl_3/(CH_3)_3N$ as a
catalyst for carbonylation of 1-pentene as shown below (13). Significant amounts of alcohol were also found in the product mixtures.
The authors also noted that isomerization of 1-pentene to 2-
pentene was rapid when trimethylamine was present. Conversions of

$$C_3H_7CH=CH_2 + CO + H_2O \xrightarrow[\substack{150\ °C \\ 600\ psi}]{RhCl_3} C_5H_{11}CHO + C_4H_9CH(CH_3)CHO \tag{13}$$

alkenes of 64-69% were obtained in the first cycle of these reactions in 24 h at 150 °C and conversions were identical with or
without the added polymer. However under these conditions,
recovery of the rhodium by the resin was not complete on cooling.
The conversion to aldehyde in subsequent reactions was 23% and 7%
in the second and third cycles. Consistent activity through 11
cycles was obtained by carrying out the reaction at a pH near 7 in
the absence of $(CH_3)_3N$ but only at the cost of having the reaction
take 10 d to achieve ca. 60% conversion. Hydroformylations using
1:1 H_2/CO were more successful and 0.5% $RhCl_3$, 600 psi CO/H_2 and
150 °C led to a 49 % conversion of 1-pentene to an equal ratio of
hexanal/2-methylpentanal in 3 h. A homogeneous catalyst under
comparable conditions in the absence of the inorganic ion exchange
resin was about 10 times more active. Loadings of catalyst onto
the ion exchange resin in these experiments were relatively low, 5
x 10^{-5} moles of Rh being used with 1 g of the ion exchange
polymer.

Linear Polystyrene Bound Transition Metal Catalysts

Bayer's group was one of the first groups to describe the use of
soluble macromolecular ligands for transition metal catalysts (49).
In their work, they used soluble polystyrenes, poly(ethylene glycol)s, poly(vinylpyrrolidinone)s and poly(vinyl chloride)s. For
example, using linear polystyrene (M_w of ca. 100,000), chloromethylation followed by treatment with potassium diphenylphosphide
could used to prepare a soluble polydiphenyl(styrylmethyl)phos-
phanes containing varying amounts of unreacted Cl and phosphino
groups as indicated below (14).
    Ligand exchange or substitution using various rhodium,
palladium and platinum complexes could then be used to prepare

$$PS-CH_2Cl \ + \ KPPh_2 \ \longrightarrow \ PS-CH_2PPh_2 \qquad (14)$$

8a: 2.59 % Cl, 0.45 % P
8b: <0.05 % Cl, 2.42 % P
8c: 1.90 % P

transition metal complexes containing this macromolecular ligand.
Separation of the soluble macromolecularly ligated homogeneous
catalyst from the low molecular weight reaction products of hydrogenation and hydroformylation reactions was accomplished by membrane filtration using polyamide membranes having a cut-off limit
of MW 10,000 or by precipitation of the diphenylphosphinated
polystyrene ligated catalyst using hexane. Recycling of both the
hydrogenation and hydroformylation catalysts was demonstrated for
six and three times respectively with no loss in activity.
Catalyst/substrate ratios for pentene hydrogenation and hydroformylation were 200/1. Complexes (**8c**)Rh(CO)$_2$Cl, (**8a**)$_2$PdCl$_2$ and
(**8c**)PdCl$_2$ were also active hydrogenation catalysts and
(**8a**)Rh(CO)(acac) was used successfully as a hydrogenation catalyst
as well.

Linear polystyrene has also been used to support asymmetric
hydrogenation catalysts containing chiral diphosphine rhodium(I)
complexes (50). Asymmetric hydrogenations of itaconic acid were
carried out, forming (R)-2-methylbutanedioic acid with e.e.'s
ranging from 20-37%. None of the polymer-bound catalysts were
more effective than (-)-DIOP-RhCl and the observed e.e.'s were
found to be dependent on the molecular weight of the polymer
chain, its microstructure and solubility.

<u>Soluble Inorganic Polymers as Supports for Transition Metal
Catalysts</u>

Ladder polyphenylsilsequioxane and linear polydiphenylsiloxane
polymers have been used to complex Cr(CO)$_6$ to form Cr(CO)$_3$ complexes which proved to be better than similar complexes of Cr(CO)$_3$
with cross-linked polystyrene as stereoselective hydrogenation
catalysts in hydrogenation of methyl sorbate to <u>cis</u>-3-hexenoate.
(51). The catalyst-polymer complexes were isolated and separated
from hydrogenation reaction products by precipitation and filtration. These polymer-catalyst complexes were readily dissolved in
THF. Solvents affected both the recyclability and activity of
these catalysts, the highest catalytic activity being seen in THF.
Cyclohexane proved to be a poor solvent both in terms of the
activity of the catalysts and in terms of the catalyst's recover-
ability. Using THF, the polyphenylsiloxane and poly(phenyl-
disiloxane) complexed Cr(CO)$_3$ catalysts were both recycled with
only modest losses in activity. Reactions were normally run at
160 °C in THF in an autoclave. Control experiments showed that
loss of catalyst activity and diminished Cr content in the
recycled polymers was not due to thermal decomposition of the
polymer-Cr(CO)$_3$ complexes. Awl speculated that the THF acts as a
donor ligand, coordinating to the Cr and gradually destroying the
arene-chromium complex during hydrogenation. After a hydrogenation reaction, the polymer-chromium complex was separated from the
methyl cis-3-hexenoate by precipitation and filtration.

The use of soluble phosphinated organometallic polymers to prepare recoverable recyclable hydroformylation catalysts has been described in several recent publications (52-54). Fellman and Seyferth found that ferrocenylenephenylphosphine polymers and oligomers such as **9** could be prepared from reaction of (1,1'-ferrocenediyl)phenylphosphine with one equivalent of 1-lithio-1'-(diphenylphosphino)ferrocene in THF. The resulting oligomeric or polymeric phosphines are very robust thermally and were stable at temperatures in excess of 350 °C. These authors found that addition of $Co_2(CO)_8$ to these tertiary phosphines resulted in formation of various polydentate phosphine cobalt carbonyl complexes which could be characterized by both ^{31}P and ^1H NMR spectroscopy and by IR spectroscopic studies of the metal carbonyl stretching vibrations. At lower Co/P ratios, a tridentate phosphine chelated cobalt complex was thought to have formed. At higher Co/P ratios complexation was thought to occur mainly in a bidentate fashion. These complexes were found to behave as homogeneous hydroformylation catalysts at 170-190 °C. When THF was used as a solvent, the polymers were readily soluble, and there was a significant change in the catalyst's selectivity with different polymers. In particular, there was a steady change in the relative amounts of alcohol and aldehyde product in hydroformylation of 1-hexene. More aldehyde product formed when the organometallic polymeric ligands were higher in molecular weight. Although only limited data were presented, there was a linear relationship between the yield of alcohol product and the log(oligomer molecular weight). The authors surmised that the higher molecular weight phosphinated polymeric ligands led to diminished amounts of species such as $HCo(CO)_3L$ from equilibria such as 15 and consequent slower aldehyde to alcohol reduction.

Garrou and Allcock have also shown that diphenylphosphinated (aryloxy)phosphazenes (e.g. **10**) can also be used to prepare polymer-bound hydroformylation catalysts (53). However, they also noted that attempts to recycle these catalysts consistently led to loss of both cobalt and phosphine functionality, apparently by reductive scission of the polymeric aryl-phosphorus bond under the reaction conditions. They suggest that this result may be indicative of a more general problem than realized with homogeneous catalysts immobilized to polymers in that immobilization strategies have typically not taken into account possible ligand degradation reactions which might occur (a review by P. E. Garrou in this volume describes this problem in greater detail).

Diphenylphosphinated poly((aryloxy)phosphazenes) such as **10** have been shown by Allcock to be generally useful as polymeric phosphine ligands for complexing transition metals (54). Osmium complexes of **10** were found to catalyze the isomerization of 1-hexene to 2-hexene. The products in this case were separated from the catalyst by vacuum transfer. Catalytic activity ceased after [HOs_3(1-hexene)(CO)$_{10}$(phosphine)] formed.

Hydroformylation catalysts based on soluble organic polymers have also been described in the patent literature (55). For example, Trevillyan has reported using a polyphenylene-bound diphenylphosphine ligand to form a tertiary phosphine cobalt carbonyl complex useful in hydroformylation of 1-octene to mainly nonanal and 2-methyloctanal.

9

(15)

10

Polyethylene-Bound Catalysts

Recently, our group has found that functionalized ethylene oligomers can be used as ligands to prepare recoverable, reusable
homogeneous catalysts from both transition metals like rhodium and
nickel and from lanthanide salts. In complimentary studies of
polyethylene functionalization, we have found that ethylene oligomers of $M_v > 1200$ are quantitatively entrapped in polyethylene
precipitates when a solution of polyethylene and a functionalized
ethylene oligomer is cooled to room temperature or when
polyethylene and the functionalized oligomer are co-precipitated
by addition of a second solvent such as methanol (56). Suitably
functionalized ethylene oligomers are obtained by reactions such
as Equation 6 above. For example, in our initial studies we
examined the use of carboxylated ethylene oligomers as ligands for
neodymium diene polymerization catalysts (57). These carboxylated
ethylene oligomers were readily obtained by anionic oligomerization of ethylene using n-butyllithium-TMEDA in heptane followed by
quenching of the living oligomer by carbon dioxide. When a neodymium salt of a carboxylated ethylene oligomer was dissolved in
toluene at 100 °C and allowed to react with triethylaluminum and
diethylaluminum chloride and butadiene, the catalyst formed in
situ polymerized butadiene stereospecifically to form high molecular weight trans-1,4-polybutadiene (57). Dilution of the viscous
product mixture with toluene allowed us to then readily filter the
solution and recover the neodymium salt as a polyethylene dispersion. We have now extended this work to include studies in which
Rh(I) and Ni(0) catalysts are prepared as polyethylene dispersions
and used at 100 °C in reactions such as hydrogenation and diene
cyclooligomerization (13).

Soluble Polyethers as Phase Transfer Catalysts

Phase transfer catalysis generally and phase transfer catalysis
using polymeric reagents is a well established application of
polymer-bound reagents and catalysts (58). While most attention
has been focused on insoluble cross-linked polymer-bound phase
transfer catalysts, soluble macromolecules which are phase
transfer catalysts themselves or solutions of phase transfer
catalysts attached to macromolecules have been studied by several
groups. For example, Smid used styrene derivatives of benzo-18-
crown-6 and benzo-15-crown-5 as monomers to prepare linear polymeric versions of both 18-crown-6 and 15-crown-5 (15). Using
alkali metal salts of picric acid, they then studied both the
selectivity and binding of these polymeric crown ethers and con-
trasted them to their low molecular weight homogeneous analogs.
Generally these polymeric crown ethers were less selective than
their low molecular weight analogs because of facile formation of
2:1 complexes between crown ether groups on adjacent monomer
units, although their ability to bind alkali metal salts was
comparable to or better than that of their non-polymeric analogs.
    Poly(ethylene oxide) is a linear version of a cyclic crown
ether. Such poly(ethylene oxide) species have been shown to serve

as catalysts in several reactions of polar reagents with non-polar substrates. The oxidation of trans-stilbene with potassium permanganate, the alkylation of potassium acetate and diethylbenzyl sodiomalonate by alkyl halides and ether syntheses from sodium phenoxide and alkyl bromides all reportedly can be catalyzed by addition of poly(ethylene oxide) (59). Poly(ethylene oxide), like 18-crown-6, dissolves potassium permanganate in benzene to give a purple benzene solution. However, unlike the reaction of olefins with 18-crown-6 solubilized potassium permanganate, the poly-(ethylene oxide) promoted oxidations did not go to completion, the highest yield of benzoic acid from stilbene being 33% in 17 h at 25 °C. Poly(ethylene oxide) was much more effective at promoting nucleophilic substitution reactions. Poly(ethylene oxide) was about as efficacious as 18-crown-6 in esterification of n-butyl bromide or n-octyl bromide. However, when poly(ethylene oxide) was compared to 18-crown-6 in alkylations of diethylbenzylmalonate anions, 18-crown-6 was an order of magnitude better at promoting nucleophilic substitution by the carbanion.

Both poly(vinylpyrrolidone) and poly(ethylene oxide) accelerate nucleophilic substitution of alkyl halides by phenoxide anions (60). These Williamson ether syntheses of phenyl alkyl ethers were up to 100 times faster when the ether syntheses were carried out in the presence of poly(vinylpyrrolidone) versus when they were performed using an equivalent amount of N-methylpyrrolidone. Poly(ethylene oxide) was as effective in these reactions as was 18-crown-6. Procedures for separating the products from the polymeric additives were not described.

The ratio of O-alkylation/(O + C alkylation) in alkylation of phenoxide anions by allylic halides can be controlled by the presence of poly(vinylmonobenzo-18-crown-6) or benzo-18-crown-6. (61). The results of these studies shown that the percentage of O-alkylated product was consistently higher when a soluble polymeric crown ether was employed.

Poly(ethylene glycol)s of molecular weight 300-600 have also been shown to be effective as soluble phase transfer catalysts in exchange of fluoride with reactive halogens in organic compounds (62). While these lower molecular weight oligomers of poly-(ethylene glycol) were effective as soluble catalysts, high molecular weight forms (e.g. PEG 4000 or PEG 6000) of poly(ethylene glycol) proved to be unsuitable. Both carboxylic acid chlorides and sulfonic acid chlorides were converted into their corresponding acid fluorides in good yield within 3 h at 20 °C in acetonitrile. Substitution reactions of the halogens in benzyl bromide or 2,4-dinitrochlorobenzene by fluoride required reflux and up to 24 h reaction times. The poly(ethylene glycol) catalysts used in these reactions were not recycled.

Summary

While linear, soluble polymers have seen scattered use in many reactions and have achieved some recognition in certain instances such as peptide synthesis, the use of soluble polymer-bound reagents and catalysts is a promising area which merits further study. While a homogeneous reagent such as a soluble macromolecular reagent cannot always be used in a continuous process,

the ease in characterization and reaction control that is possible
with a soluble polymeric reagent or catalyst makes them attractive
alternatives to their insoluble polymer-bound analogs. Since
there has been comparatively little study of these reagents, it is
reasonable to suppose that future work will proceed to both increase their utility and applicability as well as to further
develop their unique properties.

Acknowledgments

We acknowledge the support of our studies in the area of soluble
polymeric organometallic reagents and catalysts by the Petroleum
Research Fund of the American Chemical Society, the Robert A.
Welch Foundation and by the Texas A&M Center for Energy and
Mineral Resources.

Literature Cited

1. Hodge, P.; Sherrington, D. C., Eds. "Polymer-Supported
   Reactions in Organic Synthesis"; John Wiley & Sons: New York,
   New York, 1980.
2. Akelah, A.; Sherrington, D. C. Polymer **1983**, 24, 1369-1386.
3. Birr, C. "Aspects of the Merrifield Peptide Synthesis",
   Springer-Verlag: Berlin, 1978.
4. Pittman, C. U., Jr. In "Comprehensive Organometallic
   Chemistry"; Wilkinson, G., Ed.; Pergamon Press: Oxford, 1982;
   Vol. 8, pp 553-611. Bailey, D. C.; Langer, S. H. Chem. Rev.
   1981, 81, 109-148.
5. Holy, N. L. in "Homogeneous Catalysis with Metal Phosphine
   Complexes", Pignolet, L. H., Ed.; Plenum: New York, New York,
   1984; pp 443-484.
6. Regen, S. L.; Nigam, A. J. Am. Chem. Soc. **1978**, 100, 7773-
   7775. Tomoi, M.; Ford, W. T. J. Am. Chem. Soc. **1981**, 103,
   3828-3832.
7. Bergbreiter, D. E.; Blanton, J. R., submitted for publication.
8. Morawetz, H. "Macromolecules in Solution", John Wiley & Sons:
   New York, 1975.
9. Grubbs, R. H. CHEMTECH **1977**, 7, 512-518.
10. Fyfe, C. A. "Solid State NMR for Chemists", C.F.C. Press:
    Guelph, Ontario, Canada; Chapter 7 and references therein.
11. Ford, W. T.; Periyasamy, M.; Spivey, H. O. Macromolecules
    1984, 17, 2881-2886 and references therein.
12. Naaktegboren, A. J.; Nolte, J. M. R.; Drenth, W. J. Am. Chem.
    Soc. **1980**, 102, 3350-3354.
13. Bergbreiter, D. E.; Chandran, R. J. Chem. Soc., Chem. Commun.
    in press. Bergbreiter, D. E.; Chandran, R. J. Am. Chem. Soc.
    1985, 107, 4792-4793.
14. Pillai, V. N. R.; Mutter, M. Topics in Current Chem. **1982**,
    106, 119-175.
15. Kopolow, S.; Hogen Esch, T. E.; Smid, J. Macromolecules **1973**,
    6, 133-142.
16. For a general review of anionic oligomerization and polymeri-
    zation reactions cf. Young, R. N.; Quirk, R. P.; Fetters, L. J.
    Adv. Polym. Sci. **1984**, 56, 1-90.

17. Relles, H. M.; Schluenz, R. W. J. Am. Chem. Soc. **1974**, 96, 6469–6475.
18. Schuttenberg, H.; Klump, G.; Kaczmar, U.; Turner, S. R.; Schulz, R. C. J. Macromol. Sci., Chem. **1973**, A7, 1085–1095.
19. Bergbreiter, D. E. in "Metal-Containing Polymeric Systems", Sheats, J. E.; Carraher, C. E., Jr.; Pittman, C. U., Jr., Eds.; Plenum: New York, 1985, pp. 405–424.
20. Bergbreiter, D. E.; Blanton, J. R. J. Chem. Soc., Chem. Commun. **1985**, 337–338.
21. Harrison, C. R.; Hodge, P.; Hunt, B. J.; Khoshdel, E.; Richardson, G. J. Org. Chem. **1983**, 48, 3721–3728.
22. McKinley, S. V.; Rakshys, J. W., Jr. J. Chem. Soc., Chem. Commun. **1972**, 134–135.
23. Castells, J.; Font, J.; Virgili, A. J. Chem. Soc., Perkin Trans. I **1979**, 1–6.
24. Bernard, M.; Ford, W. T. J. Org. Chem. **1983**, 48, 326–332.
25. Bernard, M.; Ford, W. T.; Nelson, E. C. J. Org. Chem. **1983**, 48, 3164–3168.
26. Clarke, S. D.; Harrison, C. R.; Hodge, P. Tetrahedron Lett. **1980**, 21, 1375–1378.
27. Hodge, P.; Hunt, B. J.; Khoshdel, E.; Waterhouse, J. Nouv. J. Chim. **1982**, 6, 617–622.
28. Heitz, W.; Michels, R. Justus Liebig's Ann. Chem. **1973**, 227–230.
29. Nishikawa, H.; Tsuchida, E. J. Phys. Chem. **1975**, 79, 2072–2076.
30. Challa, G. Macromol. Chem. Phys., Suppl. **1981**, 5, 70–81.
31. Hallensleben, M. L. J. Polym. Sci., Polymer Symposia **1974**, 47, 1–9.
32. Bergbreiter, D. E.; Blanton, J. R.; Chen, B. J. Org. Chem. **1983**, 48, 5366–5368 and references therein.
33. Avny, Y.; Marom, G.; Zilkha, A. Eur. Polym. J. **1971**, 7, 1037–1046.
34. Smith, T. W.; Cheatham, R. A. Macromolecules **1980**, 13, 1203–1207.
35. Mazur, S.; Jayalekshmy, P. J. Am. Chem. Soc. **1979**, 101, 677–683.
36. Schutten, J. H.; Piet, P.; German, A. L. Makromol. Chem. **1979**, 180, 2341–2350.
37. Collman, J. P.; Kosydar, K. M.; Bressan, M.; Lamanna, W.; Garrett, T. J. Am. Chem. Soc. **1984**, 106, 2569–2579.
38. Challa, G. J. Mol. Catal. **1983**, 21, 1–16 and references therein.
39. Sherrington, D. C. Nouv. J. Chim. **1982**, 6, 661–672.
40. Morawetz, H. in "Peptides – Chemistry, Structure and Biology", Walter, R; Meienhofer, J., Eds.; Ann Arbor Sci. Publ.: Ann Arbor, Michigan, 1975.
41. Pillai, V. N. R.; Mutter, M. Acc. Chem. Res. **1981**, 14, 122–130.
42. Bayer, E.; Mutter, M.; Uhmann, R.; Polster, J.; Mauser, H. J. Am. Chem. Soc. **1974**, 96, 7333–7336.
43. Pillai, V. N. R.; Mutter, M.; Bayer, E.; Gatfield, I. J. Org. Chem. **1980**, 45, 5364–5370.
44. Whitesides, G. M.; Wong, C. H. Aldrichimica Acta **1983**, 16, 27–34.

45. Wong, C.-H.; Whitesides, G. M. Angew Chem., Int. Engl. Ed., **1985**, in press.
46. Wilson, M. E.; Whitesides, G. M. J. Am. Chem. Soc. **1978**, 100, 306–307.
47. Nuzzo, R. G.; Haynie, S. L.; Wilson, M. E.; Whitesides, G. M. J. Org. Chem. **1981**, 46, 2861–2867.
48. Kawabata, Y.; Pittman, C. U., Jr. J. Mol. Catal. **1981**, 12, 113–119.
49. Bayer, E.; Schurig, V. Angew. Chem., Int. Ed. Engl. **1975**, 14, 493–494.
50. Ohkubo, K.; Fujimori, K.; Yoshinaga, K. Inorg. Nucl. Chem. Lett. **1979**, 15, 231–234.
51. Awl, R. A.; Frankel, E. N.; Friedrich, J. P.; Swanson, C. L. J. Polym. Sci., Polym. Chem. Ed. **1980**, 18, 2663–2676.
52. Fellman, J. D.; Garrou, P. E.; Withers, H. P.; Seyferth, D.; Traficante, D. D. Organometallics **1983**, 2, 818–825.
53. Dubois, R. A.; Garrou, P. E.; Lavin, K. D.; Allcock, H. R. Organometallics **1984**, 3, 649–650.
54. Allcock, H. R.; Lavin, K. D.; Tollefson, N. M.; Evans, T. L. Organometallics **1983**, 2, 267–275.
55. Trevillyan, A. E. U.S. Patent 4 045 493, 1977; Chem. Abstr. **1977**, 87, 183995s.
56. Bergbreiter, D. E.; Chen, Z.; Hu, H. -P. Macromolecules **1984**, 17, 2111–2116.
57. Bergbreiter, D. E.; Chen, L. -B.; Chandran, R. Macromolecules **1985**, 18, 1055–1057.
58. Ford, W. T.; Tomoi, M. Adv. Polym. Sci. **1984**, 55, 49–104.
59. Hirao, A.; Nakahama, S.; Takahashi, M.; Yamazaki, N. Makromol. Chem. **1978**, 179, 915–925.
60. Yamazaki, N.; Hirao, A. Nakahama, S. J. Polym. Sci., Polym. Chem. Ed. **1976**, 14, 1229.
61. Akabori, S.; Miyamoto, S.; Tanabe, H. J. Polym. Sci., Polym. Lett. Ed. **1978**, 16, 533–537.
62. Kitazume, T.; Ishikawa, N. Chem. Lett. **1978**, 283–284.

RECEIVED September 27, 1985

3

Catalysis with a Perfluorinated Ion-Exchange Polymer

F. J. Waller

Central Research & Development Department, Experimental Station, E. I. du Pont de Nemours & Company, Wilmington, DE 19898

NAFION is a perfluorinated ion-exchange polymer (PFIEP). Because of its chemically inert backbone, the heterogeneous resin has been used as a strong acid catalyst for a wide variety of reactions in synthetic organic chemistry. The polymeric catalyst offers an advantage over homogeneous analogs because of its ease of separation from reaction mixtures. Also in many instances, there are improved yields or an increase in selectivity over other catalysts, such as poly(styrenesulfonic acid) resins.

The resin, NAFION (1a), has been used as a heterogeneous strong acid by many researchers but most extensively by G. A. Olah. It is the objective of this review to summarize the current literature with respect to the synthetic applications of PFIEP. In addition, other uses for NAFION are mentioned briefly. The references cited are representative of the publications from both scientific journals and patents. This review will deal primarily with catalysis derived from NAFION in the acid form.

NAFION has the general structure shown in Figure 1.

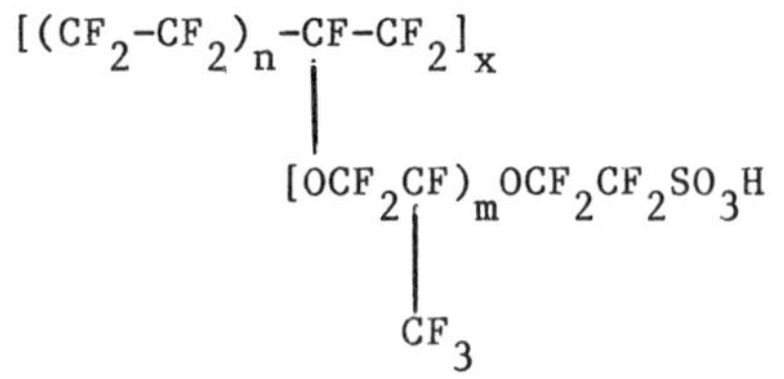

$$[(CF_2-CF_2)_n-CF-CF_2]_x$$
$$|$$
$$[OCF_2CF)_mOCF_2CF_2SO_3H$$
$$|$$
$$CF_3$$

m = 1, 2, 3,---

Figure 1. General structure of NAFION.

A series of compositions may be produced in which n can be as low as 5 and as high as 13.5. The lower value of n corresponds to an

0097-6156/86/0308-0042$07.50/0

equivalent weight of 950 and the higher value to 1800. The value
of x is about 1000. NAFION is prepared from a copolymer of
tetrafluoroethene and perfluoro[2-(fluorosulfonylethoxy)-propyl
vinyl ether (1b). The perfluorinated vinyl ether is produced by
reacting tetrafluoroethene with SO_3 to form a cyclic sultone
which subsequently rearranges to fluorocarbonylmethanesulfonyl
fluoride. The linear analog reacts with two moles of hexa-
fluoropropylene oxide to yield a compound with a terminal
1-fluorocarbonyltrifluoroethoxy group. This group loses carbonyl
fluoride on heating with Na_2CO_3 to give the perfluorinated vinyl
ether (1c). The copolymer resin in the sulfonyl fluoride form is
base hydrolyzed to the alkali form and then acidified to the
sulfonic acid form.

Polymeric perfluorinated sulfonic acids offer a variety of
advantages over their homogeneous analogs, for example
trifluoromethanesulfonic acid, foremost of those being catalyst
separation. In many instances, there are also improved yields or
an increase in selectivity. The greatest challenge is finding
applications for PFIEP where the catalysis is unique. This is an
important consideration in view of catalyst cost. If reactivity
or selectivity are only marginally better than less expensive
catalysts, the incentive to use NAFION is lost.

NAFION in the powder form can presently be purchased from two
sources: Aldrich Chemical Company and C. S. Processing Company
(2). Though literature references report polymeric perfluorinated sulfonic acids with equivalent weights of 900, 1000,
or 1200, the above two vendors sell NAFION only in the acid form
with an equivalent weight of 1100. In spite of the low number of
mequiv. per gram and incomplete accessibility of acid sites for
reaction, the utility of NAFION as a strong acid catalyst is
remarkable. It has been estimated that approximately 50% of the
acid sites are not accessible for reactions involving molecules
larger than NH_3 in non-swelling solvents (3). Therefore, coating
of PFIEP on a solid support to increase surface area or other
methods to increase porosity of the resin should enhance
activity.

Perfluorinated ion exchange polymers in the potassium or acid
form should be exchanged with nitric acid, 3N, four or five times
at approximately 75°C and then dried at ~110°C for 3-4 hours
before use. This will insure that the PFIEP is completely
converted to the acid form, free of any organics or metal ion
contamination and nearly anhydrous. Exposure to moisture will
result in a less effective catalyst if maximum Bronsted acidity
is required for a particular application.

The perfluorinated backbone of NAFION provides a resin with
chemical and thermal stability similar to that of TEFLON
fluorocarbon. Unlike TEFLON, the polymer is permeable to many
cations and polar compounds but impermeable to anions and
non-polar species. Cation size and ionic properties determine
mobility through the polymer. The polymeric fluorinated backbone
makes PFIEP highly resistant to attack from strong oxidizing and
reducing acids, strong bases and oxidizing/reducing agents. The
fluorinated resins have a greater thermal stability than their
non-fluorinated analogs and can be used at temperatures
approaching 175°C. Table I summarizes a few selected

Table I. Properties of NAFION and AMBERLYST 15

Property	NAFION	AMBERLYST 15
meq/g	0.9[*](EW* 1100)	4.7 (dry)
Maximum operating temperature	180–190°C	~150°C
Backbone Structure	Fluorocarbon Non–porous	Polystyrene Macroreticular
Surface area $(m^2 g^{-1})$	~0.8	50

* EW ≡ equivalent weight

characteristics of NAFION and AMBERLYST 15 (4), a poly(styren-esulfonic acid) resin available from Rohm and Haas. Between 180–190°C, NAFION powder in the acid form has a tendency to flow and thus fuse together. At elevated temperatures, 210–220°C, the polymer loses sulfonic acid groups (5).

PFIEP has hydrophilic [SO_3H] and hydrophobic [CF_2CF_2] regions in close proximity. These polymers have the ability to sorb relatively large quantities of water and other protic solvents despite the predominance of the hydrophobic regions. This solvation causes swelling of the solid. This morphology of the perfluorinated membrane suggests a framework where the hydrated $-SO_3^-$ groups and counterion clusters which are roughly 40A in diameter are interconnected by short channels approximately 10A in diameter. This porous ionic system is immersed in a fluoro-carbon backbone network (6). The analogy between this biphasic structure and the structure of reversed micelles has been substantiated by ^{23}Na magnetic resonance studies. This configuration minimizes both the hydrophobic interaction of water with the backbone and the electrostatic repulsion of proximate sulfonate groups (7).

The acidity, H_o, of a dried, methylene chloride swollen, per-fluorinated sulfonic acid side chain is about −6 (8). It has been suggested that NAFION has an acidity between −11 and −13 (9). The difference between the measured and a reasonable value of −11 is that the methylene chloride-swollen resin may not have been completely anhydrous. For comparison, triflic acid and AMBERLYST 15 have H_o values of −14.6 and −2.16, respectively (9–10).

However, it should be recognized that there are several additional classes of catalysts associated with metal cation ion-exchanged polymers. These include 1) a partially cation-exchanged polymer, 2) a completely cation-exchanged polymer, and 3) a cation-exchanged polymer where the metal cation is coordinated to another ligand. The general catalyst classes are shown in Figure 2. A discussion of this topic will appear elsewhere (11).

Resin [SO$_3$H]
Resin [SO$_3$H, (SO$_3$)$_z$M^z]
Resin [(SO$_3$)$_z$M^z]
Resin [(SO$_3$)$_z$M^z(ligand)$_x$]

Figure 2.   Classes of metal cation ion-exchanged resins

Synthetic organic applications.

Acylation.   Acylation of benzene and substituted benzenes
(equation 1), are conveniently carried out by refluxing a

$$XC_6H_4COCl + CH_3C_6H_5 \xrightarrow{\Delta} XC_6H_4COC_6H_4CH_3 + HCl \quad (1)$$

mixture of the benzoyl chloride, arene, and NAFION. The benzo-
phenones are isolated by filtering the hot reaction mixture to
remove the resin and followed by removing the excess arene by
distillation. Table II summarizes several examples of the
acylation of toluene with substituted benzoyl chloride (12).

Table II.   Acylation of toluene

X	Yield (%)	Isomer Distribution (o:m:p)
H	81	22.4:3.1:74.5
4-CH$_3$	83	28.7:3.1:68.2
2-F	87	16.7:2.9:80.4
4-F	87	21.5:3.4:75.1
3-Cl	82	22.6:1.1:76.3

Acetic anhydride with acetic acid has also been used to
acetylate reactive alkylbenzenes like mesitylene, equation 2.
In this case, the yield of the product was 72%.

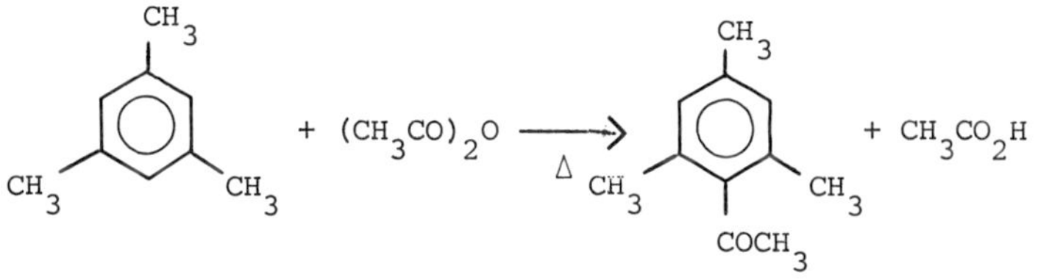

Benzoylation of p-xylene (equation 3) at 135°C for

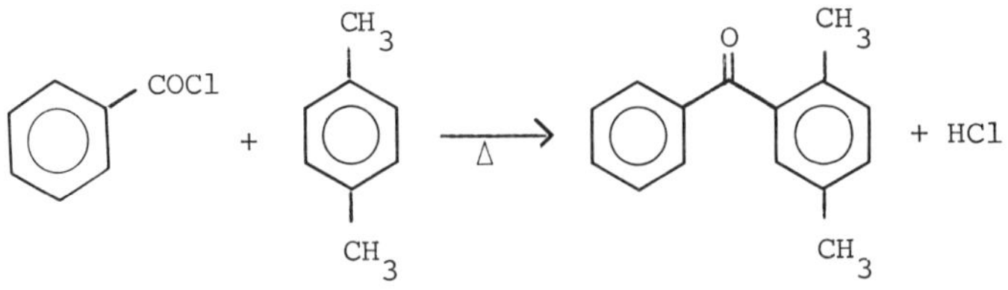

6 hours provided 2,5-dimethylbenzophenone in 85% yield (13). The
NAFION catalyst, removed by filtration, was recycled without
noticeable loss in activity. However below 100°C, the catalyst
showed less activity. An increase in the equivalent weight of
NAFION from 1100 to 1790 resulted in a marked decrease in the
rate of reaction. In 2 days, only a 16% yield of 2,5-dimethyl-
benzophenone was obtained.

The reaction of thiophene with acyclic acid anhydrides in the
presence of NAFION affords 2-acylthiophenes in moderate to good
yields (equation 4) (14). In particular, when the ratio

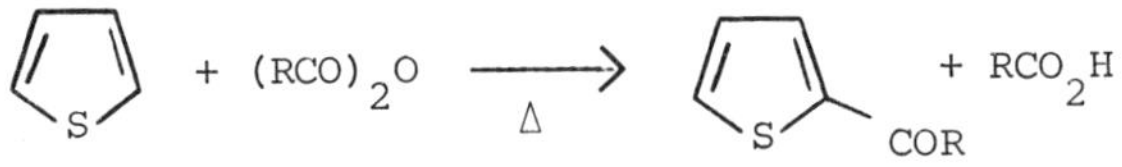

of acid anhydride to catalyst is 125 and the reaction is run in
refluxing CH_2Cl_2, ketones were obtained as shown in Table III.
Under the same set of reaction conditions, AMBERLYST 15 gave

Table III.   Acylation of thiophene

R	Yield (%)
Me	76
Et	81
n-Pr	94
n-Pr	75*

* AMBERLYST 15

lower yields. It was noted that deactivation of NAFION occurred
with repeated use. The loss of the catalytic activity seems to
be ascribable to the adsorption of polymeric materials on the
catalyst.

Rearrangements. Fries rearrangement. The Fries rearrangement of
phenol esters to hydroxyphenyl ketones, shown in equation 5,
proceeds in dry refluxing nitrobenzene with yields between 63-75%
depending upon the nature of R and R'.

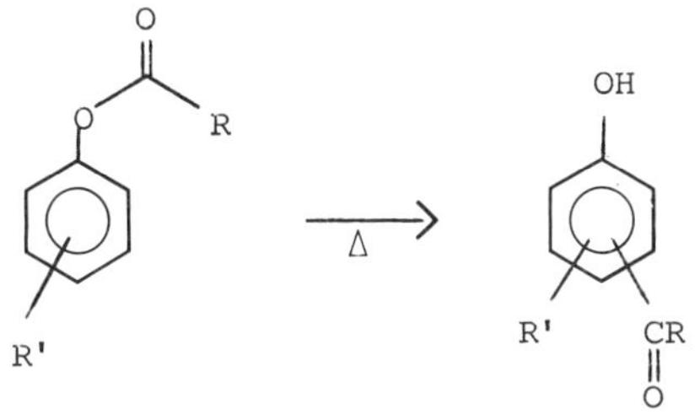

$$(5)$$

In a typical experiment, phenyl benzoate was converted to hydroxyphenyl ketone, o/p::1/2 in 73% yield (15).

Pinacol rearrangement. The second rearrangement involves the preparation of ketones from 1,2-diols (equation 6). Excellent

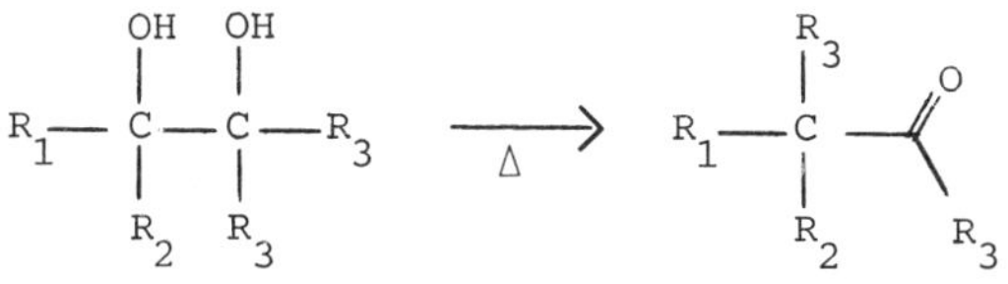

$$(6)$$

yields, 82-92%, have been reported using NAFION as the acid catalyst (16). Examples of this transformation include tetramethylethylene glycol to pinacolone, tetraphenylethylene glycol to triphenylacetophenone and dicyclohexyl-1,1'-diol to spiro[5,6]dodecane-2-one.

Rupe reaction. Rupe reaction converts alkynyl tertiary alcohols to α,β-unsaturated carbonyl compounds (17). Table IV summarizes the utility of NAFION. Yields are based upon isolated product and vary from 60 to 88%. This method suppresses the normal by-product formation from the polymerization of α,β-unsaturated compounds, giving considerable synthetic advantage.

Unsaturated alcohols to aldehydes. The last example of a NAFION-catalyzed rearrangement is shown in equation 7. Gaseous allyl alcohols, when passed over the resin at 170-190°C

$$RCH=CHCH_2OH \xrightarrow[\Delta]{} RCH_2CH_2C \underset{H}{\overset{O}{\diagup}}$$ (7)

rearrange to aldehydes (18). Table V presents examples of three isomerizations at 195°C.

Table IV. Rupe rearrangement

α-Ethynyl alcohol	α,β-unsaturated carbonyl compound	Yield (%)
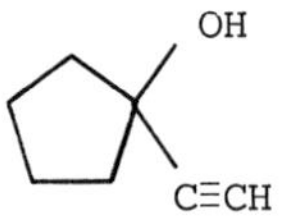	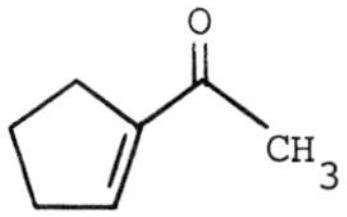	60
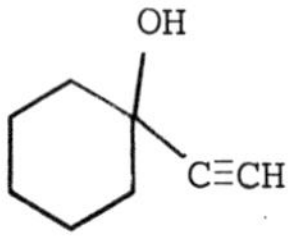	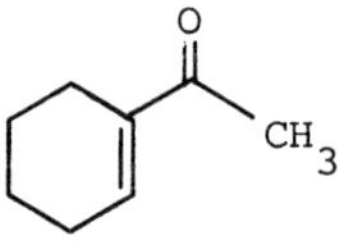	84
$CH_3CH_2 - \overset{\overset{\displaystyle CH_3}{|}}{\underset{\underset{\displaystyle OH}{|}}{C}} - C{\equiv}CH$	$CH_3CH = \overset{\overset{\displaystyle CH_3}{|}}{C} - \overset{\overset{\displaystyle O}{\|}}{C}CH_3$	83

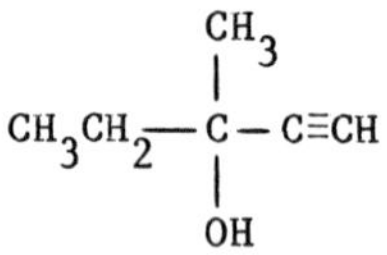

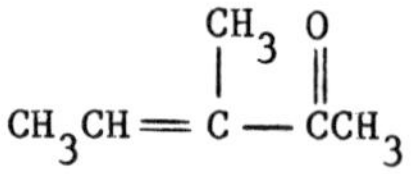

Table V. Isomerization of allyl alcohols

Alcohol	Contact Time(sec)	Yield(%)
allyl	8	60
2-methylallyl	3	88
crotyl	8	55

Ether synthesis. Cyclic ethers. As shown in equation 8, cyclic ethers are conveniently prepared by the dehydration of 1,4- and 1,5-diols at 135°C. Excellent yields ($\geq$86%) are obtained and no solvent is required (19).

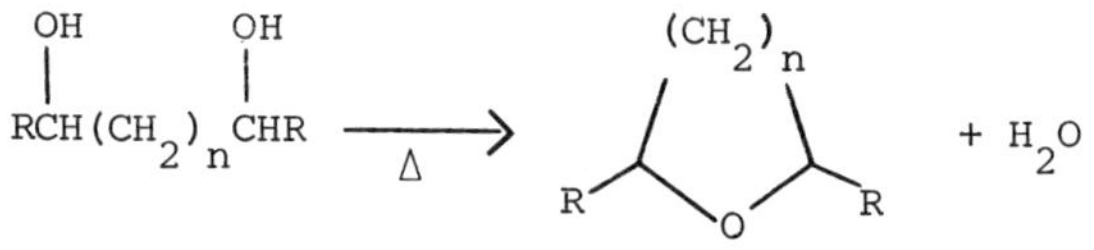

$$
\begin{array}{cc} OH & OH \\ | & | \\ RCH(CH_2)_nCHR \end{array} \xrightarrow[\Delta]{} \quad \underset{R}{\overset{(CH_2)_n}{\diagup}}\!\!\underset{O}{\diagdown}\!\!\underset{R}{\diagup} \quad + H_2O \qquad (8)
$$

1,4-Butanediol, 1,5-pentanediol, and ethylene glycol are dehydrated to give tetrahydrofuran, tetrahydropyran and 1,4-dioxane, respectively. The filtered NAFION is treated with acetone, deionized water, and dried at 150°C to give a catalyst with original activity.

Polymeric ethers. Poly(tetramethylene ether)glycol, PTMEG, can be prepared by the polymerization of THF with NAFION. Conversion of 56% at 25°C is possible after 65 hours (20). The product, PTMEG, has a molecular weight (number average) of about 1000.
   Analogous chemistry can be applied to prepare ester end-capped copolyether glycol by copolymerizing THF and propylene oxide with acetic anhydride in the presence of NAFION. The acetate end-capped copolyether glycol contains 8 mol % of propylene oxide units (21). A non-end-capped copolymer can also be prepared under similar conditions but without acetic anhydride (22).

Acetals and Ketals. There have been several reports of protecting aldehydes and ketones via acetals and ketals prepared from trimethyl orthoformate (equation 9) 1,2-ethanedithiol and 1,3-butanediol, or 1,3-propanediol.

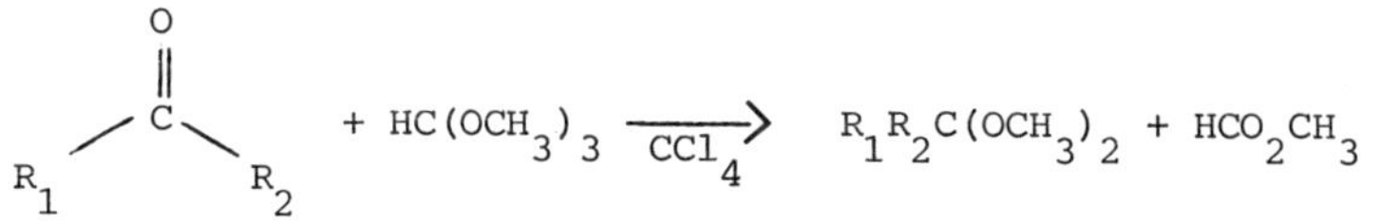

$$
\underset{R_1}{\overset{O}{\underset{\diagup}{\|}}}\!\!\overset{C}{\underset{R_2}{\diagdown}} \quad + HC(OCH_3)_3 \xrightarrow[CCl_4]{} \quad R_1R_2C(OCH_3)_2 + HCO_2CH_3 \qquad (9)
$$

Excellent yields are obtained for dimethyl acetals and ketals,
Table VI, or ethylene dithioketals (Table VII) (23). The

Table VI. Dimethyl acetal and ketal formation

R_1	R_2	Yield (%)
$-(CH_2)_5-$		98
CH_3	$n-C_5H_{11}$	83
CH_3	Ph	93
H	Ph	87

Table VII. Ethylene dithioketal formation

R_1	R_2	Yield (%)
$-(CH_2)_5-$		91
$-(CH_2)_6-$		100
$PhCH_2$	$PhCH_2$	100
Ph	CH_3	96
Ph	Ph	100

ketones and aldehydes are readily deprotected by hydrolysis which
also occurs very readily with NAFION as a catalyst. The
ethylenedithioketals are obtained by refluxing a solution of the
carbonyl compound with 1,2-ethanedithiol in benzene with
azeotropic removal of water.

2-Vinyl-1,3-dioxane has been reported to be prepared from
acrolein, 1,3-propanediol and NAFION at a rate of 16 M per hour
at 82% acrolein conversion (24).

Aldehyde diacetates. 1,1-Diacetates of aldehydes can be prepared
by using soluble protic acids. With a catalytic amount of
NAFION, the same diacetates can be obtained by vigorously
stirring equivalent amounts of the aldehyde and freshly distilled
acetic anhydride (25). The general reaction in equation 10 is
exemplified by the summary in Table VIII. The yields are good
with aromatic aldehydes and alkanals.

$$RCHO + (CH_3CO)_2O \xrightarrow[RT]{} RCH(O_2CCH_3)_2 \qquad (10)$$

Acyclic ethers. Aryl ethers can be prepared in modest yield by
reacting the phenol, methanol, and NAFION at 120-150°C for 5
hours (26). Table IX summarizes several examples. Substituting

Table VIII. Aldehyde diacetate formation

R	Reaction Time(h)	Yield(%)
Ph	1	99
p-FC$_6$H$_4$	4	73
CH$_3$	3	90
Cl$_3$C	4	67

Table IX. Aryl ether formation

Phenol	Temp(°C)	Product	Yield(%)
phenol	120	anisole	14.9
pyrocatechol	150	guaiacol	18.5
β-naphthol	125	β-naphthol monoether	13.7

AMBERLYST 15 under similar conditions produced only a 1% yield of guaiacol from pyrocatechol. Hydroquinone is converted to mono-t-butylhydroquinone and di-t-butylhydroquinone as shown in equation 11. During a two-hour reaction at 75°C, the

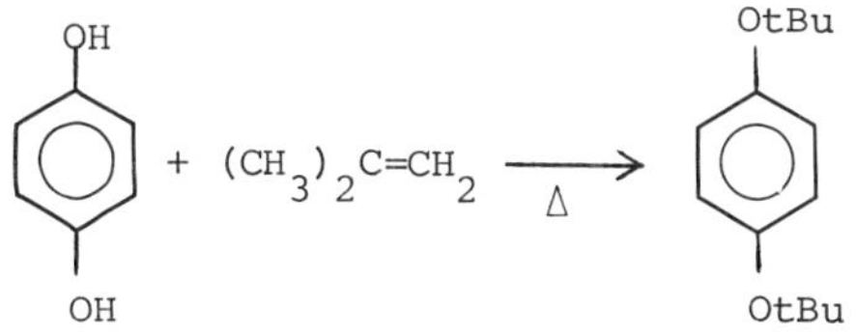

$$
\text{(hydroquinone)} + (CH_3)_2C=CH_2 \xrightarrow{\Delta} \text{(di-t-butylhydroquinone)} \qquad (11)
$$

hydroquinone and isobutylene conversion was 27 and 66%, respectively (27).

The reaction pathway changes if phenol and methanol are passed over the polymeric catalyst in the gas phase at 205°C. Not only anisole and methyl anisoles but also cresols and xylenols are formed. In a typical experiment, a product composition, in mol%, is 37.3% unreacted phenol, 37.2% anisole, 9.7% methyl anisoles, 1% dimethyl anisoles, 10.4% cresols and 4.4% xylenols. The C-methylated products were shown to occur via fast initial O-methylation of the phenol followed by an inter-molecular rearrangement of the aryl methyl ethers to methyl phenols (28).

NAFION also catalyzes the formation of diisopropyl ether when isopropyl alcohol is passed over the catalyst at 110°C. The selectivity to the ether is 92% at a 26% conversion of isopropyl alcohol (29).

The perfluorinated ion exchange resin also acts as a catalyst in the reaction between isobutylene and methanol at 80°C to yield methyl t-butyl ether. The yield based on methanol is 81.8% (30). The recovered catalyst had no decrease in the number of acid sites. However, the same reaction with AMBERLYST 15 gave a similar yield, 83.1%, but the mequiv. per gram dry catalyst decreased from 5.10 to 4.83, suggesting possible alkylation of the benzene rings in the polystyrene resin. AMBERLYST 15 is used commercially as a catalyst to manufacture methyl t-butyl ether.

Methoxymethyl ethers are prepared by exchanging alcohols with excess dimethoxymethane (equation 12) in the presence of NAFION.

$$ROH + (CH_3O)_2CH_2 \xrightarrow[\Delta]{} ROCH_2OCH_3 + CH_3OH \qquad (12)$$

The reaction has a convenient rate at a reflux temperature of 41°C (31). Table X summarizes several examples of this reaction. This method requires only filtration to remove the resin with no aqueous workup.

Table X.   Methoxymethyl ethers from dimethoxymethane

R	Reaction Time(h)	Yield(%)
$C_6H_{11}CH_2$	10	90
$C_7H_{13}CH_2$	10	90
$n-C_7H_{15}$	10	96
C_6H_{11}	16	65

Epoxide opening.   The hydration, equation 13, or methanolysis of

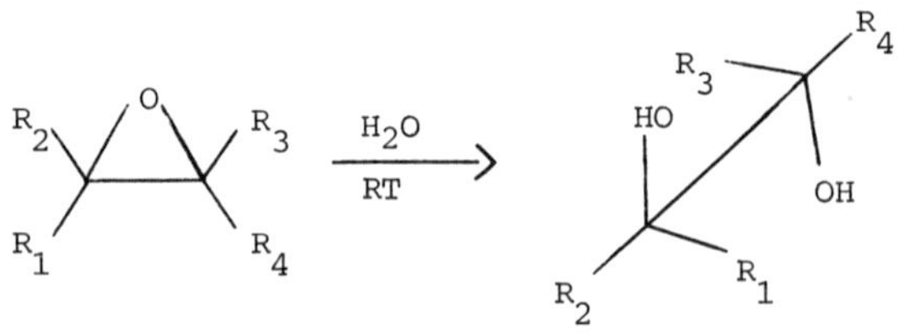

$$(13)$$

epoxides can be carried out with NAFION in good yields (66–81%) with minimal polymerization (32). Several examples are shown in Table XI. NAFION allows the reaction to be conducted under mild conditions without heating of the reaction mixtures.

The catalytic hydration of ethylene oxide with NAFION or supported NAFION at 92°C showed a 56% conversion of ethylene oxide with a 94% selectivity to ethylene glycol (33). Only a 10.1% conversion of ethylene oxide is realized when a poly-

Table XI. Epoxide ring opening

Epoxide	Nucleophile	Product	Yield(%)
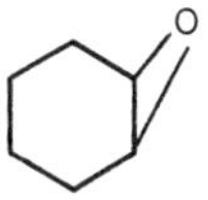	H_2O	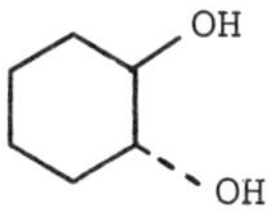	73
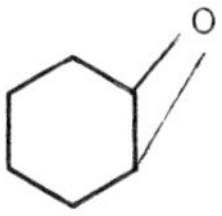	MeOH	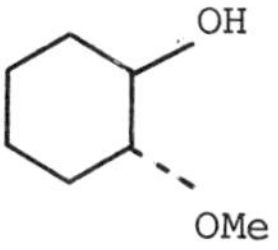	74
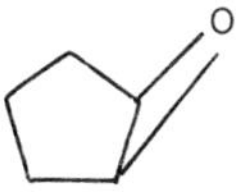	H_2O	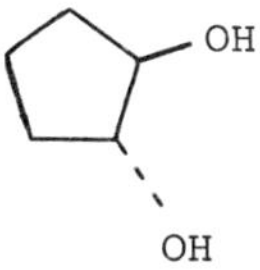	80
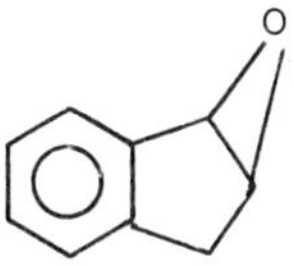	H_2O	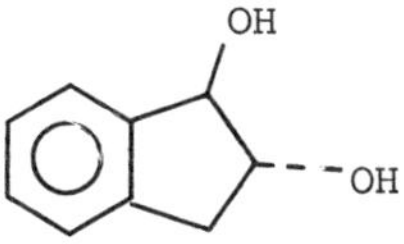	81

(styrenesulfonic acid) resin is used as a catalyst. Apparently, the high acidity of the polymeric perfluorinated sulfonic acid is preserved even when the resin is hydrated. Under the above conditions, about 90 wt % of water exists which at first thought would make the acidity of the two catalysts the same due to the leveling action of water. This phenomenon is not completely understood, but may be due to protection of some of the acid sites by the hydrophobic portions of the polymer backbone.

Esterification. Esterifications with NAFION have been carried out in both the liquid and gas phase. The reaction of acrylic acid and ethanol at 68°C with the catalyst in tubular form (approximately 25 and 35 mils inside and outside diameter, respectively) has been studied to obtain the forward rate constant in this second order reversible reaction ($\underline{34}$). A rate constant of 4.2×10^{-4} M^{-1} min^{-1} was obtained from the analysis. Utilization of sulfuric acid in the same esterification reaction gave a $k_1 = 6.15 \times 10^{-4}$ M^{-1} min^{-1} at 82°C indicating that the polymeric catalyst was as good as H_2SO_4.

When a mixture of a saturated carboxylic acid and an alcohol were passed over PFIEP at 95-125°C with a contact time

$$R_1CO_2H + R_2OH \rightleftharpoons R_1CO_2R_2 + H_2O \qquad (14)$$

of ~5 sec, high yields of the esters were obtained according to reaction 14. Table XII illustrates the scope of the reaction.

Table XII. Esterification catalyzed by NAFION

R_1	R_2	Yield(%)
CH_3	CH_3CH_2	96
$\underline{n}-C_3H_7$	CH_3CH_2	93
$\underline{n}-C_3H_7$	$\underline{i}-C_3H_7$	82
$\underline{n}-C_5H_{11}$	CH_3CH_2	98
CH_3	$\underline{n}-C_4H_9$	100
CH_3	$\underline{t}-C_4H_9$	20

Primary and secondary alcohols gave excellent yields of ester (35). However, tertiary alcohols gave poor results because the predominant pathway was dehydration followed by polymer formation.

Ethyl acetate can be formed over a NAFION catalyst from acetic acid and ethylene in the vapor phase at 135°C. Conversions were 48% based on acetic acid and after 100 hours, the activity of the catalyst was undiminished ($\underline{36}$). The reverse of the above reaction has been demonstrated by the pyrolysis of ethyl acetate to ethylene at 185°C with NAFION tubing ($\underline{37}$). The conversion of ethyl acetate is 19.4%, and ethene is the only gaseous product. Without the polymeric catalyst, the pyrolysis took place at 700°C over glass helices with 100% conversion of ethyl acetate but the exiting gaseous stream consisted of ethylene (96.5%), CO_2 (1.52%), CH_4 (1.09%), ethane (0.54%), and CO (0.33%).

Hydration. NAFION tubing of approximately 1/16 inch outside diameter has been used for the hydration of isobutylene to t-butyl alcohol at 96°C. The PFIEP tube was contained within a second plastic tube of 1/8 inch outside diameter. Liquid water was passed through the inner bore of the membrane tube while isobutene gas was passed through the void space between the two tubes. At least 84 wt % of the formed organics was t-butyl alcohol. The remaining 16 wt % consisted of isobutene dimer and trimer (38). If water is not passed through the inner bore, 83 wt % of the organics was the isobutene dimer. The remaining material was the trimer.

Hydration of acyclic olefins can be carried out readily with NAFION in a fixed bed tube reactor at 150°C. Because of thermodynamic consideration, propene conversion is 16% with a isopropyl alcohol selectivity of 97% (39).

Condensation of acetone. The condensation of phenol and acetone to bisphenol-A (equation 15) is catalyzed by polymeric per-fluorinated sulfonic acid which has been partially neutralized

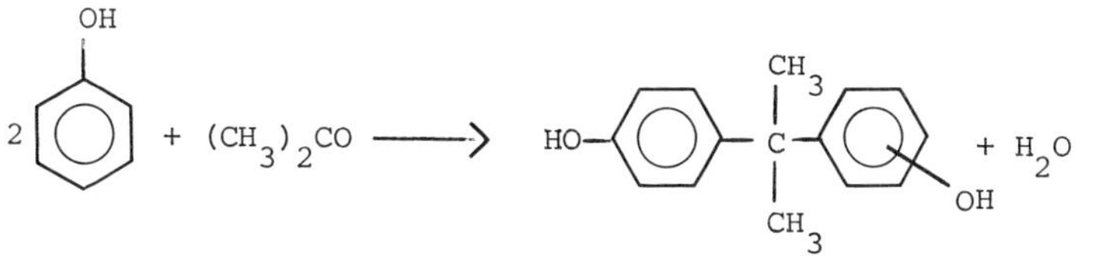

$$2 \, \text{C}_6\text{H}_5\text{OH} + (CH_3)_2CO \longrightarrow HO\text{-}C_6H_4\text{-}C(CH_3)_2\text{-}C_6H_4\text{-}OH + H_2O \quad (15)$$

with 2-mercaptoethylamine (40). When 30% of the acid sites are converted to $R_f SO_3^{\ominus} NH_3^{\oplus} CH_2CH_2SH$ and the resin is used as a catalyst, bisphenol-A was obtained in 100% selectivity at an acetone conversion of 88%. The bisphenol-A had an isomer distribution of 97.5 and 2.5% for p,p'-bisphenol-A and o,p'-bisphenol-A, respectively.

Acetone also undergoes an aldol condensation followed by dehydration to give mesityl oxide (equation 16).

$$2(CH_3)_2CO \xrightarrow{\Delta} (CH_3)_2C\text{=}CHCOCH_3 + H_2O \quad (16)$$

The reaction is limited by thermodynamic equilibrium, so typical yields for mesityl oxide, given in Table XIII, approach 20% when the reaction is catalyzed by NAFION (41).

Crossed aldol condensation of acetone and benzaldehyde gave 4-phenyl-3-penten-2-one in 27% yield after 36 hours at 60°C.

Table XIII. Acetone to mesityl oxide at 60°C

Time(h)	Yield of mesityl oxide (%)
1	10.5
6	18.1
12	19.0
24	19.7

Intramolecular condensation. The formation of anthraquinone
from o-benzoylbenzoic acid is an example of internal cyclization
(equation 17). At 150° for 3 hours, the conversion of

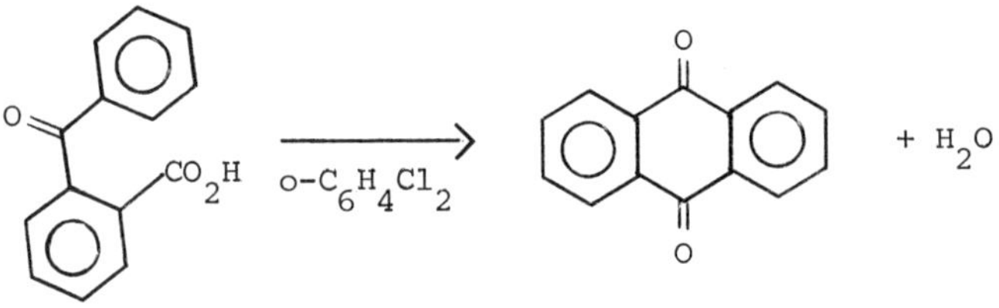

$$+ \ H_2O \qquad (17)$$

o-benzoylbenzoic was 60% with 78% anthraquinone selectivity
(42). NAFION can also intramolecular cyclize 2,5-hexanedione to
2,5-dimethylfuran and 3-methyl-2-cyclopentenone at 150°C in 29
and 2% yield, respectively (41).

Oligomerization. NAFION has been used in the cationic oligomerization of styrene. Oligomers range from dimer to hexamers
(43). With a soluble perfluorinated acid, CF_3SO_3H, a linear
dimer was the primary product in non-polar solvents. The solid,
strong acid catalyst retained catalytic activity on repeated
reaction, had higher activity than that of a conventional poly-
(styrenesulfonic acid) resin and was virtually free of solvent
effects on the reaction rate and product composition. The
perfluorinated polymeric catalyst was not very effective when
compared with CF_3SO_3H; a ten-fold excess of NAFION relative to
CF_3SO_3H was required to give oligomerizations with a similar
rate in CCl_4 at 50°C.
     NAFION has been used as an oligomerization catalyst for
decene-1 (44). A typical procedure consisted of heating the
catalyst (35 g) and decene-1 (140 g) at 120°C for 0.5 hours.
After filtering off the catalyst, and partial decene-1 removal,
the reaction mixture (114 g) consisted of C-10 (6.8%), C-20
(70.6%), C-30 (19.1%), and C-40 (3.4%). The amount of oligomers
was 75.9%. Other olefins that can be oligomerized similarly are
7-tetradecene, octene-2 and decene-5.

Nitration. NAFION resins catalyze the nitration of alkylbenzene with several different reagents: HNO_3, N_2O_4, n-$BuNO_2$ and acetone cyanohydrin nitrate (ACN). Table XIV compares the

Table XIV. Nitration of toluene

Reagent	Solvent	o	m	p
		\% isomer ratio		
HNO_3	toluene	56	4	40
N_2O_4	CCl_4	49	6	45
n-$BuNO_2$	toluene	50	3	47
ACN	toluene	47	3	50

various reagents for the nitration of toluene (45-46). The cleanest reaction occurs with n-butyl nitrate or ACN because all by-products are volatile materials. The nitro compounds are isolated simply by filtration of the catalyst without need for aqeuous basic washing or workup. In general, the yields were above 80% except for N_2O_4.

In a special case, 9-nitroanthracene can transfer a nitro group because the strong peri-interaction of the nitro group with the two neighboring hydrogens tilt the group out of the plane of the aromatic ring. Transfer nitration yields are generally low ($\leq$5%). In one example, 9-nitroanthracene in refluxing toluene with NAFION gave nitrotoluene (equation 18).

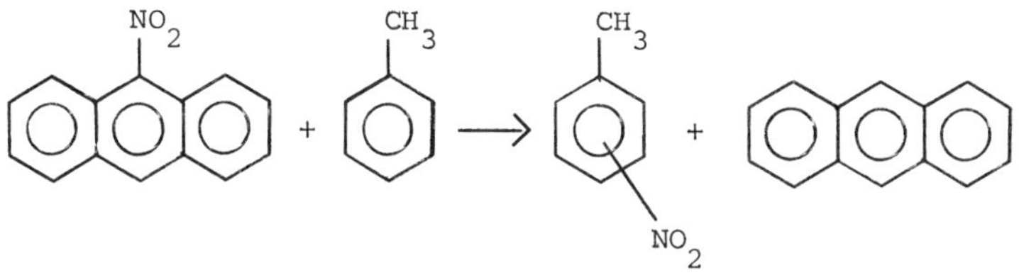

$$(18)$$

Since p-nitrotoluene is the predominant isomer, free nitronium ion itself is probably not the nitrating species (47).

Friedel-Crafts chemistry. Various alkylating agents have been used in Friedel-Crafts chemistry to prepare aromatic derivatives (48-49). NAFION is a catalyst for this general class of reaction in either the gas or liquid phase. The alkylating agents include alkyl halides, alkyl chloroformates (equation 19) and alkyl

$$ArH + ClCO_2R \longrightarrow ArR + CO_2 + HCl \qquad (19)$$

oxalates. For a comparison, see Table XV. When using alkyl
halides, polymer formation is minimized by keeping the reaction
temperature between 155-200°C.

Table XV. Alkylation of aromatics

Alkylating Agent	Aromatic	Temp(°C)	Isomer %		
			o	m	p
ClCO$_2$CH$_3$	toluene	70-72	48	26	26
ClCO$_2$Et	toluene	90	46	26	28
(CO$_2$Et)$_2$	toluene	110	48	24	28
EtCl	toluene	195	4.1	62.1	33.8

Olefin (5) and alcohols are other potential alkylating
agents. Studies have been done on the gas phase methylation and
ethylation of benzene over the polymeric perfluorinated sulfonic
acid. At 185°C, methylation of benzene with methyl alcohol gives
toluene in only 4.1% (50). On the other hand, ethylation of
benzene at 175°C gave ethylbenzene in 88% selectivity with 100%
conversion of ethylene at a weight hourly space velocity, WHSV,
of 8.0 (51).
    Naphthalene has been alkylated with propene over the
polymeric catalyst in the gas phase. At 220°C, the yield of
isopropylnaphthalene is 37% with the β-isomer being 90% (52).
    The alkylation of aromatics is often complicated by competing
side reactions such as isomerization and disproportionation of
methyl benzenes (53-55). Investigations, including a kinetic
study on the isomerization of m-xylene, have been reported in the
literature (56-57). In all cases, NAFION has been used as a
catalyst.

Diels-Alder reaction. NAFION catalysis allows the Diels-Alder
reaction to be run at a lower reaction temperature. Reactions of
anthracene with maleic anhydride, p-benzoquinone, dimethyl
maleate, and dimethyl fumarate were carried out at 60-80°C in
either refluxing CHCl$_3$ or benzene. Table XVI gives % yield of
the various adducts (58).

Table XVI. Diels-Alder adducts

Dienophile	Rxn time(h)	Yield(%)
maleic anhydride	5	91
p-benzoquinone	2	92
dimethyl maleate	15	95
dimethyl fumarate	16	94

In general, excellent results were obtained. Even if the diene
is easily polymerized, NAFION preferentially catalyzes the
Diels-Alder reaction. As an example, isoprene and p-benzoquinone
react at room temperature to give 80% of the adduct after 25
hours.

Deacylation. Acylation of substituted benzenes with aroyl
chloride has been described before. It is possible for
deacetylation and decarboxylation of aromatic substrates to take
place when unfavorable ortho steric interaction with methyl
substituents seem to provide the driving force for the reaction.
Table XVII summarizes a few examples of these reactions (59).

Other synthetic uses of NAFION.

(p-Tolyl)phosphorous dichloride. Phosphorus trichloride reacts
with toluene (60) in the presence of NAFION to yield
(p-tolyl)phosphorous dichloride and (o-tolyl)phosphorous
dichloride (equation 20). The conversion of toluene was 26%.

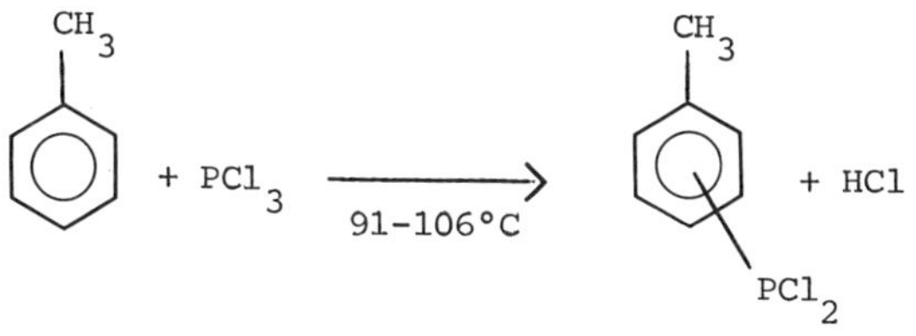

$$p/o = 17$$

Photoisomerization on NAFION. The course of cis/trans
photoisomerizations can be modified by surfaces. NAFION has been
examined as a solid acid to catalyze the photoisomerization of
ethyl cinnamate (equation 21).

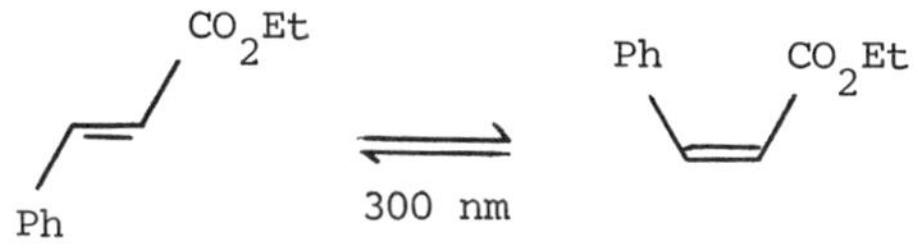

Irradiation of trans-ethyl cinnamate afforded a photostationary-
state composition of 73% cis- and 27% trans-ethyl cinnamate.
Without the solid catalyst or with FSO_3H, the cis composition was
42% or 39%, respectively. Apparently, there is selective
complexation or adsorption of the starting material by the solid
acid (61).

Table XVII. Deacylation and decarboxylation of aromatics

Ketone or Acid	Temp(°C)	Time(h)	Product(% Yield)
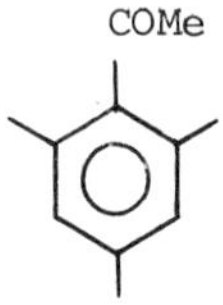 COMe	78-80	1	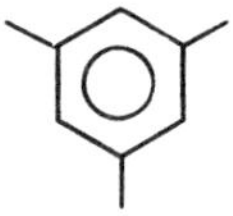 (98)
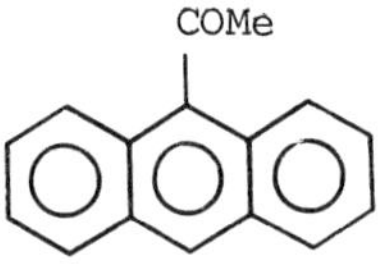 COMe	100	12	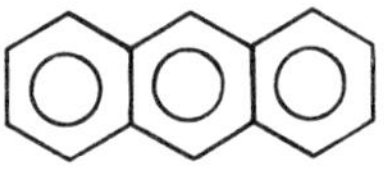 (95)
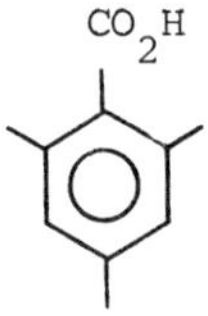 CO_2H	100	12	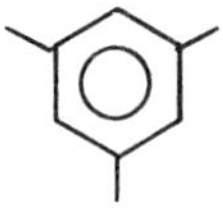 (80)
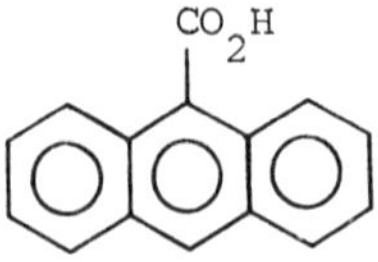 CO_2H	150	12	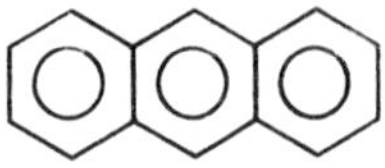 (25)

The perfluorinated polymer also functions as a photocatalyst in the photoisomerization of 3-methylene-1,2,4,5,6,6-hexamethylcyclohexa-1,4-diene (equation 22). No photoisomerization took place when the irradiation was

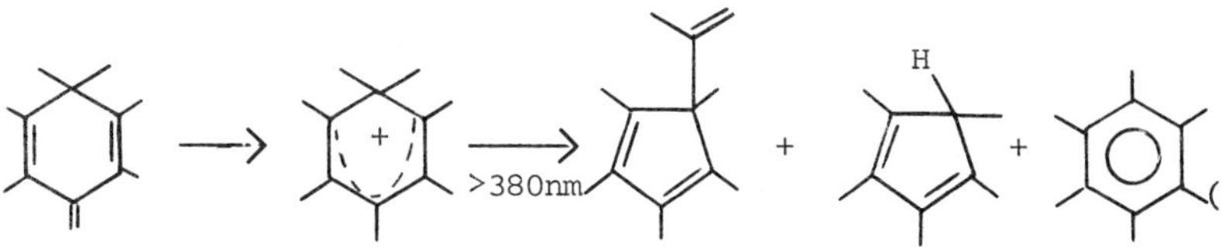

$$(22)$$

performed in the absence of NAFION (8). The heptamethylbenzenium cation is formed first with the strong acid catalyst in a dark reaction, and irradiation then converts the benzenium ion to mostly vinylcyclopentadiene.

Carbonylation chemistry. The carbonylation of formaldehyde in aqueous p-dioxane at 150°C and 3000 psig yields hydroxyacetic acid (equation 23), in about 70% yield (62).

$$CH_2O + CO + H_2O \xrightarrow{\Delta} HOCH_2CO_2H \qquad (23)$$

Principal by-products are methyl formate and polyhydroxyl aldehydes from condensation of formaldehyde. The NAFION used as a catalyst in this reaction swells controllably in the organic polar solvent. The actual products of the reaction are dehydrated forms of glycolic acid and the water-to-acid ratio influences the product yield. The hydroxyacetic acid formed is a precursor to ethylene glycol.

Immobilized reagents. NAFION can be converted into a hypohalite functionalized polymer as shown in equation 24 by treatment of

$$PFIEP[SO_3H] + ClF \xrightarrow[< -30°C]{} PFIEP[SO_3Cl] + HF \qquad (24)$$

the acid form of PFIEP with ClF below -30°C. The corresponding PFIEP[SO$_3$Br] can be formed in an analogous reaction. PFIEP[SO$_3$Br] reacts with CH$_3$Br to form PFIEP[SO$_3$CH$_3$] (63).
    The acid form of NAFION can also be reacted with chlorotrimethylsilane to form a polymer-supported silylating agent, equation 25. Incorporation of up to 0.8 mmol of the

$$PFIEP[SO_3H] + (CH_3)_3SiCl \xrightarrow{} PFIEP[SO_3Si(CH_3)_3] \qquad (25)$$

silyl group per gram of resin has been achieved (64). Unlike monomer analogs, this reagent does not fume in air. Reaction with compounds possessing an active hydrogen such as ethanol, acetic acid, ethanethiol, or diethylamine, transforms them into the corresponding trimethylsilyl derivatives.

The FITS reagent, (perfluoroalkyl)aryliodonium trifluoro-
sulfonates, can also be immobilized on NAFION ($\underline{65}$). In
particular, bis(trifluoroacetoxy)iodoperfluoroalkanes react with
PFIEP[SO_3H] as in equation 26. The immobilized reagent,

$$\text{PFIEP[SO}_3\text{H]} \xrightarrow[\text{PhH}]{R_fI(O_2CCF_3)_2} \text{PFIEP[SO}_3IR_f] \qquad (26)$$
$$\overset{|}{\text{Ph}}$$

$$R_f = n\text{-}C_mF_{2m + 1}$$

PFIEP[$SO_3IC_8F_{17}$],upon heating in thiophene at 80°C in
$\overset{|}{Ph}$

the presence of pyridine gave $\alpha\text{-}C_8F_{17}$-thiophene in 95% yield.

Supported NAFION. In order to increase the activity of the acid
sites by achieving better dispersion, NAFION has been supported
on silica gel, silica/alumina, alumina, porous glass and
Chromosorb T (fluoropolymer support). These supports can have
either low or high surface area and various pore diameters
(50–600A). Catalysts prepared in this fashion have been used in
the alkylation of benzene, isomerization of normal alkanes and
disproportionation of toluene. Table XVIII summarizes the
results on the alkylation of benzene with ethene for NAFION and
several supported catalysts ($\underline{66}$–$\underline{68}$).

Table XVIII. Alkylation of benzene with ethene

Support	–	Silica	Porous glass	Alumina
Temp (°C)	175	185	188	200
Ethene Conv (%)	100	99	95	99
PhEt sel (%)	80	91	87	85.5
WHSV	1.0	1.8	3	3

The weight hourly space velocity, WHSV, measures the activity
of the catalyst and suggests that coating NAFION on a support
increases the number of accessible acid sites.
    The primary method for dispersing PFIEP is by impregnating
supports with alcoholic solutions of soluble PFIEP in the acid
form. Alternatively, the polymeric sulfonyl fluoride precursor
is a thermoplastic and can be extruded into thin films or blended
with a powder support and extruded into various shapes ($\underline{69}$). In
addition, the sulfonyl fluoride precursor can be coextruded with
a polyethylene resin to form differently shaped parts ($\underline{70}$).
After extrusion, the sulfonyl fluoride form is converted to the
active sulfonic acid form.

Other uses of NAFION.

In addition to the synthetic applications described above, NAFION
has found other applications including coatings on electrodes
(71-72), hydrogen production (73-74), fuel cells, and membranes
for chemical devices (75-78). Some of these uses are outlined
here.

Because it is resistant to chemical attack, even by strong
oxidants at elevated temperature, perfluorinated coatings on
electrodes offer a stable environment for attaching reactants.
For example, a pyrolytic graphite electrode was coated with
NAFION (71). After electrostatically attaching metal complexes
into the membranes, chemical reactions were carried out in the
membrane or at the membrane interface with the solution. Other
studies have focused on the mechanism of charge transport through
NAFION coated on glassy carbon electrodes and containing
Cp_2FeTMA^+, $Ru(bpy)_3^{+2}$, and $Os(bpy)_3^{+2}$ where Cp_2FeTMA^+ is
[(triethylammonio)methyl]ferrocene (72).

The polymeric perfluorinated sulfonic acid has been used as a
matrix for a system which combines semiconductor CdS crystallites
and a Pt hydrogen-evolution catalyst in a photocatalytic hydrogen
generator (73-74). Upon photolysis of the platinized CdS
particles in the presence of a sacrificial electron donor, Na_2S,
the production of hydrogen gas by water reduction was observed.
The number of moles of H_2 produced with a typical NAFION/CdS
system exceeds the moles of CdS present by a factor greater than
100.

NAFION systems have also been used to construct devices for
1) separating monofunctional carboxylic acids like propionic,
butyric or valeric acid from other acids (75), 2) nitrating and
sulfonating aromatics in an annular tubular flow reactor where
the inner tube is NAFION and the outer tube is TEFLON fluoro-
carbon resin (76), 3) sulfonating aromatics in a two-compartment
cell separated by a membrane sheet (77), and 4) alkylating
aromatics with a tightly-packed bundle of parallel tubular NAFION
in a stainless steel tube (78).

Chlor-alkali membranes. Several companies are or will supply
perfluorinated membranes to the chlor-alkali industry. These
companies are summarized in Table XIX along with the type of
membrane available (79-80). Du Pont and Asahi Glass have signed
an agreement to cross-license patents (1c). Presently, the
industry workhorse in the chlorine-caustic diaphragm cell is the
asbestos diaphragm. In the membrane cell, a cation-exchange
membrane is used instead of a diaphragm. Perfluorinated
membranes will offer the chlor-alkali industry energy savings and
less pollution.

Conclusion

NAFION resins are effective, heterogeneous, strong acid catalysts
for many reactions useful in organic synthesis. Literature

Table XIX.   Chlor-alkali membranes

Company	Membrane-Type
Asahi Chemical Industry	surface carboxylation on sulfonate ionomer
Asahi Glass Company	perfluorocarboxylate
Du Pont	perfluorocarboxylate/ perfluorosulfonate
Dow	perfluorosulfonate
Tokuyama Soda Company	oxidize NAFION membrane surface

references highlighted in this review also suggest other unique applications for PFIEP. It is the author's belief that research will continue to be fruitful not only in the areas mentioned in this article but also those areas described elsewhere (11). NAFION has all the advantages of an insoluble catalyst but also has important inherent properties: a chemically inert polymeric framework and higher thermal stability. It will be interesting to see if NAFION and other PFIEP's will grow beyond their initial applications in the chlor-alkali industry and find applications in the marketplace as catalyst or unique support for catalysis.

Literature Cited

1. a. Du Pont registered trademark for its NAFION perfluorinated membranes. Henceforth in this review, NAFION will refer to Du Pont's brand of perfluorinated ion exchange polymer resins.
    b. Connolly, D. J.; Gresham, W. F. (to E. I. du Pont de Nemours & Company) U.S. Patent 3,282,875 (November 1, 1966).
    c. Chem. Eng. News, March 15, 1982, 22-25.
2. NAFION in the form of powder, film, and alcohol solutions can be purchased from C. G. Processing, Inc., P. O. Box 133, Rockland, DE 19732.
3. Waller, F. J. Brit. Polymer J. 1984, 16, 239.
4. Rohm and Haas registered trademark.
5. Olah, G. A.; Kaspi, J.; Bukala, J. J. Org. Chem. 1977, 42, 4187.
6. Gierke, T. D.; Hsu, W. Y. in "Perfluorinated Ionomer Membranes," A. Eisenberg and H. L. Yeager, Ed., ACS Symp. Ser., No. 180, American Chemical Society: Washington, D.C., 1982, Chapter 13.
7. Komoroski, R. A.; Mauritz, K. A. J. Am. Chem. Soc. 1978, 100, 7487.
8. Childs, R. F.; Mika-Gibala, A. J. Org. Chem. 1982, 47, 4204.

9. Olah, G. A.; Prakash, G. K. S.; Sommer, J. Science 1979, 206, 13.
10. Rys, P.; Steinegger, W. J. J. Am. Chem. Soc. 1979, 101, 4801.
11. Waller, F. J. accepted for publication in Catalysis Reviews - Science and Engineering.
12. Olah, G. A.; Malhotra, R., Narang, S. C.; Olah, J. A. Synthesis 1978, 672.
13. Krespan, C. G. J. Org. Chem. 1979, 44, 4924.
14. Konishi, H.; Suetsugu, K.; Okano, T.; Kiji, J. Bull. Chem. Soc. Jpn. 1982, 55, 957.
15. Olah, G. A.; Arvanaghi, M.; Krishnamurthy, V. V. J. Org. Chem. 1983, 48, 3359.
16. Olah, G. A.; Meidar, D. Synthesis 1978, 358.
17. Olah, G. A.; Fung, A. P. Synthesis 1981, 473.
18. Olah, G. A.; Meidar, D.; Liang, G. J. Org. Chem. 1978, 43, 3890.
19. Olah, G. A.; Fung, A. P.; Malhotra, R. Synthesis 1981, 474.
20. Pruckmayr, G.; Weir, R. H. (to E. I. du Pont de Nemours & Company) U.S. Patent 4,120,903 (October 17, 1978).
21. Pruckmayr, G. (to E. I. du Pont de Nemours & Company) U.S. Patent 4,153,786 (May 8, 1979).
22. Pruckmayr, G. (to E. I. du Pont de Nemours & Company) U.S. Patent 4,139,567 (February 13, 1979).
23. Olah, G. A.; Narang, S. C.; Meidar, D.; Salem, G. F. Synthesis 1981, 282.
24. Hughes, O. R. (to Celanese Corporation). U.S. patent 4,003,918 (January 18, 1977).
25. Olah, G. A.; Mehrotra, A. K. Synthesis 1982, 962.
26. Maggioni, P.; Minisci, F. GB Patent 2,085,004.
27. Malloy, T. P.; Engel, D. J. (to UOP Inc.) U.S. Patent 4,323,714 (April 6, 1982).
28. Kaspi, J.; Olah, G. A. J. Org. Chem. 1978, 43, 3142.
29. Olah, G. A. GB Patent 2,082,177.
30. Oyama, K.; Kihara, K., GB Patent 2,075,019.
31. Olah, G. A.; Husain, A.; Gupta, B. G.; Narang, S. C. Synthesis 1981, 471.
32. Olah, G. A.; Fung, A. P.; Meidar, D. Synthesis 1981, 280.
33. Kim, L. (to Shell Oil Company) U.S. Patent 4,165,440 (August 21, 1979).
34. Schreck, D. J. GB Patent 2,063,261.
35. Olah, G. A.; Keumi, T.; Meidar, D. Synthesis 1978, 929.
36. Gruffaz, M.; Micaelli, O. (to Rhone-Poulenc Industries) U.S. Patent 4,275,228 (June 23, 1981).
37. Schreck, D. J. (to Union Carbide Corporation) U.S. Patent 4,399,305 (August 16, 1983).
38. Cares, W. R. (to Petro-Tex Chemical Corporation) U.S. Patent 4,065,512 (December 27, 1977).
39. Olah, G. A. GB Patent 2,082,178.
40. McClure, J. D.; Neumann, F. E. (to Shell Oil Company) U.S. Patent 4,053,522 (October 11, 1977).
41. Pittman, C. U.; Liang, Y. J. Org. Chem. 1980, 45, 5048.
42. Nutt, M. O. (to Dow Chemical Company) U.S. Patent 4,304,724 (December 8, 1981).
43. Hasegawa, H.; Higashimura, T. Polymer J. 1979, 11, 737.

44. Marquis, E. T.; Watts, L. W.; Brader, W. H.; Darden, J. W.
    GB Patent 2,089,832.
45. Olah, G. A.; Malhotra, R.; Narang, S. C. J. Org. Chem.
    1978, 43, 4628.
46. Olah, G. A.; Narang, S. C. Synthesis 1978, 690.
47. Olah, G. A.; Narang, S. C.; Malhotra, R.; Olah, J. A. J.
    Am. Chem. Soc. 1979, 101, 1805.
48. Olah, G. A.; Meidar, D. Nouv. J. Chim. 1979, 3, 269.
49. Olah, G. A.; Meidar, D.; Malhotra, R.; Olah, J. A.; Narang,
    S. C. J. Catal. 1980, 61, 96.
50. Kaspi, J.; Montgomery, D. D.; Olah, G. A. J. Org. Chem.
    1978, 43, 3147.
51. McClure, J. D. (to Shell Oil Company) U.S. Patent 4,041,090
    (August 9, 1977).
52. Olah, G. A. GB Patent 2,090,856.
53. Olah, G. A.; Kaspi, J. Nouv. J. Chim. 1978, 2, 581.
54. McClure, J. D. (to Shell Oil Company) U.S. Patent 4,052,474
    (October 4, 1977).
55. McClure, J. D. (to Shell Oil Company) U.S. Patent 4,022,847
    (May 10, 1977).
56. Beltrame, P.; Beltrame, P. L.; Carniti, P.; Magnoni, M.
    Gazz. Chim. Ital. 1978, 108, 651.
57. Beltrame, P.; Beltrame, P. L.; Carniti, P.; Nespoli, G.
    Ind. Eng. Chem. Prod. Res. Dev. 1980, 19, 205.
58. Olah, G. A.; Meidar, D.; Fung, A. P. Synthesis 1979, 270.
59. Olah, G. A.; Laali, K.; Mehrotra, A. K. J. Org. Chem. 1983,
    48, 3360.
60. Cozens, R. J.; Hogan, P. J.; Lalkham, M. J. (to Imperial
    Chemical Industries Limited), EP Patent 24128.
61. Childs, R. F.; Duffey, B.; Mika-Gibala, A. J. Org. Chem.
    1984, 49, 4352.
62. Chem. Eng. News, April 11, 1983, 41-42.
63. Desmarteau, D. D. J. of Fluorine Chem. 1982, 21, 249.
64. Murata, S.; Noyori, R. Tetrahedron Lett. 1980, 21, 767.
65. Umemoto, T. Tetrahedron Lett. 1984, 25, 81.
66. McClure, J. D.; Brandenberger, S. G. (to Shell Oil Company)
    U.S. Patent 4,060,565 (November 29, 1977).
67. McClure, J. D.; Brandenberger, S. G. (to Shell Oil Company)
    U.S. Patent 4,065,515 (December 27, 1977).
68. McClure, J. D.; Brandenberger, S. G. (to Shell Oil Company)
    U.S. Patent 4,052,475 (October 4, 1977).
69. Peluso, S. L. Research Disclosure 1982, 221, 311.
70. Vaughan, R. J. (to Varen Technology) U.S. Patent 4,303,551
    (December 1, 1981).
71. Rubinstein, I.; Bard, A. J. J. Am. Chem. Soc. 1980, 102,
    6641.
72. White, H. S.; Leddy, J.; Bard, A. J. J. Am. Chem. Soc.
    1982, 104, 4811.
73. Krishnan, M.; White, J. R.; Fox, M. A.; Bard, A. J. J. Am.
    Chem. Soc. 1983, 105, 7002.
74. Mau, A. W.; Huang, C. B.; Katuta, N.; Bard, A. J.; Campion,
    A.; Fox, M. A.; White, J. M.; Webber, S. E. J. Am. Chem.
    Soc. 1984, 106, 6537.
75. Chum, H. L.; Sopher, D. W. (to U.S. Department of Energy)
    U.S. Patent 4,476,025 (October 9, 1984).

76. Vaughan, R. J. (to Varen Technology) U.S. Patent 3,976,704 (August 24, 1976).
77. Vaughan, R. J. (to Varen Technology) U.S. Patent 4,308,215 (December 29, 1981).
78. Vaughan, R. J. (to Varen Technology) U.S. Patent 4,316,997 (February 23, 1982).
79. Chemical Week August 25, 1982, 63-64.
80. Chemical Week November 17, 1982, 35-36.

RECEIVED January 23, 1986

]

4

The Role of Substrate Transport in Catalyst Activity

John G. Ekerdt

Department of Chemical Engineering, The University of Texas, Austin, TX 78712

This paper addresses the general subject of substrate
transport in polymer-immobilized catalyst systems.
The equations needed to interpret reaction rate data
for polymer systems are developed and their applicability is discussed. The effects of experimental
variables on observed reaction rates in the presence
of substrate transport limitations are discussed. A
simple method for estimating substrate diffusion
coefficients is presented. Methods for testing reaction rate data to determine if substrate transport is
affecting the observed reaction rates are developed
and the limitations of these methods are discussed.
Finally, examples of recent studies are reviewed and
discussed within the framework of the mathematical
formalism to demonstrate application of the formalism
and to show that carefully designed experiments are
required to establish the presence of substrate
limitations.

Numerous studies have appeared in the literature in which transition metal complexes, immobilized on functionalized-polymers, or in
which sulfonated resins have been used as catalysts. Polystyrene
is the most common polymer support and has been used in the gel-
form and macroreticular forms. Review articles on "immobilized
catalysts" have presented the advantages and disadvantages of
immobilizing the catalytic site (1-8). The primary motivation is
to "hetrogenize" a homogenous catalyst and thereby avoid costly
separations of the catalyst from the reaction mixture and, in the
case of acids, to minimize contact of the corrosive catalyst with
the reactor vessel.
    Polymer-immobilized catalysts cannot be considered to be
heterogeneous in the same sense that zeolites or metals supported
on porous oxides are heterogeneous. These latter catalysts involve
well defined rigid support structures in which the active sites
reside at specific locations within the solid particle. The

0097-6156/86/0308-0068$06.00/0
© 1986 American Chemical Society

immobilized complexes and the sulfonic acid sites are attached at specific locations along the polymer backbone. However, the polymer chains are mobile, the mobility depending on the extent of crosslinking and volume fraction of polymer versus the volume fraction of sorbed species. Gel-form polymers do not have a pore network when swollen by sorbed species, rather the polymer matrix opens up as species are sorbed into it. Macroreticular resins consist of agglomerates of microparticles of gel-form polymer, separated by macropores. In the case of macroreticular sulfonic acid resins, the active sites are generally within the interior of the gel-form microparticles (9,10,11,12).

The immobilized-catalysts are confined to a region in space defined by the dimensions of the polymer particle. Reactant(s) must diffuse from the external surface to the catalytic sites within the particle before any chemical reaction can occur. This sequential process, mass transfer with reaction, has been treated extensively for catalytic reactions in porous solids (13,14,15). A limited number of studies have shown that the mathematical formalism which is applied to heterogeneously-catalyzed reactions can be used to interpret mass transfer with reaction in immobilized catalysts which employ polymers as supports (11,16,17).

Catalytic performance for any system, heterogeneous or homogeneous, is generally assessed by measuring the activity (rate of reaction) and selectivity of the catalyst for reactions of interest. A comparison of reaction rates for catalysts prepared using different synthetic methods or different supports is used to identify the parameters which influence catalyst activity and selectivity. For studies involving heterogeneous catalysts and immobilized catalysts, reactant transport to the active sites must occur at a sufficient rate that the observed activity reveals information about the intrinsic properties of the catalytic site.

This article presents a discussion of substrate transport in polymer-immobilized catalyst systems. The mathematical formalism necessary to model a reaction system is presented and forms the basis for qualitative comments on substrate transport. The mathematical formalism also provides a common point for published reaction data and its interpretation. The mathematical models are developed from the perspective of a small scale experimental study and therefore are written for the most common research reactor for polymer-immobilized studies, a batch reactor, and for a fixed-bed reactor operating at differential conversion.

Reaction Rates

Reaction rates are never directly measured. Concentration of a reactant or product is measured as a function of time in batch reactor studies. The amount of reactant converted or product formed as a function of total reactant molar flowrate or mass of catalyst is measured for a fixed-bed reactor.

The concentration information is combined with a model which accounts for the reactor's performance to compute the reaction rate (18). The appropriate expression for a batch reactor is

$$(-r_s)_{OBS} = \frac{-dc_s}{dt} \tag{1}$$

(Nomenclature is listed at the end of this chapter.) Equation 1 assumes that the volume of the reacting system remains constant, a reasonable assumption in immobilized-catalyst studies. The rate is determined by taking the slope of a plot of concentration of substrate versus time. For reactions which involve a gas phase and a liquid phase reactant, in which the gas phase reactant is held at a constant pressure, the total moles of liquid substrate can be related to the consumption of the gas phase reactant. The rate of substrate conversion is then found by making use of the conversion variable, X, defined as

$$X(t) = 1 - \frac{n_s(t)}{n_s(0)} \tag{2}$$

which is related to the observed rate by

$$(-r_s)_{OBS} = C'(0) \frac{dX}{dt} \tag{3}$$

where

$$C'(0) = \frac{n_s(0)}{V_R} \tag{4}$$

Slopes from a plot of conversion versus time are used to determine the reaction rate.

Rates are established in fixed-bed reactors operating at differential conversion (generally $X < 0.05$) by means of

$$(-r_s)_{OBS} = (\frac{F}{W}) y_P \tag{5}$$

Equation 5 is based on the fractional amount of a product because it is generally more accurate to measure low concentrations of a product than to determine the difference in the concentration of a substrate when X is low.

Substrate Transport with Reaction Model

A fundamental issue in immobilized catalyst studies is the extent to which substrate transport influences the observed reaction rate. The issue can be resolved by making use of the mathematical formalism which has been developed for heterogeneously catalyzed reactions (13,14,15). This formalism will be presented below for both the batch reactor and the fixed-bed reactor systems. The treatment assumes that the concentration of substrate in the bulk phase is equal to the concentration immediately outside a polymer particle and that the substrate's diffusion coefficient is not a function of concentration or position in the particle. The polymer particle will be assumed to be a spherical bead. Finally, the equations will be developed for a single independent reaction. The bulk phase refers to the fluid-phase within the reactor, which is external to the polymer volume. The concentration of reactant in the bulk phase may not equal the concentration of reactant in the polymer phase due to thermodynamic partitioning of substrate between the phases. Sherrington (19) has discussed the importance of partitioning in immobilized-catalyst studies. The partitioning is expressed in terms of a distribution factor, K, which relates the equilibrium concentration of substrate in the resin to the substrate concentration in the bulk solution, viz.

$$K = \frac{[S]_r}{[S]_p} \tag{6}$$

This parameter must be measured because reaction rates are generally expressed in terms of the concentration of substrate, and one must use the correct concentration in comparing activities. The concentration in solution is the measured variable, however, the catalyst sites are exposed to K x (the bulk phase concentration).

The distribution factor has added importance when studying immobilized-catalysts in batch reactors. These studies generally involve swelling the polymer in an appropriate solvent. The distribution factor may not be unity even when the substrate distributes uniformly between the swollen-polymer and bulk phases because the total moles of substrate added resides in a finite bulk phase volume and a finite polymer phase volume, which consists of polymer and imbibed solvent (16).

The batch reactor is a transient system in which the substrate diffuses from the bulk liquid phase of volume V_b into the catalyst beads of volume $V_{CATALYST}$. The model will be developed for the situation where the substrate is added at zero time to a reactor in which the polymer beads are gel-form, completely solvent-swollen, and initially contain no substrate.

The material balance equation for radial transport and reaction within a polymer bead is

$$\frac{\partial c}{\partial t} = D \frac{1}{r^2} \frac{\partial}{\partial r} \left(r^2 \frac{\partial c}{\partial r} \right) - (-r_s) \tag{7}$$

where c, the substrate concentration within the swollen catalyst bead, is a function of time and radial position, and D is the effective diffusion coefficient. All concentrations in the rate expression, which vary with time, need to be expressed as a function of c by use of stoichiometric relations based on the overall reaction's stoichiometry. In the event more than one independent reaction is present, Equation 7 would need to be written for a sufficient number of components so that the complete composition could be described at all positions and times.

The classic approaches to reaction with substrate transport (13,14,15) convert the material balance into dimensionless form prior to solving the differential equation. This dimensionless material balance will contain a grouping of parameters, which describes the catalyst under study, referred to as the Thiele modulus. The magnitude of the Thiele modulus can be used to estimate the influence of substrate transport. The dimensionless form of Equation 7 is

$$\frac{\partial \hat{c}}{\partial \tau} = \frac{1}{R^2} \left(\frac{\partial}{\partial R} \left(R^2 \frac{\partial \hat{c}}{\partial R} \right) \right) - \Phi^2 (-\hat{r}_s) \tag{8}$$

where

$$\hat{c}(R,\tau) = \frac{c(r,t)}{c_{bo}} \tag{9}$$

$$\tau = \frac{Dt}{a^2} \tag{10}$$

$$R = \frac{r}{a} \tag{11}$$

$$\Phi = a \left\{ \frac{(-r_s(c_{bo}))}{Dc_{bo}} \right\}^{\frac{1}{2}} \qquad \text{(Thiele modulus)} \tag{12}$$

$$(-\hat{r}_s) = \frac{(-r_s(c))}{(-r_s(c_{bo}))} \tag{13}$$

The substrate concentration at the catalyst surface is related to the surrounding bulk phase concentration, which is varying with time, by using the distribution factor

$$c(a,t) = K \cdot c_b(t) \tag{14}$$

The solution to Equation 8 is subject to the initial conditions

$$\hat{c}(R,0) = 0 \quad ; \quad 0 \leq R < 1 \tag{15}$$

$$\hat{c}(1,0) = K \quad ; \quad R = 1 \tag{16}$$

which are written on the assumption that the substrate is instantly dispersed throughout the bulk solvent phase at zero time. The solution to Equation 8 must also satisfy the boundary condition at the catalyst surface,

$$\frac{\alpha}{3} \frac{\partial \hat{c}(1,\tau)}{\partial \tau} = \frac{-\partial \hat{c}(1,\tau)}{\partial R} \qquad \tau \geqslant 0 \tag{17}$$

where

$$\alpha = \frac{V_b}{V_{CATALYST} \cdot K} \tag{18}$$

The boundary condition at the catalyst bead surface arises from a material balance on the substrate over the bulk phase in which the rate of loss of substrate from the bulk phase equals the total flux into the catalyst beads at any dimensionless time, τ.

The rate in a batch reactor must be based on the total amount of substrate in the system. There is no way to measure the concentration (or moles) of substrate inside the polymer beads; it must be computed. Substrate is depleted from the bulk phase for two reasons: Initially the polymer phase contained no substrate, and substrate reacts inside the polymer beads. Total moles of substrate are easily determined in systems which involve addition of a gas phase reactant at constant pressure. The reaction rate is developed using the conversion of substrate. Equation 2 uses the total moles of substrate in the system. A material balance is performed over the batch reactor at any time t to give the moles of substrate in the bulk and polymer phases. The derivative of this balance may be used to calculate reaction rates by making use of Equation 3. The derivative of the total mole balance with respect to dimensionless time is

$$\frac{dX}{d\tau} = \frac{-V_R c_{bo}}{n_s(0)} \frac{d}{d\tau} \left\{ (\frac{\alpha}{\alpha K+1}) \; \hat{c}(1,\tau) \right.$$

$$\left. + (\frac{3}{\alpha K+1}) \int_0^1 \hat{c}(R,\tau) R^2 dR \right\} \tag{19}$$

where

$$V_R = V_{CATALYST} + V_b \tag{20}$$

The first term in the bracket of Equation 19 refers to the moles of substrate in the bulk phase and the second term refers to the moles of substrate in the catalyst beads. Equation 19 is the most general description of the slope of a plot of experimentally determined conversion versus time for reaction in a solvent-swollen polymer-immobilized catalyst. Numerical methods may be required to solve Equation 8; the solution to Equation 8 is needed to evaluate the integral in Equation 19.

In the absence of substrate transport limitations, the substrate concentration is independent of radial position in the polymer. The rate measured under these circumstances would be an intrinsic reaction rate. The substrate material balance is greatly simplified because Equation 8 no longer needs to be solved. A specific example of this can be found in the literature (16).

The approach toward describing reaction and substrate transport in a fixed-bed reactor system is analogous to the development presented for a batch reactor. The substrate concentration only depends on radial position in the polymer support, therefore all derivatives with respect to time are equal to zero. The appropriate dimensionless equation is

$$0 = \frac{1}{R^2} \frac{d}{dR} (R^2 \frac{d\hat{c}}{dR}) - (\Phi')^2 (-\hat{r}_s') \qquad (21)$$

where

$$\Phi' = a \left\{ \frac{(-r_s'(c_{bo}))}{Dc_{bo}} \right\}^{\frac{1}{2}} \qquad \text{(Thiele modulus)} \qquad (22)$$

$$(-\hat{r}_s') = \frac{(-r_s'(c))}{(-r_s'(c_{bo}))} \qquad (23)$$

The solution to Equation 21 is subject to the boundary conditions

$$\hat{c}(R) = K \qquad R = 1 \qquad (24)$$

$$\frac{d\hat{c}(R)}{dR} = 0 \qquad R = 0 \qquad (25)$$

The rate is ultimately found by integrating the concentration profile over the catalyst

$$r_{OBS} = 4\pi \int_0^a (-r'(c)) r^2 dr \qquad (26)$$

The effective diffusion coefficient will be discussed in a later section. The models above assumed that D was independent of concentration and radial position. This is not an appropriate assumption for all immobilized-catalyst systems. Bischoff (20) has presented a generalized method for dealing with situations in which the diffusivity is a function of concentration. Dooley et al. (11) applied Bischoff's approach to describe the reaction in the micro-particles of a macroreticular resin. For the case where a reaction is conducted in a fixed-bed reactor the observed rate is given by

$$r_{OBS} = \frac{3}{a} [2 \int_{c(o)}^{c(a)} D(c) (-r_s(c)) dc]^{\frac{1}{2}} \qquad (27)$$

where $c(o)$ is the substrate concentration at the center of a polymer bead. The solution to Equation 27 requires that this center point concentration be specified, and this is not known a priori. It may be necessary to assume values for $c(o)$ until the solution to Equation 27 matches the experimentally observed rate.

Studies in macroreticular resins are also complicated by the fact that diffusion proceeds by different physical processes in the macropores and in the microparticles. An effective diffusion coefficient cannot be defined on the basis of total particle area and one must write a material balance for substrate transport in the macropores and a material balance for substrate transport with reaction in the microparticles. It is difficult to generalize this approach and the interested reader should refer to a specific application in a macroreticular resin (11) and to general discussions of the approach for heterogeneous catalysts (15,21,22).

Qualitative Features of Substrate Transport with Reaction

The mathematical model for substrate transport with reaction can be used to devise a general approach to predict reaction rates in catalyst particles. The observed rate is defined in terms of the intrinsic rate and an effectiveness factor

$$r_{OBS} = \eta \; (-r_{s,INTRINSIC}) \tag{28}$$

The intrinsic rate is based on the assumption that the concentration throughout the polymer particle is given by the concentration at the external surface. The effectiveness factor, η, is a measure of the extent to which the concentration is not uniform, and at the surface value, throughout the polymer particle. For simple nth-order reactions (n = 0, 1, 2) analytical relations have been developed in which

$$\eta = f \; (\Phi) \tag{29}$$

For example, when n = 1

$$(-r_s) = kC \tag{30}$$

$$\Phi_1 = a \; [\frac{k}{D}]^{\frac{1}{2}} \tag{31}$$

$$\eta = \frac{3}{\Phi_1} \; [\frac{1}{\tanh \; \Phi_1} - \frac{1}{\Phi_1}] \tag{32}$$

The effect of the magnitude of Φ_1 on η is easily seen from Equation 32. At sufficiently small Φ, η is unity, and for Φ_1 greater than 15, the effectiveness factor is inversely proportional to the Thiele modulus.

Aris (15) has discussed the functional dependence in Equation 29 for nth-order kinetics and Langmuir-Hinshelwood kinetics. Immobilized-catalyst kinetics are similar to Langmuir-Hinshelwood kinetics because of the equilibria between substrates and the active sites. Aris has shown that as long as the apparent reaction order is zero-order or positive, there are two asymptotic limits to Equation 29

$$\eta \propto \frac{1}{\Phi} \qquad \text{large } \Phi \tag{33}$$

$$\eta = 1 \qquad \text{small } \Phi \tag{34}$$

Intrinsic reactions are realized when Equation 34 is true, and mass

transfer-limited reactions are realized when Equation 33 is true.
The exact functional dependence for intermediate values of Φ depends
on reaction kinetics; Equation 32 is only valid for first-order
reactions.
The asymptotic behavior shown in Equation 33 can be used to
make general comments about the effect of mass transfer on the
experimental (observed) reaction rate. When the Thiele modulus is
large,

$$r_{OBS} \propto \frac{1}{a} \tag{35}$$

$$r_{OBS} \propto D^{\frac{1}{2}} \tag{36}$$

$$E_{OBS} \neq E_{TRUE} \tag{37}$$

where E_{TRUE} is the Arrhenius activation energy for the reaction.
The observed activation energy, for simple $\underline{n}$th-other kinetics, is
effectively one-half the true activation energy.
The results in Equations 35, 36 and 37 can be used to test for
substrate transport limitations. Examples from the literature
using these simple tests will be presented later. Caution must be
exercised in applying the simple tests individually because they
are written for a situation where all other factors are equal, and
this may not be the case for polymer-immobilized catalysts.
More definitive criteria have been proposed to test for the
absence of substrate transport limitations ($\underline{15,18,23}$). They involve
calculating the magnitude of a dimensionless quantity, using experimentally measured variables, and comparing the magnitude to theo-
retically predicted values of the dimensionless quantity. The
predicted values are based on concentration profiles within a porous
support which would result in negligible substrate transport limitations. Butt and Aris compare the different criteria in their
texts and the arbitrary nature of a negligible limitation. The
most common criterion is one proposed by Weisz ($\underline{23}$). The Weisz
modulus is

$$\Phi_w = \frac{(-r'_s)a^2}{D \cdot c_{bo}}$$

<6.0 (zero order)	(38a)
<0.6 (first order)	(38b)
<0.3 (second order)	(38c)

All of the parameters in the Weisz modulus can be measured experimentally. The inequalities in Equations 38 are for effectiveness
factors greater than 95%; it is best to satisfy these inequalities
by at least an order of magnitude ($\underline{18}$). Given the general form of
the solution to Equation 21 for Langmuir-Hinshelwood kinetics,
which are effectively zero order or greater ($\underline{15}$), it seems
reasonable to propose that the inequalities in Equation 38 provide
a safe estimate of negligible substrate limitations for more
complex kinetic expressions.

Diffusion in Polymers

The polymer-immobilized catalyst system requires the substrate molecules to diffuse through the solvent-swollen polymer (16,17) or substrate/product-swollen polymer (11) structure to reach active complexes contained therein. The polymer will be referred to simply as swollen in this section. Diffusion of solute molecules in swollen crosslinked polymers has been related to diffusion in the pure solvent by Mackie and Meares (24) and Paul et al. (25). The subject has been reviewed by Muhr and Blanshard (26) and they have presented additional correlations beyond the one presented below. Paul et al. have shown that the diffusion coefficient in the polymer depends on the volume fraction of polymer, volume of polymer/total swollen volume, and the properties of the swelling solvent. The functional relationship may be represented as

$$D = D_o f(V_{Ro})$$ (39)

where D_o is the diffusion coefficient of the solute in the pure solvent at infinite dilution, and V_{Ro} represents the volume fraction of swollen polymer. In general, the diffusion coefficient within the swollen polymer decreases as the polymer volume fraction increases. The functional dependence is specific to the polymer-solvent-solute system.

Mackie and Meares (24) have modeled diffusion in polymers for the situation where the size of the diffusing molecules are similar to the size of the polymer segments. The mobile polymer segments were considered to act as a physical obstruction to diffusion and the solute was considered to be restricted to the free sites. They arrived at the following functional dependence for Equation 39,

$$f(V_{Ro}) = \left\{ \frac{1 - V_{Ro}}{1 + V_{Ro}} \right\}^2$$ (40)

Paul et al. (25) observed that for polymer volume fractions less than 0.8, the functional dependence of the diffusion coefficients on the polymer volume fraction was, generally, in accordance with Equation 40. Muhr and Blanshard (26) provide additional supporting data on different polymers than those reported by Paul et al. Roucis and Ekerdt (27) measured the diffusion of cyclic hydrocarbons in benzene-swollen polystyrene beads; their diffusion coefficients satisfy the general form of Equation 40. The effective diffusivities of organic substrates in crosslinked polystyrene reported by Marconi and Ford (17) also follow trends predicted in Equation 40. In the absence of experimental data, it appears that Equation 40 provides a reasonable, and the simplest, means to estimate D for use in detailed modeling or in estimation methods such as Equation 38. Equation 40 was used by Dooley et al. (11) in their study of substrate diffusion and reaction in a macroreticular sulfonic acid resin which involved vapor phase reactants.

Vrentras and Duda have developed an alternate model for describing diffusion in polymers which is referred to as the free-

volume theory. A recent article describing polymer-solvent-solvent
systems (28) details this theory for situations which are similar
to a substrate reacting in a solvent-swollen polymer. The theory
contains a large number of coefficients which must be obtained from
pure component properties and binary transport data for the two
solvents. The free-volume theory requires that the polymer chains
move relative to each other in addition to the proposed local
segmental motion present in Mackie and Meares' theory. This macroscopic chain motion is not possible in crosslinked polymers.
Furthermore, the general applicability of the free-volume theory to
the types of polymers which are employed as catalyst supports
remains to be demonstrated.
      The constancy of the effective diffusion coefficient in the
substrate transport with reaction model, Equation 26 versus Equation
27, depends on two factors. First, if the volume fraction of
polymer is not constant with time, radial position in the gel, or
extent of reaction, then D is influenced by the relation given in
Equation 40. Dooley et al. (11) present an example of this in their
study. Second, if the substrate's diffusion coefficient in the
solvent alone is dependent on substrate concentration at the range
of concentrations in the reaction system, then D is influenced
similarly according to Equation 39. The assumption of a constant
diffusion coefficient in the substrate transport with reaction model
must always be justified.

Rate Studies of Immobilized Complexes

The preceding discussion has shown that the measured reaction for
an immobilized catalyst may not be an accurate representation of
the intrinsic properties of the catalyst. One of the major goals
in the development of an immobilized catalyst is to prepare a fixed-
bed catalyst which has the inherent selectivity of the homogeneous
complex without sacrificing losses in activity because substrate
transport is slow relative to the intrinsic reaction rate. Catalytic activity and selectivity are invariably altered when a
transition metal complex is attached to a polymer through a chelating ligand. There are numerous instances where substrate diffusional limitations are suspected of altering activity because the
observed rate was influenced by variables that can be argued to
increase diffusional resistances in the catalyst (8,29-36).
Without a complete presentation of the rate data, it is not
possible to discount substrate diffusion as a contributing factor
in these or other studies (37). However, careful attention to the
effect of the variables identified as important in the mathematical
model should assist in a better appreciation of substrate transport
effects in designing future studies.
      There are many factors which influence the activity and selec-
tivity of an immobilized metal complex, mass transfer of the
substrate is but one. Additional factors are associated with the
chemical properties of the complex and include altered ligand
environment about the complex (29,38); steric constraints imposed
by the presence of the polymer structure in the vicinity of the
attached complex (8,16,17,30,38); reduced activity due to thermodynamic exclusion of the reactant molecules from the solvent-swollen

polymer support phase (17,33); and local differences in the chelating ligand to complex molar ratios (30,39,40). The role of
substrate transport is the most easily identified and quantified of
the above effects; however it has only been treated in detail in a
limited number situations (11,12,16,17).
It was pointed out earlier that observed activation energies
can be expected to be as much as a factor of 2 lower than the
intrinsic activation energy (Equation 37) when severe substrate
transport limitations are present. De Croon and Coenen (35)
studied cyclohexene hydrogenation over rhodium immobilized on a
phosphinated crosslinked polystyrene gel-form resin and reported
that the immobilized catalyst had an activation energy of 11.7
Kcal/g-mole and that the homogeneous rhodium analog had an activation energy of 22 Kcal/g-mole. They associated this factor of two
difference to substrate diffusional limitations. Roucis and Ekerdt
(16) reported an activation energy of 10.0 Kcal/g-mole for cyclohexene hydrogenation for a similar immobilized rhodium complex.
This latter study established that the activation energy was based
on observed rates which were intrinsic by making use of the criteria
proposed by Weisz (23). Roucis and Ekerdt went on to argue that De
Croon and Coenen's rate data, that gave the similar activation
energy, were also intrinsic. These two studies demonstrate that it
is extremely risky to compare the activity of an immobilized complex
directly to the activity of the homogeneous analog because the
immobilized complex is in a very different environment and should
be expected to be inherently altered.
The diffusion coefficient for a substrate in the pure solvent
decreases as the substrate molecular size increases. The diffusion
coefficient of the substrate in the swollen polymer will be corres-
pondingly lower on the basis of Equation 39. Studies of reaction
rates of substrates of differing sizes have appeared in which the
ratio of the reaction rate of a larger substrate relative to a
smaller substrate in the immobilized catalyst was compared to the
homogeneous analog (31,32,36). The fact that the larger substrate
reacted more slowly in the immobilized catalyst has been used by
these authors to signify transport limitations were present. The
larger substrates do diffuse more slowly; however, this observation
alone, of rate depending on reactant size, is insufficient to claim
transport limitations are responsible.
The rate ratio analysis must also account for entropy effects
associated with the limited mobility of the substrate relative to
the active site within the polymer. Roucis and Ekerdt (16) measured
the relative rates for hydrogenation of cyclooctene and cyclohexene
over an immobilized rhodium complex and found that the rate for
cyclooctene was less than that for cyclohexene for the homogeneous
rhodium complex, $RhCl(PPh_3)_3$, and in 1, 2, and 3 % divinylbenzene
polystyrene resins. Furthermore, they found that the relative rate
differed with divinylbenzene content. The relative rates were shown
to be intrinsic rate ratios (16). Roucis and Ekerdt proposed that
steric constraints due to the presence of the polymer support in
the vicinity of active rhodium affected activity. All the rhodium
sites were exposed to the same substrate concentration; the sites
displayed a different inherent activity toward the two substrates.
Selectivity effects may be related to diffusional limitations

in the other studies (31,32,36). The cyclohexene and cyclooctene
rate ratios of Grubbs et al. (31,32) were estimated to represent
intrinsic activity differences (16). There is insufficient
information to estimate the intrinsic limit for the largest
substrate tested, Δ^2-cholestene. The magnitude of the Weisz
modulus should be estimated for any rate selectivity study prior to
assigning intrinsic differences to diffusional effects.

A common approach in testing for transport limitations in
heterogeneous systems involves measuring the reaction rate over
different particles sizes of the same catalyst. Marconi and Ford
discuss this approach (17). The absence of an effect of particle
radius on the measured rate is used to establish the absence of
diffusional limitations. The differently sized catalyst particles
must be made from the same preparation, as done by Greig and
Sherrington (41), so as not to introduce uncertainties arising from
the reproducibility of a synthesis procedure.

Substrate limitations have been documented and quantitatively
described (11,16,17). Dooley et al. (11) present an excellent
description of modeling a reaction in macroreticular resin under
conditions where diffusion coefficients are not constant. Their
study was complicated by the fact that not all the intrinsic variables could be measured independently; several intrinsic parameters
were found by fitting the substrate transport with reaction model
to the experimental data. Roucis and Ekerdt (16) studied olefin
hydrogenation in a gel-form resin. They were able to measure the
intrinsic kinetic parameters and the diffusion coefficient independently and demonstrate that the substrate transport with reaction
model presented earlier is applicable to polymer-immobilized catalysts. Finally, Marconi and Ford (17) employed the same formalism
discussed here to an immobilized phase transfer catalyst. The
reaction was first-order and their study presents a very readable
application of the principles as well as presents techniques for
interpreting substrate limitations in triphase systems.

Summary

There are a number of factors which may influence the activity or
selectivity of a polymer-immobilized catalyst. Substrate diffusion
is but one. This article has reviewed the mathematical formalism
for interpreting reaction rate data. The same approach that has
been employed extensively in heterogeneous systems is applicable to
polymer-immobilized systems. The formalism requires an
understanding of the extent of substrate partitioning, the
appropriate intrinsic kinetic expression and a value for the
substrate's diffusion coefficient. A simple method for estimating
diffusion coefficients was discussed as were general criteria for
establishing the presence of substrate transport limitations.
Application of these principles should permit one to identify
experimental conditions which will result in the intrinsic reaction
rate data needed to probe the catalytic properties of immobilized
catalysts.

Nomenclature

a   Swollen catalyst bead radius (length)

c   Substrate concentration within the polymer catalyst (mole/vol)

$\hat{c}$ Dimensionless substrate concentration within the polymer catalyst

c_b Bulk phase substrate concentration (mole/vol)

c_{bo} Initial bulk phase substrate concentration for a batch reactor; bulk phase substrate concentration for a differential fixed-bed reactor (mole/vol)

c_s Substrate concentration (mole/vol)

D   Effective substrate diffusion coefficient ($length^2$/time)

D_o Substrate diffusion coefficient in the pure solvent ($length^2$/time)

E_{OBS} Measured Arrhenius activation energy (energy/mole)

E_{TRUE} True (intrinsic) Arrhenius activation energy (energy/mole)

F   Total molar feedrate (mole/time)

k   First-order reaction rate constant ($time^{-1}$)

K   Substrate partition factor (dimensionless)

n_s Total moles of substrate within the reactor (mole)

r   Radial coordinate (length)

R   Dimensionless radial coordinate

r_{OBS} The observed (measured) rate of reaction (moles/time·vol)

$(-r_s)$ Reaction rate (moles substrate reacted/time·vol)

$(-r_s)'$ Reaction rate (moles substrate reacted/time·mass of catalyst)

$(-r_s')$ Reaction rate (moles substrate reacted/time·volume of catalyst)

$(-\hat{r}_s)$ Dimensionless reaction rate defined by Equation 13

$(-\hat{r}_s')$ Dimensionless reaction rate defined by Equation 23

$(-r_{s,INTRINSIC})$ Intrinsic rate of substrate reaction
                     (mole/time·vol)

$[S]_p$ Concentration of substrate in bulk phase (mole/vol)

$[S]_r$ Concentration of substrate in resin phase (mole/vol)

t Time (dimensional)

V_b Bulk phase volume (vol)

$V_{CATALYST}$ Total swollen catalyst volume (vol)

V_R Total reactor volume (vol)

V_{Ro} Volume fraction of polymer in the swollen polymer resin
           (dimensionless)

W Total mass of catalyst in the reactor (mass)

X Substrate conversion (dimensionless)

y_p Product mole fraction (dimensionless)

α Parameter defined by Equation 18 (dimensionless)

η Effectiveness factor (dimensionless)

Φ Thiele modulus (dimensionless)

Φ' Thiele modulus defined in Equation 22 (dimensionless)

Φ_1 Thiele modulus for a first-order reaction (dimensionless)

Φ_w Weisz parameter defined in Equation 38 (dimensionless)

τ Dimensionless time

Literature Cited

1. Murrell, L. L. In "Advanced Materials in Catalysis" (J. J.
    Burton, and R. L. Garten, Eds.); Academic Press: New York,
    1977.
2. Yermakov, Y. I. Cat. Rev.-Sci. Eng. 1976, 13, 77.
3. Pittman, C. U., Jr.; Evans, G. O. CHEMTECH 1973, Sept., 560.
4. Heinemann, H. CHEMTECH 1971, May, 286.
5. Bailar, J. C., Jr. Cat. Rev.-Sci. Eng. 1974, 10, 17.
6. "Polymer-supported Reactions in Organic Synthesis" (P. Hodge
    and D. C. Sherring, Eds.); John Wiley: New York, 1980.
7. Collman, J. P.; Hegedus, L. S. In "Principles and
    Applications of Organotransition Metal Chemistry"; University
    Science Books: Mill Valley, CA, 1980.

8. Ciardelli, F.; Braca, G.; Carlini, C.; Sbrana, G.; Velentini, G. J. Mol. Catal. 1982, 14, 1.
9. Reinicker, R. A.; Gates, B. C. AIChE J. 1974, 20, 933.
10. Wesley, R. B.; Gates, B. C. J. Catal. 1974, 34, 288.
11. Dooley, K. M.; Williams, J. A.; Gates, B. C.; Albright, R. L. J. Catal. 1982, 74, 361.
12. Diemer, R. B., Jr.; Dooley, K. M.; Gates, B. C.; Albright, R. L. J. Catal. 1982, 74, 373.
13. Satterfield, C. N. In "Mass Transfer in Heterogeneous Catalysis"; M.I.T. Press: Cambridge, MA, 1970.
14. Petersen, E. E. In "Chemical Reaction Analysis"; Prentice Hall: Englewood Cliffs, NJ, 1965.
15. Aris, R. In "The Mathematical Theory of Diffusion and Reaction in Permeable Catalysts"; Vol. 1, Oxford University Press: Oxford, 1975.
16. Roucis, J. R.; Ekerdt, J. G. J. Catal. 1984, 86, 32.
17. Marconi, P. F.; Ford, W. T. J. Catal. 1983, 83, 160.
18. Butt, J. B. In "Reaction Kinetics and Reactor Design"; Prentice Hall: Englewood Cliffs, NJ, 1980.
19. Sherrington, D. C. In Ref. 6, p. 170.
20. Bischoff, K. B. A.I.Ch.E. J. 1965, 11, 351.
21. Carberry, J. J. A.I.Ch.E. J. 1962, 8, 557.
22. Smith, J. M.; Wakoa, N. Ind. Eng. Chem. Fund. 1964, 3, 123,
23. Weisz, P. B. Z. Phys. Chem. N.F. 1957, 11, 1.
24. Mackie, J. S.; Meares, P. Proc. Roy. Soc. London 1955, A232, 498.
25. Paul, D. R; Garcin, M.; Garmon, W. E. J. Appl. Poly. Sci. 1976, 20, 609.
26. Muhr, A. H.; Blanshard, J.M.V. Polymer 1982, 23, 1012.
27. Roucis, J. B.; Ekerdt, J. G. J. Appl. Poly. Sci. 1982, 27, 3841.
28. Vrentas, J. S.; Duda, J. L.; Ling, H.-C. J. Polym. Sci. Polym. Phys. Ed. 1984, 22, 459.
29. Strukul, G.; D'Olimpio, P.; Bonevento, N.; Pinna, F.; Graziani, M. J. Mol. Catal. 1977, 2, 179.
30. Pittman, C. U., Jr.; Smith, L. R.; Hanes, R. M. J. Amer. Chem. Soc. 1975, 97, 1742.
31. Grubbs, R. H.; Kroll, L. C. J. Amer. Chem. Soc. 1971, 93, 3062.
32. Grubbs, R. H.; Kroll, L. C.; Sweet, E. M. J. Macromol. Sci.- Chem. 1973, A7, 1047.
33. Guyot, A.; Graillat, C. H.; Bartholin, M. J. Mol. Catal. 1977/78, 3, 39.
34. Whitehurst, D. D. CHEMTECH 1980, Jan., 44.
35. De Croon, M.H.J.M.; Coenen, J.W.E. J. Mol. Catal. 1981, 11, 301.
36. Kim, T. H.; Rase, H. F. Ind. Eng. Chem., Prod. Res. Dev. 1976, 15, 249.
37. Joó, F.; Beck, M. T. J. Mol. Catal. 1984, 24, 135.
38. Manassen, J.; Dror, Y. J. Mol. Catal. 1977/78, 3, 337.
39. Grubbs, R. H.; Sweet, E. M. J. Mol. Catal. 1977/78, 3, 259.
40. Pittman, C. U., Jr.; Honnick, W. D.; Yang, J. J. J. Org. Chem. 1980, 45, 684.
41. Greig, J. A.; Sherrington, D. C. Polymer 1978, 19, 163.

RECEIVED August 28, 1985

5

Stability of Polymer-Supported Transition Metal Catalysts

Philip E. Garrou

Central Research–New England Laboratory, Dow Chemical USA, Wayland, MA 01778

During the last 15 years chemists have been enamoured with the idea of anchoring transition metal catalysts to organic polymers. Such studies have sought to produce heterogenized catalyst systems that are as active and selective as their homogeneous counterparts while having the distinguishing characteristic of being easily separable from the reaction media. As polymer supported catalysts are virtually certain to be more expensive than their homogeneous analogues, it is vital that they be recycled. Most researchers have been aware of the need to recycle but unfortunately very few studies have determined an activity vs. time relationship for such immobilized catalyst systems. It is difficult to accurately estimate the operating cost (cost of catalyst per unit of product) of a given polymer immobilized catalyst system. Such costs depend not only on the primary cost of the transition metal species and the functionalized support, but also on the catalysts activity and the length of its working life. Such factors can only be determined by experimentation for each individual case. While the cost/unit metal is about the same for polymer immobilized homogeneous catalysis and conventional heterogeneous catalysts, the cost of the matrix will obviously be higher for the immobilized systems (1). One must therefore show some unique activity vs. selectivity. conventional heterogeneous catalysts in order to warrant use of the polymer immobilized systems. It must be remembered, however, that use of such materials in industrial chemical processing necessitates knowledge of the long term stability of such catalyst systems since reactor downtime to unload, regenerate or replace the catalyst can dramatically impact the "operating cost" of the catalyst system. While typical publications in the field describe systems that "lose little activity upon recycling" one look at the data presented in Figure 1 reveals that even minimal activity losses of 3% per day could severely limit the useful lifetime of immobilized catalyst systems (2). Another such evaluation is depicted in Figure 2. This plot depicts the relative price for sale of a typical hydroformylation product vs the projected catalyst lifetime at a constant loss of 1 ppm Rh catalyst in the effluent stream (3). It is clear that 60+ days (1,440 hours) of continuous usage would be needed to approach the full economic potential of this supported catalyst system.

0097–6156/86/0308–0084$06.75/0
© 1986 American Chemical Society

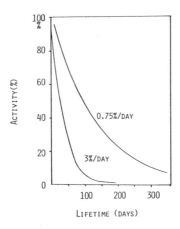

Figure 1. Activity vs. lifetime of immobilized catalyst systems. Reproduced with permission from reference 2. Copyright 1980 Springer-Verlag Heidelberg.

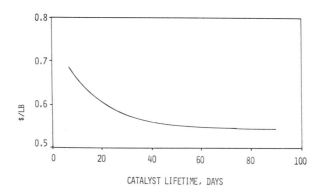

Figure 2. Relative price for sale of a typical hydroformylation product vs. projected catalyst lifetime at a constant loss of 1 ppm Rh catalyst in the effluent stream.

The basic question that has to be answered for each individual reaction examined is whether or not the products produced are of high enough value to warrant the use of immobilized catalysts. It is likely that the commercial impact of such catalyst systems will be greatest in the pharmaceutical area where the selectivities inherent in homogeneous catalysis are needed, the value of the products is high and the ability to easily separate the catalyst could pay for itself after a limited number of recycles. Impact in this area has not been obvious, probably because the polymer immobilized catalyst systems have not been in the hands of the typical synthetic organic chemist.

As we have mentioned, analysis of the modes by which catalyst deactivation occurs is important to understanding where such catalysis could best be applied. It has become clear that deactivation can occur by any or all of the following mechanisms: (a) "leaching" of the catalyst from the support due to the lability of the transition metal-functionality bond; (b) chemical instability of the support (both backbone and functionality) under reaction conditions; (c) production of metal crystallites in the polymer matrix under reductive conditions.

Lability

Metal loss may occur if the ligands by which the complexes are bound to the polymer undergo reversible dissociation during the reaction. For example, the well known Wilkinson catalyst, $RhCl(PPh_3)_3$, is known to dissociate a tertiary phosphine ligand in order to become catalytically active. If the phosphine bound catalyst depicted in Equation 1 underwent such

$$\bigcirc-\textcircled{\scriptsize O}-PPh_2RhCl(PPh_3)_2 \;\rightleftharpoons\; \bigcirc-\textcircled{\scriptsize O}-PPh_2 + RhCl(PPh_3)_2 \qquad (1)$$

dissociation, loss of Rh to the solution would obviously result. If such catalysts were used in a continuous flow reactor, rhodium would be slowly drained from the catalyst system. Increasing the ligand to metal ratio will inhibit dissociation reactions but will also obviously exhibit an inhibiting effect on the rate of the reaction.

Chemical Instability

It has generally been assumed that the bonds that link the catalyst to the polymer support are chemically stable under the reaction conditions one employs. Until recently, the literature offered little information in this regard, since lifetime studies are needed to properly evaluate stability. Recent publications have pointed out the chemical instability of the phosphorus-carbon bond of tertiary phosphine functionalized supports and the chemical reactivity of various nitrogen functionalized polymeric support materials under reaction conditions. If such chemical stability problems are present, the consequences are indeed serious. While a typical "leach" situation would necessitate a periodic reloading of the metal complex, cleavage of polymer functionality would necessitate replacement of both the metal complex and the polymer.

Metal Crystallite Formation

Clearly the most observed problem of polymer immobilized hydrogenation catalysts is the decomposition of the anchored organometallic complexes to metal particles. Interpretation of literature observations are complex since catalysts are not deactivated, but rather reveal catalytic activity usually only observed for heterogeneous catalysts. Crystallites can sometimes be detected by visual inspection for darkening. However, electron microscopy is a more reliable detection method. The presence of small amounts of metal in homogeneous catalyst systems has recently been reviewed by Laine (4). The presence of active "invisible" metal colloids in homogeneous catalyst systems has been the topic of much recent activity by Crabtree and co-workers (5). In general, reports that observe the presence of an induction period and/or unusual catalytic activity such as arene hydrogenation, conversion of syn-gas to hydrocarbons or nitrile reduction to amines, should be treated as suspect since such results are usually indicative of colloid or metal crystallite formation. In general it appears that the polymer matrix enhances this decomposition problem. Reports of crystallite formation, under reaction conditions where the corresponding homogeneous systems are known to be stable, continue to appear in the literature. It has been shown that such catalyst decomposition is retarded at high P/M ratios, but again this usually retards the rate of the reaction.

What follows is a review of immobilized catalyst research that has specifically addressed the question of catalyst stability. The literature coverage is selective, but comprehensive enough to present an accurate picture on the current state of research in this area.

Phosphorus Functionality

Haag and Whitehurst (6,7) first described the use of weak base anion exchange resins containing phosphorus and nitrogen functionality as supports for organometallic complexes in the early 70's. Because of its dominant role in numerous catalytic reactions (8,9) trivalent phosphorus has been the the predominant link by which metal complexes have been attached to a variety of organic polymers (10). Crosslinked phosphinated polystyrene has been the predominant support due both to its commercial availability and to the ease with which it can be prepared in the academic laboratory (11). Due to ease of preparation and better air stability, 1 and 2 have been the ligands most frequently studied although 3 and more recently 4 are available (12,13).

Lability of the Metal-Phosphorus Bond

By far the most complete study on elution or leaching of metal complexes under "industrial" flow conditions was reported by the

Mobil group (14) who examined olefin hydroformylation catalyzed by Rh complexes supported on $P-C_6H_4P(n-Bu)_2$. The catalyst resins contained 2-3 meq/g phosphorus and 0.2 meq/g Rh. Catalytic reactions were carried out in a tubular down-flow reactor. They observed that the rhodium concentrations in solution over the phosphine resins were proportional to the percent loading of the metal, indicating an equilibrium was established. As solvent, substrate or product polarity increased the rhodium concentration in solution increased. A higher concentration of rhodium was also noted when there was a higher CO partial pressure and/or the temperature was decreased as shown in Table 1.

A most important glimpse of the dynamics of a working catalyst was obtained by examining the catalyst concentration in the reactor bed as a function of time. Figure 3 reveals a depletion of Rh with time as a function of the distance down the bed. It is clear that depletion occurs in the direction of flow. For propylene hydroformylation under the given reaction conditions, 48% of the Rh was lost in   30 days. It was assumed that all of the rhodium loss was due to dissociation of the rhodium-phosphorus bond. These systems were not examined for cleavage of the phosphorus from the resin, nor was there any mention of special attempts to exclude the last traces of O_2 from the catalyst beds, both of which, as we shall discuss below, are potential problems.

British Petroleum group (15,16) also examined the use of rhodium complexes bonded to 1 and 5 for the hydroformylation of hexene-1.

Reactions were carried out in either stirred autoclaves or a single pass trickle flow reactor at 80-90 °C and   650 psig of CO/H_2. They noted that strict exclusion of O_2 from the reaction mixture lowered the Rh concentration in the effluent from 8 to 2 ppm in autoclave studies. A continuous flow pilot plant test of a catalyst consisting of 2.2% Rh loaded onto 1 revealed up to 90 ppm Rh in the products after 24 hours on stream. When such experiments were repeated under rigorously oxygen-free conditions the Rh content of the products fell to   1 ppm. Higher concentrations of oxygenodes/hydrocarbons 1:2 v/v, resulted in 7 ppm Rh in the liquid eluate. Rhodium catalysts supported on 5 were insoluble in hydrocarbons but exhibited significant solubility in aldehydes which facilitated the loss of rhodium from the reactor. It was concluded that even ppm levels of O_2 in the feedstocks would give rise to Rh elution due to loss of functionality because of rhodium catalyzed tertiary phosphine oxidation to tertiary phosphine oxide.

Chemical Reactivity of the Metal-Phosphorus Bond

As noted above one source of chemical reactivity is the observation of phosphine oxidation which lowers the availability of ligating functionality. Another mode of deactivation is actual rupture of the phosphorus-carbon bond. In order to understand the chemical instability of transition metal phosphine complexes on resins, we must first examine some recent data on the chemical stability of tertiary phosphines during homogeneously catalyzed

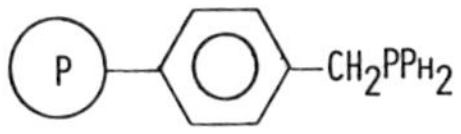

$$\text{(P)}-\langle\text{O}\rangle-CH_2PPH_2 \qquad \text{(P)}-\langle\text{O}\rangle-PE^+_2$$

$$\underset{\sim}{1} \qquad\qquad \underset{\sim}{3}$$

$$\text{(P)}-\langle\text{O}\rangle-PPH_2 \qquad \text{(P)}-(CH_2)_4-P(N-BU)_2$$

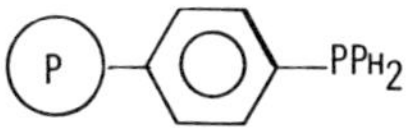

$$\underset{\sim}{2} \qquad\qquad \underset{\sim}{4}$$

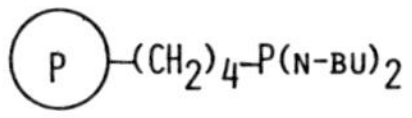

Table 1. Effect of Variables on Rh Concentration in Solution

Feed	H_2(psig)	CO(spig)	Temp($^\circ$C)	Loading Rh(%)	(mx10^6)
Benzene	1,000	1,000	100	0.08	1.5
	1,000	1,000	100	0.64	8.0
1-Hexene[a]	750	750	85	0.06	4.0
[b]	750	750	85	1.02	13.0
2-Ethylhexanal	1,000	1,000	100	0.08	4.0
	1,000	1,000	100	0.64	23.0

a) 25% converted to heptanals   b) 80% conversion to heptanals

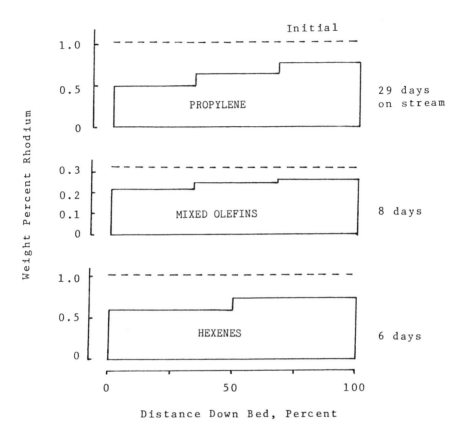

Figure 3. Depletion of Rh with time as a function of the distance
down the bed. Reproduced with permission from reference 14.
1977 Elsevier Science Publishers, B.U.

5

reactions. Gregorio and co-workers (17) first reported the slow decomposition of PPh_3 to $PPh_2(n-Pr)$ during rhodium catalyzed propene hydroformylation. Later studies by Tanaka and co-workers (18) in Japan, Abatjoglou and co-workers at Union Carbide (19,20) and our work with Allcock and Lavin (3,21) confirmed such results and pointed towards oxidative-addition of the phosphorus-carbon bond as a plausible reaction mechanism. Primary degradation products from R_3P were shown to be RH and RCHO. Secondary products as shown in Scheme 1 are derived from hydrogenation of RCHO and subsequent hydrogenolysis or homologation of RCH_2OH. ^{31}P NMR studies have observed R_2PH when the hydroformylation reaction is run in the absence of olefin substrate and R_2PR' when the reaction is run in the presence of olefin. Such tertiary phosphine degradation has now been observed for Co, Rh, Ru, Pd, Ni, and Os tertiary phosphine complexes (23,24). In addition to hydroformylation, such degradation has been observed in carboxylation, hydrogenation and dehydrogenation reactions (23,24). Aryl group scrambling catalyzed by group 8-10 transition metals have also been recently described (19,21) with species such as 6 implicated as intermediates.

It thus appears clear that given the proper circumstances, tertiary phosphines may become significantly involved in the reaction chemistry of the compounds of which they are a part. With such chemistry in mind, it was thus not surprising that examination of the hydroformylation reaction catalyzed by by $Co_2(CO)_8$ supported on 2 or 7 revealed both loss of Co and loss of phosphorus (3,21).

A decline in catalytic activity with use was detected for reactions catalyzed by either species. Polymers 2 and 7 in the absence of cobalt both revealed excellent stability at 190°C (hydroformylation temperatures). This is illustrated by the TGA curves shown in Figure 4. Curve A shows an onset of decomposition for phosphinated polyphosphazene of 400°C, slightly better than that of phosphinated polystyrene (curve B, 20% crosslinked; curve C, 2% crosslinked). Loss of phosphorus was observed over a period of 45 hours for a catalyst derived from 2 (2% DVB crosslinked). The data depicted in Figure 5 reveal benzene, toluene, benzyl alcohol, diphenylphosphine and triphenyl phosphine as cleavage products. If one recalls the previously discussed homogeneous results it should be clear that the PPh_3 is derived from a phosphido intermediate such as 8.

Cleavage, as depicted, would give a PPh_3 cobalt complex. Such PPh_3 would only build up in solution to a small degree since it would compete with the resin for bonding to the Co and would itself be susceptible to the phosphorus-carbon bond cleavage reaction.

In a similar fashion elemental analysis of recovered $7/Co_2(CO)_8$ revealed a 7% drop in carbon and phosphorus, while the nitrogen content remained constant. Such data indicate that a loss of the pendant phosphine groups is not occuring by a random chain scisson or depolymerization. PPh_3 and Ph_2PH can also be be detected in the reaction solution after catalysts derived from 7 have been filtered

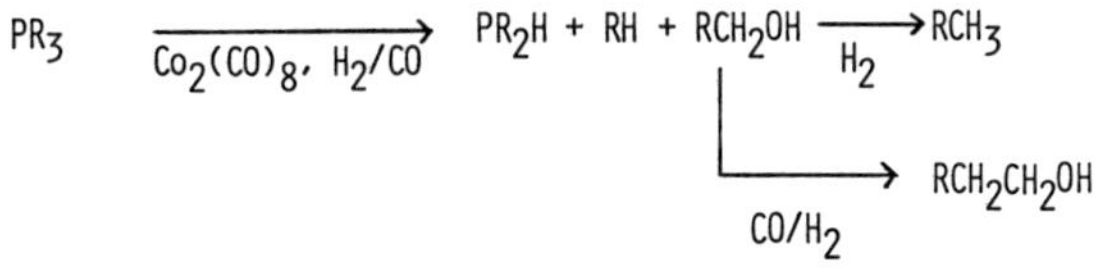

$$PR_3 \xrightarrow[\text{Co}_2\text{(CO)}_8, \ H_2/\text{CO}]{} PR_2H + RH + RCH_2OH \xrightarrow[H_2]{} RCH_3$$

$$\xrightarrow[\text{CO}/H_2]{} RCH_2CH_2OH$$

Scheme 1

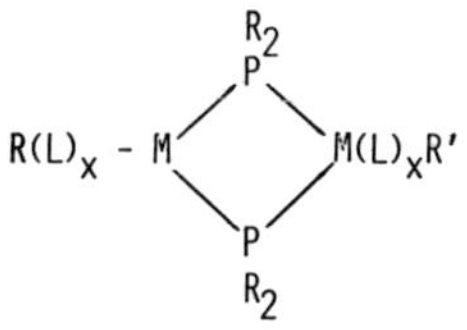

$$R(L)_x - M \underset{\underset{R_2}{P}}{\overset{\overset{R_2}{P}}{\Big\langle}} M(L)_xR'$$

6

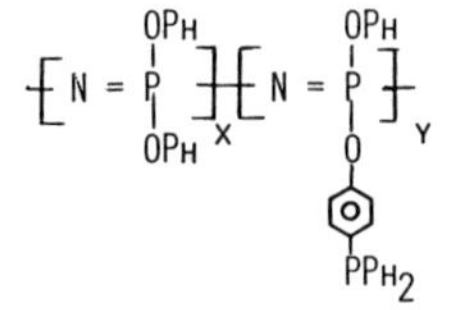

$$\left[N = \overset{\overset{OP_H}{|}}{\underset{\underset{OP_H}{|}}{P}} \right]_x \left[N = \overset{\overset{OP_H}{|}}{\underset{\underset{O}{|}}{P}} \right]_y$$

7

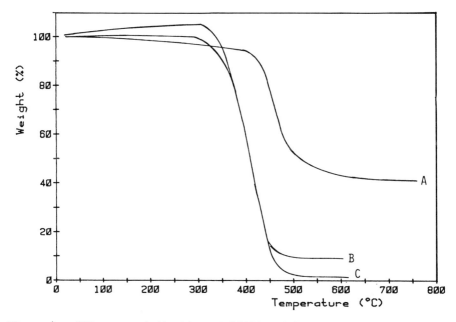

Figure 4.   TGA curves indicating stability of Polymer 2 and 7 at 190 C
in the absence of cobalt. Curve A shows the onset of decomposition for
phosphinated polyphosphazene of 400 C, slightly better than that of
phosphinated polystyrene (Curve B, 20% cross-linked; Curve C, 2% cross-
linked.

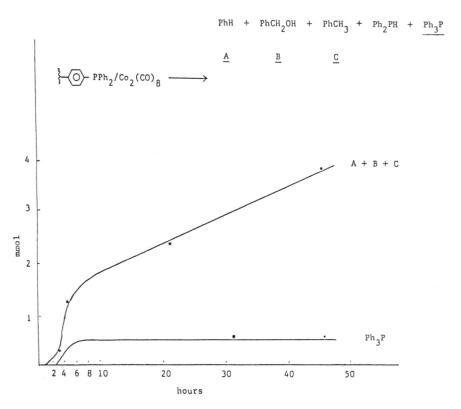

Figure 5.   Data revealing benzene, toluene, benzylalcohol, diphenylphosphine, and triphenylphosphine as cleavage products.

off, in agreement with results obtained for catalysts derived from **2**. In a similar fashion close analysis of the hydroformylation catalysts derived from $Co_2(CO)_8$ and **9** (mw 161,000)(**22**) revealed an 18% loss of phosphorous from the polymer and substantial lowering of the molecular weight.

Formation of Metal Crystallites

Collman and co-workers (**25**) first reported the use of **2** to support transition metal species in 1972 when they prepared complexes of Rh, Co, and Ir. They found that complexes of **2** when exposed to oxygen or 1 atmosphere of H_2 decomposed to metal particles and that such catalysts were active for typical "heterogeneous" reactions such as the hydrogenation of arenes. They compared the activity of $Rh_6(CO)_{16}$ on **2** to that of 5% Rh on Al_2O_3.

Wilkinson's catalyst, $(PPh_3)_3RhCl$, is a very useful hydrogenation catalyst and has found great utility in the hands of synthetic organic chemists. It is not poisoned by sulfur functionality, it is selective for mono- and di-substituted olefins, it does not catalyze hydrogenolysis and such catalysts containing chiral phosphine ligands permit asymmetric induction. The major drawback to its use is that it has to be removed from the reaction product by chromatography. This is time consuming, costly and wasteful of rhodium.

Various groups have sought through the years to immobilize Wilkinson-like rhodium catalysts on phosphinated resins to overcome such problems. Strukul and co-workers (**26**) have studied rhodium (I) supported on phosphinated polystyrene (2 or 20%) divinylbenzene coploymers. Catalysts at various phosphorous and rhodium contents were examined for the hydrogenation of cyclohexene. They observed an induction period and catalyst activity dependent on the P/Rh ratio, the catalysts becoming less active at higher P/Rh. After a few recycles catalysts having low P/Rh ratios displayed activity insensitive to the presence of the ligand, in contrast to fresh catalyst or homogeneous analogues. In similar fashion acetonitrile poisoned fresh catalyst but had no effect on recycled catalyst. These results were interpreted in terms of catalysis by rhodium metal.

In a related study Guyot and coworkers (**27**) examined similar catalysts varying the crosslinking by 2, 25, and 60% divinyl benzene incorporation. After rhodium fixation the catalysts we tested for hydrogenation of 1-hexene in ethanol. The activity increased with decreasing rhodium content. Upon treatment with H_2 the catalyst turned grey-black. The catalysts were also examined for benzene hydrogenation activity. The activities were always low for fresh catalyst and increased after hydrogen treatment or with age. This happened more quickly for the macroporous resins then for the gel resins. If the H_2 treatment was carried out at 100 °C the reduction to metal was also facile for the gel resins. There is also a relationship to the Rh/P ratio, the activity for benzene hydrogenation developing more quickly for higher Rh/P ratios. Similar results were obtained for similar Rh and Ir catalysts supported on 1 (27,27).

Gates and co-workers (28) allowed $Rh_6(CO)_{16}$ to react with 2 (1.9-5.4% crosslinked) at 25°C and produced catalysts active for cyclohexene hydrogenation. When the $Rh_6(CO)_{16}$ was allowed to react at 50 °C with the functionalized polymer, the resulting product was grey-black and TEM (transmission electron microscopy) revealed 25-40A° aggregated metal particles on the catalyst surface. When oxygen was allowed to contact the catalysts the rhodium aggregation occurred more quickly. Only with polymers having unusually high P/Rh ratios, i.e. 40/1, was this phenomenon slowed down. Similar results were obtained for $Ir_4(Co)_{12}$ and analogous polymer supports (29). Although it could be shown by IR that Ir_4 carbonyl clusters were the predominant species present in the functioning hydrogenation catalysts, their concentration did not correlate with reaction rates. It was thus questioned whether catablytic activity was due to undetectably small amounts of metal aggregates.

Pittman and coworkers (30) studied the reaction of butadiene with carboxylic acids to give octadienyl esters catalyzed by Pd loaded phosphine functionalized styrene–divinylbenzene resin beads (1% DVB). Although the resins were recycled they noted a significant decrease in activity. The Pd content dropped, for example, from 1.74 to 1.29% after 2 recycles and to 1.05% after 5 recycles. In addition to Pd loss they noted Pd metal deposited in the resin matrix. Higher P/Pd ratios appeared to inhibit metal agglomeration.

Nitrogen Functionality

Relatively little research has been done on nitrogen functionalized resins. Figure 6 depicts the commercial resins 10, 11 and 12, their maximum recommended operating temperature and the resin type.

Moffat (31,32) was among the first to study catalysts derived from such nitrogen containing ligands when he examined 2-4% crosslinked poly-2-vinylpyridine, 12, as a support for cobaltcarbonyl hydroformylation catalysts. He found that the soluble cobalt concentration at reaction temperature was 100 ppm. When comparable homogeneous reactions were run with equivalent concentrations of cobalt the rates were comparable. The rate of metal release and uptake by the polymer matrix was found to be an important consideration. Changes in cobalt concentration paralleled changes in temperature or solvent. For example, rapid change of the temperature from 182 to 124 °F, revealed that the equilibrium concentration of cobalt in solution was reestablished in 1 to 3 minutes. When n-heptane was pumped into an autoclave containing 13 mmol of cobalt on 5 grams of 12 at 191 °F, equilibration of the cobalt between solid and solution occured before the first sample could be removed in less than 1.5 minutes.

The Mobil group (6,7,14) pioneered the use of Rh/10 or 11 catalysts for hydroformylation. When the amine resin catalysts were analyzed versus time on line in a packed bed reactor, redistribution of the rhodium was observed reminiscent of the previously described

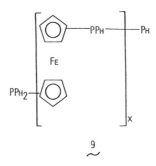

9

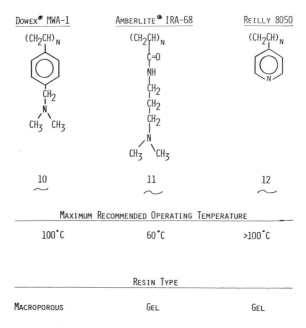

Figure 6. Comercial resins 10, 11, and 12, along with their maximum recommended operating temperatures and their resin type. Reproduced with permission from reference 33b. Copyright 1985 Elsevier Science Publishers, B.U.

phosphorus studies. Rhodium concentration in solution was shown to be 170×10^{-6} molar for a 2% Rh loading at 25% conversion of olefin to aldehyde. Using pure benzene as feed it was shown that both increased temperature and decreased CO pressure led to reduced Rh concentration. Thus at 1000 psig H_2 and 100 °C an increase in the CO pressure from 250 to 1000 psig effected an increase in soluble Rh from 9 to 11 $\times 10^{-6}$ molar. At 2000 psig of 1:1 H_2:CO the equilibrium concentration of Rh decreased from 11 to 9 $\times 10^{-6}$ molar when the temperature was raised from 100 to 125 °C. It was again shown that more polar solvents or products increased the Rh leach.

Our work at Dow has also sought to examine the commercial viability of Rh/10, 11, 12 catalysts for the hydroformylation of a variety of olefins (33). The results that follow were obtained during laboratory and pilot plant studies on the hydroformylation of dicyclopentadiene (DCPD) to dicyclopentadiene dimethanol (DCPDDM) using various Rh carbonyl sources immobilized on 10, 11, and 12.

Thermal Stability

The thermal stabilities of the functionalized resins were eva luated by thermogravimetric analyses (TGA). The temperature employed for the hydroformylation catalyst life studies was 130 °C. TGA evaluations at this temperature indicated that none of the three resins lost weight during 12 hour isothermal studies. For 11, although the recommended operating temperature is only 60 °C, a TGA study carried out for 48 hours revealed no weight loss. Although the resins probably do have differing thermal stabilities, under the defined experimental conditions each resin was "stable".

Catalyst Deactivation as a Function of Amine Resin

Catalyst deactivation as a function of the relative rate of formation of alcohol (DCPDDM from DCPD) is depicted in Figure 7. It is clear that deactivation was a problem with rhodium catalysts derived from 11 and 12. We will show later that much of that deactivation was caused by the chemical instability of the catalysts under reaction conditions.

The studies using 10 as the catalyst support revealed no significant deactivation, as shown in Figure 7, until the temperature was raised above 130° C. (See Figure 8.) For batch 26, the temperature was increased from 130 to 140° C and then to 150° C at Batch 36. Deactivation obviously occurs at the higher temperatures.

Deactivation Due to Resin Chemical Stability

The resins were evaluated to identify any changes that had occured during the lifetime studies, with particular attention being given to the amine functionality.

Deactivation of 12 was studied by infrared spectroscopy and determination of elemental nitrogen. There was a decrease in the weight percent nitrogen from 12.3 to 7.5% for the new and used resins, respectively. Infrared analyses were performed after 30- and 50-batch runs. In addition to the bands due to the pyridine ring, there were bands due to primary hydroxyl, and aliphatic C-H. The

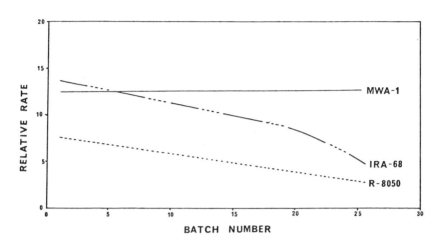

Figure 7.   Catalyst deactivation as a function of the relative rate of formation of alcohol (DCPDDM from DCPD).   Reproduced with permission from reference 33b.   Copyright 1985 Elsevier Science Publishers, B.U.

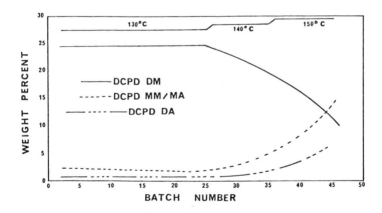

Figure 8.   Catalyst deactivation as a function of weight percent. Reproduced with permission from reference 33a.   Copyright 1985 Elsevier Science Publishers, B.U.

infrared spectrum of catalyst reused 50 times revealed a higher concentration of methylene groups than the one of a catalyst used 30 times. In addition, when comparing the absorbance at 810 cm^{-1} (2 adjacent protons on an aromatic ring), one observed intensities that represent 55% more pyridine on the unused catalyst. A possible explanation of the deactivation of 12 lies in the reduction of pyridine to a saturated piperidine moiety which then undergoes reductive alkylation with a DCPD aldehyde intermediate followed by hydrogenolysis of the hemiaminal. Such a process is depicted in Scheme 2.

Several types of analyses were performed on fresh and used 11 rhodium-loaded catalysts in order to determine any chemical sources of deactivation. Infrared analysis of the used resin revealed a decrease in the intensity of the Amide II band; the amide carbonyl band did not decrease; there was a new weak band at 1725 cm^{-1}; a new band due to a primary hydroxyl and a decrease in the intensity of the C-H stretch for the N-CH$_3$ groups. Elemental analysis data for nitrogen for three lifetime studies were 15.37, 15.83, 16.12% and 5.83, 6.39, 8.09% for new and used catalysts, respectively. In comparing the new and used catalyst, there was 60% less nitrogen, which could be due to cleavage of functional nitrogen or increasing the molecular weight of the pendant group.

TGA/mass spectral studies confirmed the presence of a DCPD moiety bound, not adsorbed, to the resin backbone. Scheme 3 shows some possible reactions that explain the chemical deactivation of 11. No single pathway in the figure can explain all of the data from the analyses of the new and used 11/ rhodium-loaded catalyst. The ester in Pathway A would satisfy the presence of a 1725 cm^{-1} band and a DCPDDM fragment. The Hoffman elimination of Pathway B is one reaction that could take place at the tertiary amine after quaternization by an anionic rhodium hydrido cluster. The reaction of the substituted amide with DCPDMA is not without precedence and is required for the decrease in the Amide II band relative to the amide carbonyl. This would also lead to a DCPD DM fragment bonded to the resin. Pathway D depicts an amide reduction followed by: (1) cleavage to yield an aldehyde which is then reduced to the alcohol, and (2) hydrogenolysis to the secondary amine. It should be recognized that the catalyst deactivation is slow, taking about 150-200h. When similar studies were conducted on 12, no loss of activity was observed (Figure 7). Consistent with these observations, no differences in infrared analysis or elemental analysis were observed between fresh and used catalyst resins. These results clearly show that the MWA-1 backbone was far more stable to chemical degradation/reaction under operating hydroformylation conditions.

Metal Leach

The reaction profile depicted in Figure 9 reveals the correlation between aldehyde and rhodium concentration in solution vs time. The correlation between aldehyde concentration and Rh concentration is shown replotted in Figure 10. The catalyst life study data unmistakably points to aldehyde as being responsible for the Rh leach.

Metal leach places restrictions on how to construct a process.

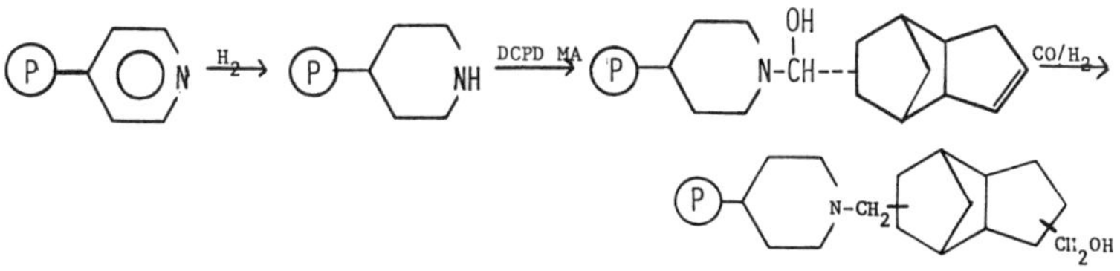

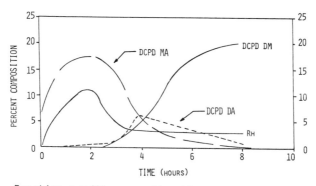

Scheme 2

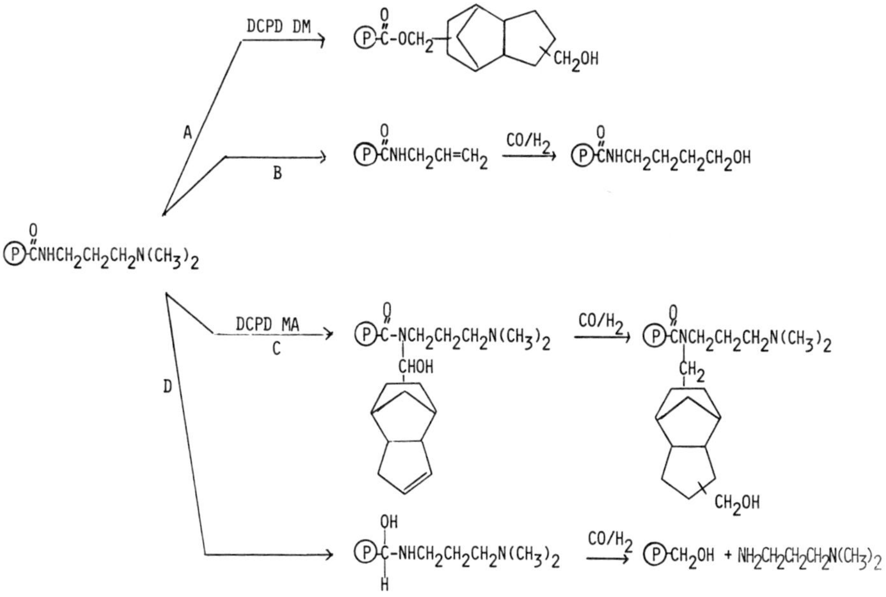

Scheme 3

Figure 9.   Reaction profile revealing the correlation between aldehyde and rhodium concentration solution vs. time.   Reproduced with permission from reference 33a.   Copyright 1985 Elsevier Science Publishers, B.U.

Using a conventional continuous flow reactor, after a period of time the Rh would wash down and out of a fixed bed reactor. Such migration down the catalyst bed was reported by Mobil workers. Our work concluded that the only way to reduce rhodium leach in such processes is to operate batch-wise. One to three ppm of Rh was always observed in the product stream even under optimal conditions in batch operation.

Cyclopentadienyl (Cp) Ligands

Cyclopentadienyl groups attached to cross-linked polymers have been examined as ligands for catalytic systems involved in olefin-hydrogenation (34,35,36,37), -isomerization (34,35), -disproportionation (35) and hydroformylation (34,35,36,37) as well as acetylene cyclotrimerization (35) and Fischer-Tropsch (38) reactions. Various synthetic methods of attachment have been devised (13,14,15,16). It was felt that catalyst attachment via the Cp ligand would provide kinetically stable systems in which no separation or
detachment of the metal species would take place. In addition it was hoped that 15 and 16 would provide better mechanical, chemical and thermal stability than polystyrene which is known to slough off polymer, has a ceiling temperature (T_c) of 150°C and has poor thermal stability. Catalysts derived from 13 - 16 were found to be no more stable than those of other functionalities as described below.
    Grubbs and coworkers (35) while examining Rh and Co catalysts derived from 14 reported the loss of infrared CO stretches and visual darkening of the catalysts after use for hydrogenation of olefins, aldehydes or ketones, cyclohexene disproportionation to benzene and cyclohexane or the cyclotrimerization of a wide variety of acetylenes. Stille (36) using a rhodium catalyst prepared from 14 observed activity for the hydrogenation of benzene that increased with reuse, a phenomenon usually associated with metal crystallite formation. Rhodium catalysts of 15 and 16 used to hydroformylate octene-1 revealed a loss of carbonyl adsorptions and a loss in catalytic activity upon reuse (37).
    Vollhardt and co-workers (38) observed Co catalysts of 14 were inactive in alkyne cyclizations (vs. homogeneous CpCo(CO)$_2$) but active for methanation and Fischer-Tropsch reactions. Turnover of CO was reported to be 0.01 mmol/molCo/hr. Stille later reported (36) a failure to reproduce these results using "the same" catalytic species under similar conditions. This author tends to agree with the observations of Collmann and Hegedus (39) that "In the absence of other evidence, formation of hydrocarbons from CO and H$_2$ is diagnostic of metal particles arising from degradation of the homogeneous precatalysts".
    It should also be pointed out that disagreement exists over the results of hydroformylation of olefins using Cp supported CpCo(CO)$_2$. Both activity (34,36) and a lack of activity (35) have been reported for very similar catalysts under similar conditions. Results in this

author's laboratory (40) have determined that peroxides are the cause of these irreproducible results. If the olefin substrates are not vigorously treated to remove the last traces of peroxide contamination, activity is observed, presumably due to oxidation of the Co off the support. When one uses rigorously cleaned olefins no activity is observed.

Miscellaneous Functionality

Farona (41) found that polystyrene tricarbonylmolybdenum, like its homogeeous counterpart, catalyzed the Friedel-Crafts reaction. The polymer anchored tricarbonylmolybdenum, although showing decreased activity with respect to the homogeneous system, was reused several times with little loss of activity. To be more exact 1.01 g of 17 was refluxed in 50 ml of toluene under N_2 for two hours. The toluene solution, filtered away from the beads and evaporated to dryness yielded 4.2 mg of toluenetricarbonylmolybdenum, which represents 0.15% exchange of the $Mo(CO)_3$ groups from poystyrene to toluene. Such intermolecular transfer reactions are commonplace (42). Such results are indicative of behavior similar to that previously discussed for the Rh catalysts on phosphorus or nitrogen functionalized resins. Such equilibration would presumably drain catalyst from the resin if the reactions were conducted in a continuous flow system.

Rollman (43) synthesized a series of polymer-bonded metalloporphyrins. Functionalized tetraphenylporphyrines were attached to porous polystyrene resins via amine, carbonyl and ester linkages and metal ions (Co,Ni,Cu,Zn) were then incorporated into the structures. The porphyrin polymers that contained both oxidation (Co) and proton-acceptor sites (amine and carboxylate groups) were effective catalysts for the oxidation of thiols to disulfides. When the catalyst was exposed to a refinery stream containing only 180 ppm mercaptan sulfur, deactivation occurred. Deactivation was thought to occur via oxidation of the porphyrins by free radicals known to be present in such catalytic systems.

Conclusions

It is clear that if industry wishes to use a polymer immobilized catalyst, a good deal of research will be needed to determine the durability of the various functionalized supports under the particular reaction conditions. Investigation of the mechanism of breakdown, when it occurs, will hopefully indicate ways in which improved immobilized systems might be prepared.

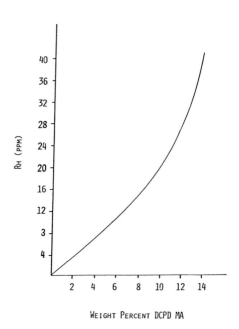

Figure 10. The correlation between aldehyde and Rh concentration replotted
from Figure 9.   Reproduced with permission from reference 33a.   Copyright
1985 Elsevier Science Publishers, B.U.

LITERATURE CITED

1. Whitehurst, D. D., CHEMTECH, **1980**, 44.
2. J. Falbe, "New Syntheses with Carbon Monoxide"; Springer-Velag: New York, **1980**.
3. Dubois, R. A.; Garrou, P. E.; Hunter, D. L., unpublished results.
4. Laine, R. M., J. Molec. Catal., **1982**, 14, 137.
5. Anton, D. E.; Crabtree, R. M., Organometallics, **1983**, 2, 855.
6. Haag, W. O.; Whitehurst, D., USP 4 111 856, **1974**.
7. Haag, W. O.; Whitehurst, D., USP 4 098 727, **1978**.
8. Parshall, G. W. "Homogeneous Catalysis"; Wiley-Interscience: New York, **1980**.
9. Heck, R. F., "Organotransitionmetal Chemistry"; Academic Press: New York, **1980**.
10. Hodge, P.; Sherrington, D. C., "Polymer-Supported Reactions in Organic Synthesis"; Wiley: New York, **1980**.
11. Lieto, J.; Milstein, D.; Albright, R. L.; Minkiewicz, J. V.; Gates, B. C., CHEMTECH, **1983**, 46.
12. McKinley, S. V.; Rakshys, J. W., USP 3 708 462, **1973**
13. Kim, B.; Kodomari, M.; Regen, S. L., J. Am. Chem. Soc., **1984**, 49, 3233.
14. Lang, W. H.; Jurewicz, A. T.; Haag, W. O.; Whitehurst, D. D.; Rollman, L. D., J. Organometal. Chem., **1977**, 134, 85.
15. Allum, K. G.; Hancock, R. D.; Howell, I.V.; Pitkethler, R. C.; Robinson, P. J., J. Catal., **1976**, 43, 322.
16. Allum, K. G.; Hancock, R. D.; Howell, I. V.; Pitkethler, R. C.; Robinson, P. J., J. Organometal. Chem., **1975**, 87, 189.
17. Gregorio, G.; Montrasi, G.; Tampieri, M.; Calaveri d'Oro, P.; Pagoni, G,; Andreeta, A., Chem. Ind. (Milan), **1980**, 62, 389.
18. Sakakura, T.; Kobayashi, T. A.; Hayoshi, T.; Kawabata, Y.; Tanaka, M.; Ogata, I., J. Organomet. Chem., **1984**, 267, 171.
19. Abatjoglou, A. G.; Bryant, D. R.; Organometallics, **1984**, 3, 923.
20. Abatjoglou, A. G.; Bryant, D. R., Organometallics, **1984**, 3, 932.
21. Dubois, R. A.; Garrou, P. E.; Lavin, K. D.; Allcock, H. R., Organometallics, **1984**, 3,649; Dubois, R. A.; Garrou, P. E., submitted for publication.
22. Fellman, J. D.; Garrou, P. E.; Withers, M. P.; Seyferth, D.; Traficante, D. D.; Organometallics, **1983**, 2, 818; Fellman, J. D.; Garrou, P. E., Unpublished Data.
23. Garrou, P. E.; Dubois, R. A.; Jung, C. W., CHEMTECH, **1985**.
24. Garrou, P. E., Chem. Rev., 1985, 85, 171.
25. Collman,J. P.; Hegedus, L. S.; Cooke, M. P.; Norton, J.R.; Dolcetti G.; Marquardt,D. N., J. Am. Chem. Soc., 1972, 94, 1789.
26. Strukul, G.; D'Olimpio, P.; Bonivento, M.; Pinna, F; Graziani, M., J. Molec. Catal., **1977**, 2, 179.
27. Guyot, A.; Graillat, C.; Bartholin, M., J. Molec. Cat., **1977**, 3, 39; Bartholin, M.; Graillat, C.; Guyot, A., J. Molec. Cat., **1981**, 10, 361; Bartholin, M.; Graillat, C.; Guyot, A., J. Molec. Cat., **1981**, 10, 377.

28. Jarrell, M. S.; Gates, B. C.; Nicholson, E. D., J. Am. Chem. Soc.
    1978, 100, 5727.
29. Lieto, J.; Rafalko, J. J.; Minkiewicz, J. V.; Rafalko, P. W.;
    Gates, B. C. "In Fundamental Research in Homogeneous Catalysis"
    Vol. 3; Tsutsui, M., Ed.; Plenum: New York, 1979, p.637.
30. Pittman, C. U.; Wuu, S.K.; Jacobson, S. E.; J. Catal., 1976,
    44, 87.
31. Moffat, A. J., J. Catal., 1970, 19, 322.
32. Moffat, A. J., J. Catal., 1970, 18, 193.
33. Hunter, D. L.; Moore, S. E.; Dubois, R. A.; Garrou, P. E.,
    Applied Catalysis, 1985, 19, (a) 259-273; (b) 275-285.
34. Brintzinger, H.H.; von Vicarıster, H.; GB 2,000,153A; 1978.
35. Chang, B. H., Grubbs, R. H., Brubaker, R. H., J. Organomet.
    Chem., 1979, 172, 81.
36. Sekiya, A.; Stille, J. K., J. Am. Chem. Soc., 1981, 103, 5096.
37. Verdet, L.; Stille, J. K., Organometallics, 1982, 1, 380.
38. Perkins, P.; Vollhardt, K.P.C., J. Am. Chem. Soc., 1979, 101,
    3987.
39. Collman, J. P.; Hegedus, L. S., "Principles and Applications of
    Organometallic Chemistry"; 1980, p. 450.
40. Dubois, R. A., private communication.
41. Tsonis, C. P.; Farona, M. F., J. Organometal. Chem., 1976, 114,
    293.
42. Garrou, P. E., "Advances in Organometallic Chemistry", vol. 23,
    1984, p. 95.
43. Rollman, L. D., J. Am. Chem. Soc., 1975, 97, 2132.

RECEIVED November 12, 1985

Polymeric Photosensitizers

D. C. Neckers

Center for Photochemical Sciences, Bowling Green State University, Bowling Green, OH 43403

This paper reviews polymeric photosensitizers from the first case indicating that the concept was possible to polymer-rose bengal. The review focuses on the chemistry of the appended small molecule, and the role of the polymer. The most important polymeric photosensitizer to date is polymer rose bengal – a heterogeneous source of singlet oxygen; and the most extensive review is of this system.

Immobilized photosensitizers derive ultimately from the work of Kautsky and deBruijn (1). In determining the mechanism of "Photodynamic Action" (2), they carried out the following experiment:

Two dyes were separately absorbed at the expanded inner surfaces of a gel. One surface contained the photosensitizer adsorbate, and the other surface the acceptor sorbate. When mixed, the particles are in contact with one another, but the sensitizer and the acceptor are not. The critical experiment for the detection of diffusable, activated oxygen involved irradiating the sensitizer dye, and observing photochemical bleaching of the acceptor dye. The sensitizer was trypaflavine and the acceptor was leuco-malachite green.

Kautsky concluded that an excited state of oxygen was involved in the bleaching of the particle immobilized methylene blue and he also reasoned that the metastable oxygen was singlet oxygen (3). Though 30 years of erroneous conclusions and false premises led to a number of other suggested mechanisms for dye sensitized photooxidation, in 1964 singlet oxygen was confirmed as the

0097-6156/86/0308-0107$07.25/0
© 1986 American Chemical Society

reactive intermediate produced when low energy triplet dyes are
irradiated in air or pure oxygen.

The first uv sensitizer immobilized to a synthetic polymer
for the purpose of actually carrying out a photosensitized
process was reported by Moser and Cassidy (4). These workers
reported poly(acrylophenone) photoreduced to the corresponding
α-phenethyl alcohol. They concluded that the photoreduction
chemistry of poly(arylophenone) was like that of the
photoreduction chemistry of acetophenone. Peter Leermakers (5)
reported that poly(phenylvinylketone)--a "brittle, plastic
mass"--was used for heterogeneous energy transfer to three
dienes: norbornadiene, cis-piperylene and myrcene. This work
suffered from incompleteness, however, and when we repeated the
work according to Leermakers' experimental protocol (5), we
observed that the polymer was significantly soluble and also
contained oligomeric sensitizers. Our conclusion was that
Leermakers and James (5) did not consider an authentic heterogeneous energy transfer process, but were instead looking simply at
energy transfer reactions of dissolved small acetophenone analogs
in solution.

Our own interest in polymeric energy transfer donors derived
from Merrifield's work (6). Our first three papers (7) belie the
fact that we were considering polymeric uv sensitizers at its
inception, too. Thus we synthesized acetyl co(styrene-
divinylbenzene) only to discover (P)-AlCl₃ (7). And we made "(P)
-benzoyl" only to discover what we (8) and several others (9)
already knew--that acetophenone and benzophenone triplets were
virtually identical in chemistry to alkoxy radicals. Hydrogen
abstraction by alkoxy radicals on benzylic centers is a well
known chemical process. (P)-Benzoyl (10) presented to the world
not an effective heterogeneous energy transfer donor, but a procedure for enhancing the photocrosslinking of polystyrene. Asai
and Neckers (11) resolved this problem by fluorinating the polymeric backbone, but the results were unspectacular.

Enter Rose Bengal

Our love affair with rose bengal is made clear by our recent
publications (12). (P)-Rose bengal(13) came about for three
rather obvious reasons:

1. The rose bengal triplet energy is about 40 kcal/mol.
This excited state hasn't the energy to abstract hydrogen from a
Merrifield poly(styrene-co-divinylbenzene) backbone. Rose bengal
triplets are completely quenched by oxygen at concentrations over
10^{-3}m.

2. Heterogeneous energy transfer from immobilized dyes to
oxygen had precedent in Kautsky's experiments. Long wavelength
visible light--particularly in the red--is not absorbed as
extensively by polymer beads as is uv radiation. In addition,
dye chromophores have huge extinction coefficients. A little
incident radiation goes a long way in activating the dyes.

3. Rose bengal has two nucleophilic centers--the C-2'
carboxylate and the C-6 phenoxide. One or both might displace
chloride from chloromethyl functions in Merrifield's "Bio-Beads".

These are chloromethylated poly(styrene-co-divinylbenzene) beads.

In fact, as the (P)-rose bengal patent discloses (13), approximately 40 different dyes were immobilized (or at least we tried to immobilize 40 different dyes) to Merrifield resin. We also reported immobilization of rose bengal to cellulose, DMAE cellulose, paper, nylon, and polyethylene in that patent application. However, of the immobilized dyes tested in 1971, rose bengal immobilized to Merrifield resin was the most effective. In fairness, though, the test protocol employed in these early experiments (14) was naive, and many factors discovered by my recent coworkers (15) make it clear that we should reinvestigate other immobilized dyes. About the only thing the early experiments established was that (P)-rose bengal was a good source of singlet oxygen in non-polar solvents. The quantum yields reported in the early studies were incorrect, since the actual quantum efficiency of energy transfer depends on oxygen pressure.

Rose bengal presents an interesting historical case in its own right. Related to Baeyer's fluorescein (16) and Emil Fischer's eosin (17), it was synthesized originally by Gnehm (18), the first president of the ETH in Zurich. Gnehm was employed early in his career by the Swiss chemical firm Offenbach and Geschwanden, the predecessor in the 19th century to Chemical Industry Basel--CIBA (19).

Gnehm's synthesis of rose bengal was made possible because CIBA held the patent on the process for forming perchlorophthalic anhydride (20) from phthalic anhydride by gas phase chlorination using antimony pentachloride. The name "Rose Bengal" is a commercial name and results because the dye is very similar in color to the cinnabar paste Bengali women used at that time as a center forehead spot to symbolize marriage--Indian happiness wart (21).

Rose bengal shows up first as a singlet oxygen sensitizer in von Tappeiner's 1904 paper (2) describing the effect of fluorescent materials on enzymes and paramecia; this was the first report of something von Tappeiner referred to as photodynamic action, i.e. light, dyes and oxygen killed paramecia. Rose bengal's role as a sensitizer in contemporary mechanistic studies in singlet oxygen chemistry is largely the result of Schenck and his coworkers (22).

Rose bengal is a xanthene and related to both the triphenylmethanes as well as to fluorescein. Rose bengal differs from fluorescein, which is still among the most fluorescent compounds known, in a number of important ways. In principle only one is important: It intersystem crosses to the triplet with an efficiency of 78% in polar solvents and fluoresces with a quantum yield of less than 10% in these solvents. Fluorescein effectively does not intersystem cross to the triplet. It emits from the singlet state with very high efficiency--an altogether useful property in its own right.

The fundamental structures of the important xanthenes are shown in Table I. The numbering system is outlined in the figure in the table.

Polymer based rose bengals were first reported by Neckers and his coworkers in 1973 (13). (P)-rose bengal was the first-- and it remains the only--important polymer based photosensitizer. Three properties make it useful:

1. Quantum yields of singlet oxygen formation from polymer rose bengals are essentially the same as are quantum yields of singlet oxygen formation in solution (15).

2. Rose bengal immobilized to polystyrene beads is not bleached. Rose bengal in solution bleaches with continued radiation.

3. Energy transfer from rose bengal immobilized within a polystyrene bead is a slower process than is energy transfer from a rose bengal immobilized on the surface of a polystyrene bead. There is an influence of oxygen pressure on the observed efficiency of singlet oxygen formation from immobilized rose bengals.

4. Self quenching of excited state dye by ground state dye with rose bengals immobilized to soluble polystyrenes occurs, but it is not observed with the beads.

Photochemistry and Photophysics of Immobilized Rose Bengal

The absorption spectrum of rose bengal in MeOH is shown in Figure 1,a. The fluorescence excitation spectrum of a (P)-rose bengal is virtually identical. The total emission spectrum of rose bengal taken at 77°K in MeTHF shows the expected fluorescence and phosphorescence, Figure 2,a. The fluorescence relates to the absorption spectrum as the expected mirror image relationship and the triplet energy corresponding to the O-O band is at 740 nm, corresponding to a triplet energy of 40 kcal/mole, Figure 2a.

The absorption spectrum of the rose bengal chromophore is almost entirely due to the xanthene portion of the molecule. Changes in the so-called A ring have little or no effect on any of the spectroscopic properties. Changes in the extent of ionization at C-6 do affect the spectrum, however, and this is shown nicely by the spectra of rose bengal in MeOH and the bis piperidinium salt in methylene chloride, Figure 1b. The relative peak heights of the shorter wavelength λmax ($\approx$520 nm) to the longer wavelength λmax (558 nm) are 1:3 in the fully dissociated form, but almost 1:1 in the piperidinium salt which is not as dissociated in the solvent in question. Hydrogen bonding also affects the relative position of the absorption maxima causing a red shift in the longer of the two maxima. Aggregates of these dyes are also important. Rose bengal solutions do not follow Beer's law even at moderate concentrations, deviating negatively because of aggregation effects. The dimers absorb uniquely from the monomers (23), which exist only in dilute solution. The equilibrium constant for their formation at 25°C is 4 x 10^{-3} M^{-1}. One can assume the dimers to be stacked (Figure 3) though no direct evidence exists to support this structure.

The fluorescence spectra of rose bengal monomers and dye dimers (24) (a) belie the extent of dissociation at C-6 (Figure 2b,c) and are (b) sensitive to proximity effects. The latter is

Table I.
The Commercial Xanthene Dyes

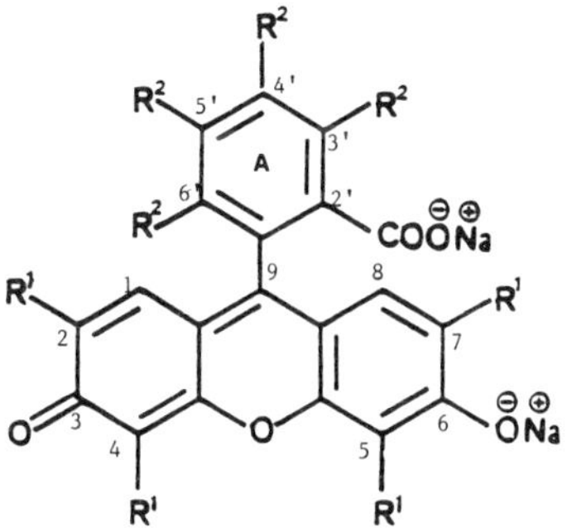

		R^1	R^2	$\lambda_{max}(nm)$	ϕ_F	ϕ_{ST}	$\phi 1_{O_2}$
1	uranine*	H	H	491	0.93	0.03	0.1
2	eosin	Br	H	514	0.63	0.3	0.4
3	erythrosin	I	H	525	0.08	0.6	0.6
4	rose bengal	I	Cl	548	0.08	0.76	0.76

* Uranine is the disodium salt of fluorescein.

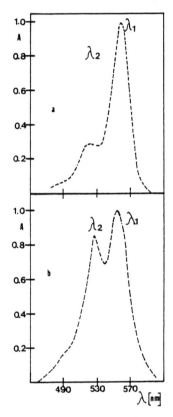

Figure 1. Electronic absorption spectra of rose bengal derivatives
1a. rose bengal absorption spectrum in MeOH
1b. rose bengal dipiperidine salt absorption spectrum in CH_2Cl_2

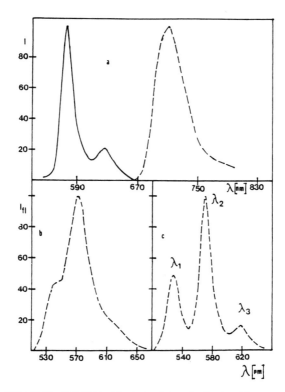

Figure 2.   Emission spectra of rose bengal derivatives.
2a.   Total emission spectra of rose bengal at 77 K.
2b.   Fluorescence emission of rose bengal, di trimethyl ammonium
      salt in EtOH at ambient temperature.
2c.   Fluorescence emission of rose bengal, di trimethyl ammonium
      salt in EtOH recorded at 77 K. (λ_{ex}=520 nm).

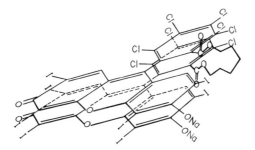

Figure 3.   Stacked rose bengals in rationally synthesized dimers.

Table II. Do the Dyes Interact? Absorption Spectra in Visible
Region for Rose Bengal Dimer Forms in Ethanol

Rose Bengal Dimer Form	λ_1	$A_{\lambda 1}/A_{\lambda 2}$	ϵ_1	$C[\text{mol}/2]$
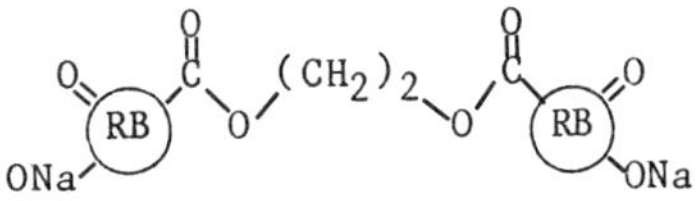 Ia	559.0	3.24	1.02×10^5	0.48×10^{-5}
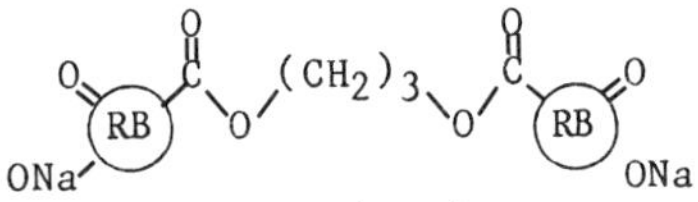 Ib (Na+)	565.5	2.55	0.858×10^5	0.539×10^{-5}
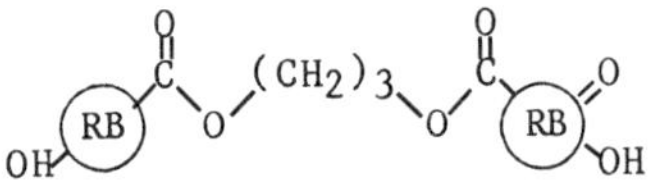 Ib (H+)	565.0	2.51	0.893×10^5	0.704×10^{-5}
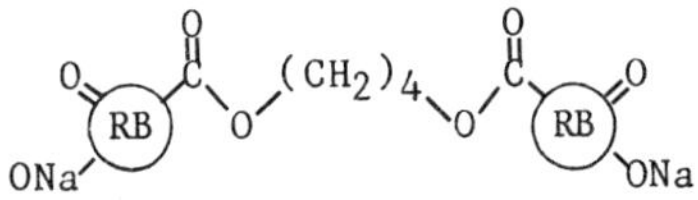 Ic	560	2.85	0.896×10^5	1.35×10^{-5}
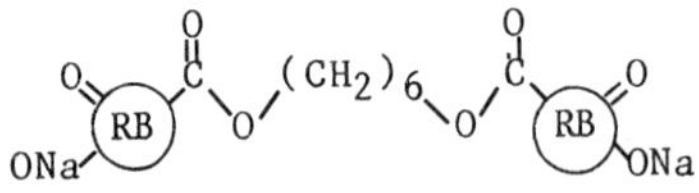 Id	553	2.07	0.77×10^5	1.95×10^{-5}
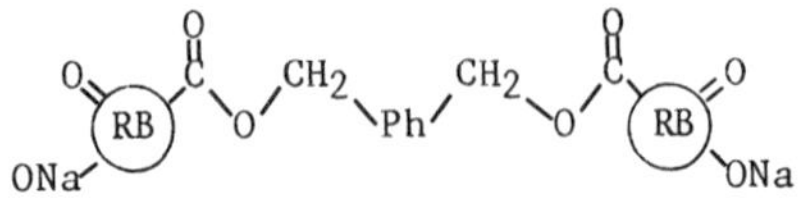 II	561	2.58	0.682×10^5	1.85×10^{-5}

Note: Molar absorption coefficients were calculated for 0.5
molecular weight.

shown from the absorption spectra (Table II) of rationally synthesized rose bengal dimers in which the dye chromophores are separated by methylenc units 2 to 6 carbons long (12). In the emission spectra of these dimers, the relative ratio of I_1/I_2 is a sensitive function of the length of the methylene chain separating the two chromophores (Figure 4). This is also the result of the stacking of the dyes as a function of the stereochemical mobility of the chain of carbon atoms separating the two chromophores. The longer methylene chains allow for more rose bengal/rose bengal interactions.

There are several effects of dye/dye interactions on photochemical energy transfer.

1. Dye triplets can be quenched by appropriately positioned, proximate ground state dye molecules of the appropriate excited state energies. This is so-called self quenching.

2. Dye aggregates absorb at different wavelengths, emit at lower wavelengths and have lower triplet energies. This translates into absorption from dimers, or emission from excimers.

3. Dye aggregates may be quenched by O_2 differently than are site isolated dyes.

Dye/Dye Interactions and Their Effect on Singlet Oxygen Formation.

The overall effect of one dye interacting with another dye on a polymeric support will likely be to decrease the efficiency of energy transfer to oxygen, i.e. the quantum efficiency of singlet oxygen formation.

We have studied the effect of dye/dye interactions on singlet oxygen formation both in soluble polymers and with polymer beads. Rose bengal is our probe.

The immobilization of rose bengal to non-crosslinked poly(styrene-co-chloromethylstyrene) is shown for a 3:1 co-polymer with differing concentration of rose bengal in Table III.

Table III. The Yields, %-CH_2Cl which Correspond to the Number of Rose Bengals Added to the Reaction Mixture of $\textcircled{P}$-CH_2Cl, and the cm^{-1} of C=O in $\textcircled{P}$-RB

Polymer†	Yield [g]	% I in Polymer	% of CH_2Cl group	C=O cm^{-1}
$\textcircled{P}$-RB-51*	0.85	----	1.65	1725.0
$\textcircled{P}$-RB-102	0.95	----	3.30	1725.0
$\textcircled{P}$-RB-152	0.77	----	5.00	1725.0
$\textcircled{P}$-RB-305	1.03	----	10.0	1741.4
$\textcircled{P}$-RB-450	1.15	76.2	15.0	1728.1
$\textcircled{P}$-RB-610	0.90	70.3	20.0	1738.3
$\textcircled{P}$-RB-1520	1.05	76.2	50.0	1739.8

† mg. rose bengal per gram of polymer

Three features of the data are important.
   1. The reaction occurs with 75% + 2 yield.
   2. The infrared stretching frequency of the carboxyl of the
immobilized dye increases as the dyes come into closer proximity
(higher loading).
   3. Dye loading influences the λ_1/λ_2 ratio in both the
absorption spectrum and the emission spectra.
   Photooxidation using the polymers of Table III as sensitizers in CH_2Cl_2 solution is given as a function of polymer
loading by rose bengal in Figure 5.
   The results divide into two separate regimes. The lightly
loaded regime (a) is not particularly interesting since effects
in this regime are solely the result of O_2 mobility effects--and
these depend on solution viscosity. In the more highly loaded
regime (b) the effects are those of self quenching. Beyond a
loading of 340 mg/g, site/site interactions quench the dye
excited state in solution and the quantum yield of singlet oxygen
decreases beyond the loading point. Based on simple statistical
calculations, this suggests that if the polymer immobilized dyes
are closer than 50 styrene units, "site/site" quenching (11)
occurs in solution, and the efficiency of singlet oxygen formation decreases.
   Site/site interactions in soluble polymers show up in the
absorption spectra of the immobilized dyes (Figure 6), and this
influence is significantly observed in the color of the polymers.
Thus, lightly loaded polymers are purple--more heavily loaded
polymers are deep red.
   The efficiency of photochemical energy transfer from rose
bengal immobilized to Merrifield resin beads has now been studied
in great detail by Paczkowski, Paczkowska and Neckers (26). The
rate of photooxidation of 2,3-diphenyl-p-dioxene, a chemical
quencher of singlet oxygen, can be compared for the various
Poly-RBs (i.e. soluble polymer rose bengals as the source of O_2^1)
or (P)-RBs with the rate of photooxidation of 2,3-diphenyl-p-
dioxene for rose bengal in MeOH ($\phi O_2^1 = 0.76$). The important
experimental parameter which must be controlled when using this
method to study the energy transfer process from the triplet
state of the sensitizer to oxygen is to assure that it be a zero
order reaction. The singlet oxygen quenching process by 2,3-
diphenyl-p-dioxene must also be a zero order reaction. As
Gottschalk and Neckers have shown (27), this depends on the lifetime of singlet oxygen in the solvent in question. Below the
defined crucial quencher concentration, the Schaap/Thayer relative actinometric method is not a valid measure of the quantum
yield of O_2^1 formation.
   Figure 7 shows observed efficiency of singlet oxygen
formation (using (P)-RB-2.0 as a sensitizer) as a function of the
flow of oxygen in the cell.
   The rate of 2,3-diphenyl-p-dioxene photooxidation using RB
dissolved in MeOH as the sensitizer essentially does not change
as a function of the rate of flow of oxygen. The rate of

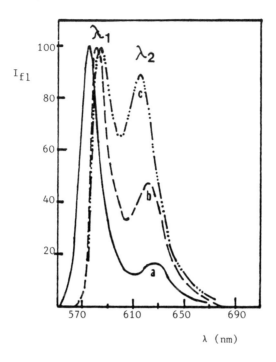

λ (nm)

Figure 4. The fluorescence emission spectra of rose bengal dimers in EtOH recorded at 77°K (λ_{ex} = 520 nm).

Compounds numbered in Table II.

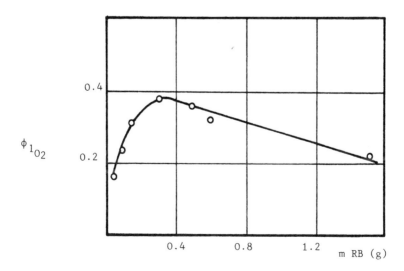

Figure 5. Quantum yield of singlet oxygen formation as a function of the amount of RB attached to a polymer.

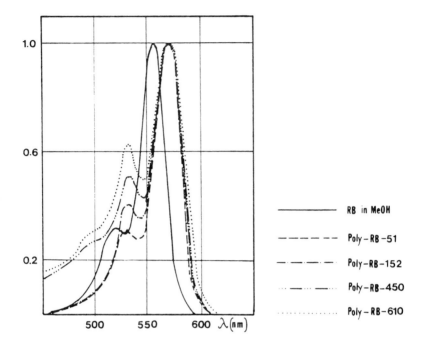

Figure 6.   Electronic spectra in the visible region of Poly-RB
samples.

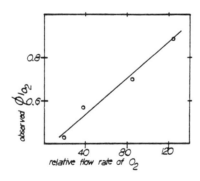

Figure 7.   Observed efficiency of singlet oxygen formation (for
(P)-RB-2.0) as a function of the flow of oxygen.

2,3-diphenyl-p-dioxene photooxidation using (P)-RB as sensitizer, however, increases as the oxygen flow increases. It is clear that there must be a barrier to penetration of oxygen into the (P)-RB beads which is O_2 flow dependent, and said penetration by O_2 into the bead occurs at a rate which is slower than is the rate of thermal diffusion in fluid solution. Higher rates of oxygen flow influence the quantum yield of singlet oxygen formation in the heterogeneous polymer sensitizer case, but have no influence on the rate sensitized by RB in solution, because in solution the quantum yield of quenching by O_2 is diffusion controlled and not influenced by solution viscosity. For purposes of quantum yield studies all solvents have the same viscosity. For a (P)-rose bengal, however, two different phases exist: a solution of the 1_{O_2} acceptor in CH_2Cl_2 solution outside the bead and another within the swollen polymer beads. The bead/solution interface and the solvent/acceptor system in the bead have a much higher effective overall viscosity. Since diffusion of oxygen into the beads from the solution is a function of the effective viscosity in the bead's interior, the rate of oxygen flow must be much higher to force O_2 to quench all of the rose bengal triplets formed in the interior of the bead. The pertinent observation is that there is a critical oxygen flow rate below which not all RB^3 in the interior of the beads are not quenched by O_2.

Rose bengal loading on Merrifield beads does not affect the quantum yield of singlet oxygen formation significantly, Table IV. The highest observed value is 0.91 (for (P)-RB-0.4); other values are only 15% lower, e.g. (ϕO_2 = 0.76 for (P)-RB-0.2). Both these values are unexpectedly higher than those observed by Schaap, Thayer, Blossey, and Neckers (14) and their coworkers ($\phi_{1_{O_2}}$ = 0.43). The original value obtained for the quantum yield[2] of singlet oxygen formation from (P)-rose bengal must have been obtained under conditions in which the energy transfer process from the sensitizer to oxygen was not zero order. Under the conditions of the original measurement, the concentration of oxygen in the bead was therefore not sufficient to quench all of the rose bengal triplets. Even for the highest loaded polymer beads (282 mg RB/g), the observed quantum yield of singlet oxygen formation is unexpectedly high (0.82). This values suggests, importantly, that no self-quenching processes occur even with the most heavily loaded Merrifield beads. This is easy to explain by considering the results obtained with soluble Poly-Rbs (25) and by comparing these results to those obtained with insoluble beads. The quantum yield for singlet oxygen formation observed for Poly-RB-450 with loading about 337 mg of rose bengal per 1 g of polymer is slightly lower than is the maximum value obtained in solution. Self-quenching processes are most distinct when the loading of the soluble polymers is in the range of 330-400 mg of rose bengal for 1 g of polymer. This loading value is much

Table IV. Quantum Yield for Singlet Oxygen Formation for (P)-RB
           in CH_2Cl_2 with Various Rose Bengal Loading

Type of (P)-RB*	Amount of RB used in reaction with 1 g of polymer beads (mg)	Amount of RB attached to 1 g of polymer beads (mg)	ϕ^1O_2
(P)-RB-0.1	25.0	9.2	0.82
(P)-RB-0.2	50.0	12.5	0.82
(P)-RB-0.4	100.00	33.4	0.91
(P)-RB-0.8	200.00	95.5	0.88
(P)-RB-1.6	400.00	127.8	0.77
(P)-RB-2.0	500.00	138.1	---
(P)-RB-3.0	750.00	179.8	----
(P)-RB-4.0	1000.00	213.0	0.76
(P)-RB-2/4**	2000.00	284	0.82

*Cl = 0.75 meq/g   12% of the chloromethyl groups react with rose
                          bengal.
       Notation: grams of RB per 4 g. of polymer beads
              **4 g RB for every 2.0 g beads.

higher than is the loading even of the most highly loaded polymer
beads ((P)-RB-2/4). In Merrifield beads, no self-quenching is
observed because the dyes are effectively "site-isolated".
    We observed that if an amine is used as the cation at C-6,
it has significant influence on the quantum yield of singlet
oxygen formation from the polymer-based reagent. The observed
efficiency of singlet oxygen formation from (P)-rose bengals in
which different ammonium ions are substituted at C-6 is shown in
Table V.

Table V. Quantum Yield for Singlet
Oxygen Formation for CH_2Cl_2 Soluble
Poly-RB Containing Various Cations
at C6 Phenolate Group

Polymer-RB	$\phi\, ^1O_2$
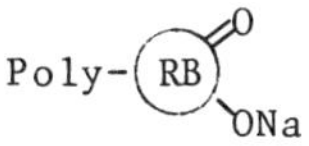 Poly-(RB) ONa	0.38*
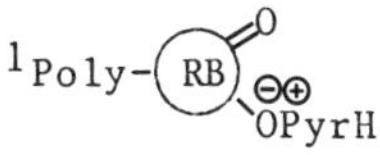 ^{1}Poly-(RB) $\ominus\oplus$ OPyrH	0.41
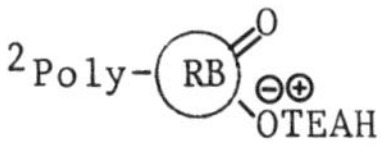 ^{2}Poly-(RB) $\ominus\oplus$ OTEAH	0.61

1. Pyr - pyridine
2. TEA - triethylamine

For the sodium and pyridinium salts the quantum yield is lower,
while for triethylammonium salts the quantum yield of $^1O_2$
formation is much greater and comparable to the value observed by
Lamberts and Neckers for rose bengal (C-2'benzyl ester,
triethylammonium salt) in CH_2Cl_2 solution (0.67).

The efficiency of singlet oxygen formation for $\left(P\right)-\left(RB\right)$ ONHR / O
is actually a function of the K_B of the amine. Salts of stronger
bases give higher quantum yields of singlet oxygen formation in
CH_2Cl_2 (Figure 8).

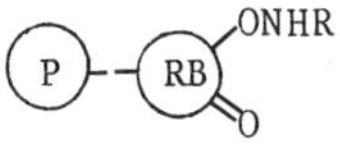

We attribute the effect of the amine at C-6 as deriving from
the extent of dissociation of the ammonium salt at C-6, and it is
an equilibrium effect which depends on the relative
concentrations of the protonated phenolate group and the
dissociated salt form. Strongly basic amines remove the proton
completely and form the ammonium salt. Weakly basic amines do
not cause complete ionization of the C-6 phenol. This same
effect has been observed for monomeric models in solution. (We
can ignore free amine quenching; because the concentration of the
amine is very low.)

As we pointed out in the introduction, in the case of 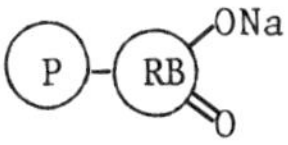
$\left(P\right)-\left(RB\right)$ ONa / O   the hydrophobic polymer serves to carry an
essentially polar dye into non-polar solvents, but it does so
only by virtue of the polymer backbone. Since rose bengal C-2'
benzyl ester sodium salt is not soluble in non-polar solvents

CH_2Cl_2, we could say that the hydrophobic polymer in non-polar solvents produces a heterogeneous sensitizer with the bead swollen by the solvent but with the dye essentially not knowing what to do. It is an insoluble component of the swollen gel. The polymer chain produces essentially an ideal dispersion of the insoluble rose bengal moieties in the non-polar solvent. For this reason

we consider (P)-(RB)⟨ONa / O⟩ to be an essentially pure heterogeneous

sensitizer, and the photooxidation process is indeed heterogeneous (Scheme 1).

The situation differs when the rose bengal moiety itself is soluble in non-polar solvent as it is in the case of the appended rose bengal C-6 triethylammonium salt. When the C-2'(polystyryl) ester is converted to the C-6 ammonium salt, the solubility of the appended moiety in non-polar solvents is increased, the efficiency of singlet oxygen formation is now limited by (a) the effective isolation of the polymer appended dye and (b) by the possibility of formation of the dissociated form, i.e. by equilibrium between the protonated phenol group and dissociated

salt form. For (P)-(RB)⟨ONHR / O⟩ the efficiency of singlet oxygen

formation is not dependent on the rate of oxygen flow because the system is more like a homogeneous solution than a heterogeneous

process. (P)-(RB)⟨ONHR / O⟩ shows very similar behavior to the

corresponding homogeneous sensitizers containing rose bengal in this regard, and the shape of the curve shown in Figure 3 is the

same for different rates of oxygen flow, whereas (P)-(RB)⟨ONa / O⟩

drastically changes its efficiency of singlet oxygen formation as a function of oxygen flow.

There are two probable explanations for this behavior. Since both the polymer chain and the appended rose bengal are soluble in CH_2Cl_2, an increase in swelling of the beads may result, allowing more efficient oxygen diffusion into the beads. Alternatively, the triplet lifetime of the dissociated rose bengal-trialkylammonium salt may be longer than is the lifetime of the sodium salt in these same solvents. Under these conditions, the rate of oxygen flow can be lower in the case of the ammonium salt polymer than in the case of the sodium salt polymer and still allow all of the rose bengal triplets to be quenched.

Though (P)-(RB)⟨ONHR / O⟩ 's show the typical properties observed

for homogeneous photosensitizers, the crosslinked polymer chain does not allow the soluble sensitizer molecule to dissolve in the solvent to form an authentically homogeneous solution. The polymer based ammonium salts are not pure heterogeneous photosensitizers, nor are they purely homogeneous sensitizers.

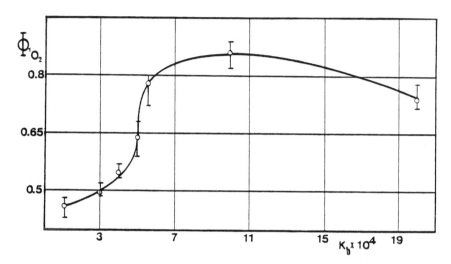

Figure 8. Relationship between basicity of amines used and quantum yield of singlet oxygen formation for P -rose bengals in which different ammonium ions are substituted at C-6.

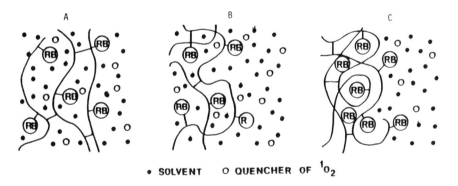

Scheme 1. The effect of pendent ion (C-6 ammonium or C-6 Na$^+$) on the availability of sensitizer moieties in solution.

A. soluble pendent
B. less soluble pendent
C. insoluble pendent

In this case a soluble dye is appended to an insoluble, but
swollen, polymer bead. We choose to call this type of
photosensitizer a semi-heterogeneous photosensitizer.

The results obtained for (P)-(RB)^{ONa} used as heterogeneous

photosensitizers in MeOH are also important. This solvent system
was not studied by Neckers, Schaap, and their coworkers because
they reasoned that a polymer bead needed to be swollen by the
solvent in which it was dispersed in order for it to function
effectively as a heterogeneous sensitizer. Only the most lightly

loaded (P)-RBs ((P)-(RB)^{ONa} - 0.1, 0.2 with loading 9.2 mg RB/g

and 12.5 mg RB/g, respectively) were viable photosensitizers when
used in MeOH, and quantum yields were low ($\phi_{1_{O_2}}$ = 0.21). When
loading by dye is increased, the efficiency of singlet oxygen

formation decreases (for (P)-(RB)^{ONa} - 0.4 with loading 33.4 mg

RB/g; $\phi_{1_{O_2}}$ less than 0.05). (P)-(RB)^{ONa} in MeOH is a

semi-heterogeneous photosensitizer (rose bengal C-2' benzyl ester
sodium salt is soluble in MeOH), but the hydrophobic polymer
chain does not swell in MeOH, and distance between rose bengal
molecules is controlled by polymer environment only. There is no
ideal dispersion of rose bengal molecules, and self-quenching
processes (intermolecular aggregate formation) decrease the
efficiency of singlet oxygen formation (28). A more precise
explanation of those observations is possible through analysis of
certain spectral properties of (P)-RBs.

Spectral Properties of Heterogeneous Photosensitizers.

Direct measurement of the absorption spectra of heterogeneous
photosensitizers is impossible because of light dispersion from
the polymer surface. Schaap, Thayer, Blossey and Neckers
reported the use of diffuse reflectance techniques to approximate
absorption spectra (14), but this technique does not produce
"true" absorption spectra. The sample for reflectance is
prepared in a MgO tablet. The (P)-RB is not, therefore,
dispersed in a solvent and the spectra obtained are the spectra
of (P)-RB without ideal dispersion of the rose bengal moieties in
solution. It is impossible to observe the true spectrum in the
absence of perturbing effects using reflectance techniques.

The excitation spectra of (P)-RB in a frozen non-polar
solvent (MTHF) do characterize the absorption spectra. These
spectra can be compared with those of monomeric models for (P)-RB
under identical conditions and with the spectral properties
observed for soluble Poly-RB.

Figure 9 shows the fluorescence excitation spectra for rose

bengal and for (P)-(RB)^{OTEA} in MTHF recorded at 77°K (λ_{ob} = 605
nm).

The general structures of the fluorescence excitation spectra for rose bengal and for $\text{(P)}-\text{(RB)}$ OTEA are similar. Table VI shows the fluorescence excitation data for (P)-RBs, absorption spectra data for monomeric models of (P)-RBs, and absorption spectra data for soluble Poly-RBs. The results are also similar.

Table VI. Spectral Data of (P)-RBs--Their Monomeric Models and Poly-RBs

*Excitation fluorescence spectra of $\text{(P)}--\text{(RB)}$ OR in MTHF			Absorption ** spectra of (RB) BE OR in CH_2Cl_2		Absorption spectra of Poly-(RB) OR in CH_2Cl_2		
R	λ_1	λ_2	λ_1	λ_2	λ_1	λ_2	A** of 1 mg/2
Na	570	522	--	--	572	532	22.35×10^{-3}
TMA	567	522	566.5	528.0	--	--	--
iso-BA	573	523	563.0	529.0	--	--	--
sec-BA	573	522	563.0	529.0	00	00	00
tert-BA	510	522	559.0	530.0	00	00	00
TBA	566	522	--	--	--	--	--
TEA	571	523	564.0	527.0	573	531	11.21×10^{-3}
DEA	573	522	561.0	530.0	--	--	--
Pip	572	520	552.5	526.5	573.5	533	--
NH_3	573	524	--	--	--	--	--
Pyr	575	525	--	--	--	--	9.55×10^{-3}

*   Excitation spectra are not corrected
** Calculated absorption for 1 mg of Poly-RB dissolved in 1 L of solvent

From this observation one can predict that the absorption properties of a heterogeneous photosensitizer will be comparable to the absorption spectrum of its soluble equivalent. Thus if one knows the absorption properties of a monomeric model, takes into regard corrections for solvent and microenvironment, and measures the excitation fluorescence (or phosphorescence) spectrum of the polymer at low temperatures, these spectra should be essentially comparable.

As the data of Table VI show, the absorption maxima for soluble Poly-RBs are shifted about 10 nm to the red when compared with monomeric models. The longest wavelength maximum of fluorescence excitation spectra for (P)-RBs and the longest wavelength maximum of absorption for soluble Poly-RBs occur at virtually the same wavelength.

Soluble Poly-RBs (triethylammonium salt, pyridine salt) show a decrease in absorption for 1 mg of polymer compared with the absorption observed for Poly-(RB)$\overset{ONa}{\underset{O}{\diagup}}$. Similar behavior is

observed for monomeric models of Poly-RBs (12,29).
Table VII shows fluorescence and phosphorescence data for (P)-RBs
with different loadings (in MTHF). The data in the Table and the
information in Figure 10 give still more information about the
differing behaviors of (P)-rose bengals as a function of that
cation at C-6. Rb loading essentially does not change the
position of the fluorescence and the phosphorescence maximum. The
maximum fluorescence emission for rose bengal C-2'benzyl ester,
Na+ salt is at 575 nm (see Figure 11). An increase of more than
20 times in rose bengal loading on a polymer chain shifts the
maximum of fluorescence emission only 3 nm, and the shape of the
fluorescence spectrum does not change. Thus the loading of (P)
-RB, does not influence the energy level of the singlet state in
non-polar solvents. The total shift of the fluorescence emission
maximum from the rose bengal C-2'benzyl ester sodium salt (13-16
nm) is likely connected with a different type of microenvironment
(30,31), in which the rose bengal is located. Similar behavior
was also observed for soluble Poly-RBs with loading less than 400
mg RB for 1 g of polymer carrier.

Table VII. Fluorescence and Phosphorescence Data for

(P)--(RB)$\overset{O}{\underset{ONa}{\diagup}}$ with Different Loading of Rose Bengal

Type of (P)--(RB)$\overset{O}{\underset{ONa}{\diagup}}$	Loading mg RB / 1g (P)	MTHF		EtOH
		Fl λ max [nm]	Ph λ max [nm]	Fl λ max [nm]
(P)-RB-0.1*	9.2	588	758	602-603
(P)-RB-0.2	12.5	585	753	603-604
(P)-RB-0.4	33.4	589	758	608-604
(P)-RB-0.8	95.5	585	759	608-609
(P)-RB-1.6	127.8	591	759	612-613
(P)-RB-2.0	138.1	590	760	612-613
(P)-RB-3.0	179.8	591	759	613-614
(P)-RB-4.0	213.0	592	760	615-616
(P)-RB-2/4	284.0	--	--	620

* Refer to Table IV for notation.

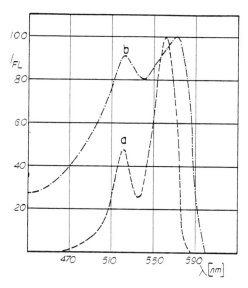

Figure 9. Fluorescence excitation spectra for rose bengal (curve a) and for (P)-(RB)⟨OTEA (curve b) in MTHF at 77°K (λ_{ob} 605 nm).

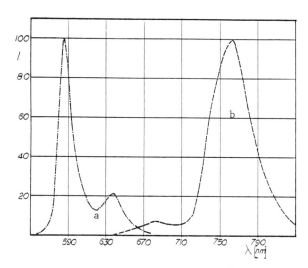

Figure 10. The total emission spectra for (P)-RB-0.1 recorded in MTHF at 77°K; fluorescence spectrum (curve a) phosphorescence spectrum (curve b).

A similar situation is observed with the phosphorescence spectra. Rose bengal C-2'benzyl ester, sodium salt shows a maximum phosphorescence emission at 719-720 nm in MTHF. The most lightly loaded (P)-(RB) shows a maximum at 758 nm. The total shift is about 40 nm in the case of the polymer. This relatively large shift results either from a different type of microenvironment of the dye in the polymer or emission from aggregate forms of the dye. The same spectral observations were obtained for soluble Poly-RBs with variable loading.

The fluorescence spectra of (P)-RBs were also measured for beads frozen in polar solvents (EtOH). In EtOH the beads are not swollen; there is no solvent between the polymer chains; and the rose bengal molecules are isolated by polymer chain only. Even for the most lightly loaded (P)-(RB), a large shift of fluorescence emission maximum is observed (see Figure 12). When the loading is increased, this shift is increased also. Figure 12 shows the relationship between the shift in the fluorescence emission maximum and the degree of loading. It is a linear relationship and may be used for calculating the degree of loading of the bead with dye.

Since the microenvironment is the same for all of the (P)-RBs, the red shift of the fluorescence emission is the result of emission from aggregates. The phosphorescence emission maximum for (P)-RBs observed in EtOH essentially does not change position when compared with the phosphorescence emission obtained for (P)-RBs in MTHF. Since there is no difference between the phosphorescence spectra obtained in MTHF and in EtOH, it suggests that self-quenching processes must be taking place from the singlet state. As was stated earlier, as powders in MeOH only the most lightly loaded (P)-RBs (.01,0.2) give detectable singlet oxygen formation. When dye loading is increased, the efficiency of singlet oxygen formation in MeOH is decreased. Since this loading essentially influences the fluorescence properties of the (P)-RBs, we can assume that self-quenching processes occur from the singlet state of excited rose bengal molecules, and it is this type of self-quenching process which decreases the efficiency of singlet oxygen formation.

Summary

The effective use of heterogeneous energy transfer donors, or polymer based photosensitizers, is defined by polymeric rose bengals--the only heterogeneous photosensitizer which works well and has been carefully studied. Based on the high extinction coefficient and intersystem crossing efficiency of the dye, rose bengal polymers produce singlet oxygen yields in non-polar solvents which are a function of the flow rate of oxygen and of the counter ion at C-6, but not of the polymeric support or the degree of loading with the dye.

Energy transfer self quenching effects between dye molecules

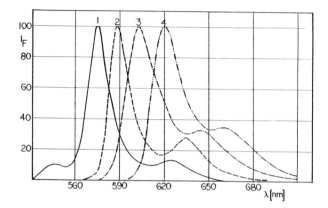

Figure 11. Fluorescence spectra recorded at 77°K. (P)- rose
bengals of increasing loading (left to right).

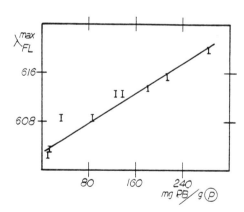

Figure 12. The relationship between the shift in the
fluorescence maximum and the degree of loading for (P)-rose
bengals (recorded in EtOH at 77°K).

is minimized in rose bengals immobilized to Merrifield resins
such that the dyes function as though they were isolated from one
another in the excited state. This is not so with the same dyes
immobilized either in small molecule models, or immobilized on
soluble polymers studied in solution. In these latter cases,
interaction between the dyes are observed both in the ground state
and the excited state.

Polymer rose bengal beads are most easily analyzed by
fluorescence excitation spectroscopy. The emission spectra of
immobilized dyes are presented as a function of solvent, of C-6
counter ion and dye-dye aggregation in this review.

Acknowledgments

This work has been supported by the National Science Foundation,
Division of Materials Research. The author gratefully acknowledges
this support. The work would have been impossible without the
extensive synthetic efforts of Dr. J.J.M. Lamberts and the pho-
tophysical skills of Dr. J. Paczkowski (Institute of Technology and
Chemical Engineering, Bydgoszcz, Poland). These individuals not
only contributed their technical expertise, but much of the thinking
to making the rose bengal system by far the most well understood
polymeric energy transfer device. The author cannot thank these two
co-workers enough. At various times other co-workers, mentioned in
the references, have contributed to the rose bengal affair. The
efforts of these individuals, as well as my colleague, Prof. J. C.
Dalton, are also acknowledged with gratitude.

Literature Cited

1. Kautsky, H.; deBruijn, H. Naturwiss. 1931, 19, 1043; Kautsky,
    H.; de Bruijn, H.; Neuwirth, R.; Baumeister, W. Chem. Ber.
    1933, 66, 1588.
2. von Tappeiner, H.; Jodlbauer, A. Dtsch. Arkiv. Klin. Med.
    1904, 80, 427.
3. Kautsky, H. Trans. Faraday Soc. 1939, 35, 216.
4. Moser, R. E.; Cassidy, H. G. J. Polym. Sci. 1964, B2, 545.
5. Leermakers, P. A.; James, F. J. Org. Chem. 1967, 32, 2898.
6. Merrifield, R. B. J. Am. Chem. Soc. 1963, 85, 2149; see
    Blossey, E. C.; Neckers, D.C. "Solid Phase Synthesis";
    Dowden, Hutchinson and Ross, 1975.
7. Neckers, D. C.; Kooistra, D. A.; Green, G. W. J. Am. Chem.
    Soc. 1972, 94, 9284; Blossey, E. C., Turner, L. M.; Neckers,
    D. C. Tetrahedron Lett. 1973, 1823; Blossey, E. C.; Turner, L. M.;
    Neckers, D. C. J. Org. Chem. 1975, 40, 959.
8. Blossey, E. C.; Neckers, D. C. Tetrahedron Lett. 1974, 323.
9. Walling, C. "Free Radicals in Solution", J. Wiley: New York,
    1957.
10. Neckers, D. C. Chemtech 1978, 8, 108; Neckers, D. C. Nouv.
    J. Chim. 1982, 6, 645.
11. Asai, N.; Neckers, D. C. J. Org. Chem. 1980, 45, 2903.
12. Lamberts, J. J. M.; Neckers, D.C. J. Am. Chem. Soc. 1983,
    105, 7465; Lamberts, J. J. M.; Neckers, D. C. Z.

Naturforsch. 1984, 39b, 474; Lamberts, J. J. M.; Schumacher. D. R.; Neckers, D.C. J. Am. Chem. Soc. 1984, 106, 5879. Lamberts, J. J. M.; Paczkowski, J.; Paczkowska, B.; Neckers, D. C. Photochemistry and Photobiology, in press.

13. Neckers, D. C.; Blossey, E. C.; Schaap, A. P. U.S. Patent 4 315 998, 1982; Blossey, E. C.; Neckers, D. C.; Thayer, A. C.; Schaap, A. P. J. Am. Chem. Soc. 1973, 95, 5820.
14. Schaap, A. P.; Thayer, A. L.; Blossey, E. C.; Neckers, D. C. J. Am. Chem. Soc. 1975, 97, 3741.
15. Paczkowski, J.; Paczkowska, B.; Neckers, D. C.; Macromolecules in press.
16. von Baeyer, A. Chem. Ber. 1871, 4, 555, 658.
17. Fischer, E. Chem. Ber. 1874, 7, 1211.
18. Gnehm, R. Chem. Ber. 1874, 7, 1742; see also Schultz's Tables (1881).
19. Graf, G., Ciba-Geigy, Basel, Private communication.
20. Gnehm, R. Ger. Off. 1885, 32654.
21. Courtesy of Dr. Mary Fieser; Harvard University.
22. Gollnick, K.; Schenck, G. O. Pure Appl. Chem. 1964, 9, 507.
23. Rohatgi, K. K.; Mukhopadhysy, A. K. Photochem. and Photobiol. 1971, 14, 551.
24. Paczkowski, J.; Neckers, D. C. Macromolecules 1985, 18, in press.
25. Paczkowski, J.; Neckers, D. C. Macromolecules 1985, 18, 1245.
26. Paczkowski, J.; Paczkowska, B.; Neckers, D. C. Macromolecules, in press.
27. Gottschalk, P.; Neckers, D. C. Tetrahedron Lett., in press.
28. Martin, M. M.; Lindquist, L. J. Lumin. 1975, 10, 381.
29. Lamberts, J. J. M.; Schumacher, D. R.; Paczkowski, J.; Neckers, D. C. "Photochemistry in Organized Media"; Fox, M. A., Ed.; American Chemical Society: Washington, D.C., 1985.
30. Yuzhakow, V. T. Russian Chem. Revs. 1979, 48, 1076.
31. Arbeola, L. J. Chem. Soc., Faraday Trans. 1981, 79, 1735.

RECEIVED October 15, 1985

7

Polymer-Bound Oxidizing Agents

Richard T. Taylor

Chemistry Department, Miami University, Oxford, OH 45056

The attachment of an oxidizing agent to a polymer can be carried out through two different strategies. In one instance a ligand can be bonded to the polymer and the oxidizing agent bound by electrostatic forces which may be of a chelation or an ion-exchange type. Alternatively, covalent attachment of the oxidizing agent to the polymer is often appropriate. The chemistry of the polymer-bound reagent resembles that of the non-polymeric analog. To the extent that diffusion between phases is important, the reactivity of the polymeric reagent is depressed. It is accelerated under instances where intraresin effects begin to minimize site-site interactions and generate reactive isolated species.

The binding of oxidizing metal ions to polymeric ligands include examples of Cu, Co, Fe, Mo and V. The various Cr(VI) reagents are especially noteworthy in their extent of development. While polymer-bound perbenzoic acid can be recycled after a variety of reactions, the arsenic and selenium analogs operate in a catalytic sense. The seleninic acid species can be used for alcohol and hydroquinone oxidation as well. Polymer-bound amine oxides and sulfoxides operate through an oxygen transfer mode while electron exchange polymers (such as quinones) can act as redox couples. Singlet oxygen can be generated through use of a polymeric photosensitizer or via decomposition of a polymeric endoperoxide. Halogenation-oxidation can be promoted by polymeric analogs of N-bromoamides, or perbromides on ion-exchange media. Hypervalent iodine compounds, attached covalently or by ion-exchange, can act as halogenating agents (ICl_4^-, RIX_2, etc.) or as oxidizing agents (periodates, RIO_2).

The development of polymeric oxidants has proceeded parallel to the development of polymeric reagents over the past decade. As methods

0097-6156/86/0308-0132$06.75/0
© 1986 American Chemical Society

for the functionalization of polymers have improved, oxidizing reagents became a natural research goal. Many previous general reviews have included sections on such oxidizing agents (1-5).

This present work examines the present development of a variety of polymer-bound oxidizing agents. Since the review is not meant to be exhaustive, it focuses on reagents which are chemically bound to a synthetic polymer. Reagents supported on inorganic supports such as celite, silica, graphite, clay, etc. are not included (6,7), nor are enzymes.

The nature of the polymeric attachment to the oxidizing reagent can be categorized into two classes, based on the mechanism of attachment. A cationic or anionic oxidizing agent can associate with the polymer via electrostatic forces. In this case, the polymer would contain chelating units or would carry charge (ion-exchange resin). In another strategy, the oxidant is covalently bound to the polymer. The strategies can be used for a variety of reagents, and both have been shown to be adaptible to similar oxidants. The rationales for carrying out such an immobilization procedure for a given oxidant are varied. Principal among these is recovery of the spent reagent. Many oxidants are environmentally hazardous, and as such, attachment to a polymer allows control of wastes. In the same way an expensive reagent can be recovered for recycle. The ultimate form of recycle is, of course, catalysis and many of the following systems have been shown to operate catalytically.

On a more fundamental level, changes in reactivity can occur from polymer-binding of a reagent. In most cases the polymer is insoluble in the reaction medium; the reaction becomes heterogeneous. Normally such reactions are retarded and the selectivity of the reagent increases. There are instances, however, where reactivity increases. Normally this event is interpreted in terms of polymer-bound species which are "site-separated" and in an unassociated state are more reactive.

The principal support for these polymeric species remains polystyrene-co-divinylbenzene. This resin is readily available in various degrees of crosslinking with corresponding variation in physical properties. The methods for functionalization are simple and provide access to most of the necessary functional groups. On the other hand, it is clear that switching to other polymeric systems can provide distinct advantages in some oxidations. Hence an increasing amount of study is headed in that direction.

The polymer-bound oxidizing agents to be considered in this review are divided according to the nature of the oxidizing functionality. The reagents to be considered are:

a) the polymer-bound oxidizing metals with special emphasis on chromium reagents,
b) polymer-bound peracid species,
c) polymer-bound oxygen transfer agents, $^1O_2$ sources,
d) electron exchange polymers,
e) polymeric halogenating agents, and
f) polymeric hypervalent iodine sources.

Cationic Metal Oxidants

The oxidizing abilities of transition metals at a variety of oxidation states are well-known. The easiest method by which such species

can be adapted to a polymeric basis is by the preparation of a polymeric ligand system. Both monodentate and polydentate ligands have
been attached to polymers. A variety of metals have been coordinated with polymeric ligands in attempts to optimize oxidizing properties. Most of the metals which have been examined in this way have
proven to operate in a catalytic fashion.

At first examination, one would expect that a polydentate
ligand which can saturate the metal coordinatively might bind the
metal most effectively. On the other hand, unsaturated metals would
produce a species with solvent molecules as ligands in some sites so
that exchange and reaction can occur. The following examples are
grouped by ligand structure, and concentrate on oxidations of
organic species.

Amine complexes. The effect of immobilization of the Cu(II)
complexes of polymeric amines has been examined and reviewed by
Challa et al. (8). The polymeric ligands were varied by using the
following species: polymer-bound dimethylbenzylamine (formed by
treatment of chloromethylated polystyrene with dimethylamine), the
copolymer of styrene and 4-vinyl-pyridine and the copolymer of
styrene and N-vinylimidazole. The reaction examined was the Cu(II)
oxidative dimerization of 2,6-disubstituted phenols (Equation 1).
An air oxidation of Cu(I) to Cu(II) renders this reaction catalytic

$$2 \quad \text{(ring)} - OH + O_2 \xrightarrow{Cu(II)} O = \text{(rings)} = O + H_2O \quad (1)$$

in copper complex. It was demonstrated that the active agent in
this oxidation is a binuclear copper complex. Since the polymer is
flexible, the polymeric attachment serves to increase entropically
the local concentration of the active species. As a result, the
polymeric ligands showed hightened reactivity versus the monomeric
species. Increasing the degree of functionalization also increased
the local concentration and hence the reactivity. Copper complexes
of vinylpyridine copolymers (9) and vinylamine polymers (10) have
been used by Chanda, O'Driscoll and Rempel to study catalytic
oxidations of thiosalts as well.

Multidentate amine complexes. In a more general study, Drago et al.
(11,12) were able to attach bis(cyanoethyl)amine to iodomethylated
polystyrene (Equation 2). Reduction affords the bound dipropylene-
triamine system.

$$PS - \text{(ring)} - CH_2I \rightarrow PS - \text{(ring)} - CH_2N \begin{smallmatrix} CN \\ CN \end{smallmatrix} \rightarrow PS - \text{(ring)} - CH_2N \begin{smallmatrix} NH_2 \\ NH_2 \end{smallmatrix} \quad (2)$$

A variety of transition metal ions, including Mn(II), Co(II), Ni(II), Cu(II), and Zn(II) were bound to this system. The catalytic oxidation of 2,6-dimethylphenol in air is catalyzed by the Co(II) complex. The reactivity of the catalyst is reduced from that of the homogeneous analog. Some variation in product distribution (benzoquinone vs. dimer) is observed with changes in loading. Benzoquinone is formed by a second oxidation of the phenoxy radical and is thus promoted by a higher concentration (load) of Co(II).

Related polydentate ligands are the polymer-anchored bis(phosphonomethyl)amino and bis(2-hydroxyethyl) amino species, studied by Suzuki (13). These ligands have been attached to both microreticular and macroreticular polystyrene-co-divinyl benzene and coordinated to both oxo-vanadium (V) and oxo-molybdenum(VI). Using the catalytic epoxidation of (E)-geraniol as a model system (with t-BuOOH) it was found that the macroreticular oxo-vanadium(V) catalyst was the most reactive, particularly with the phosphorous ligand (Equation 3).

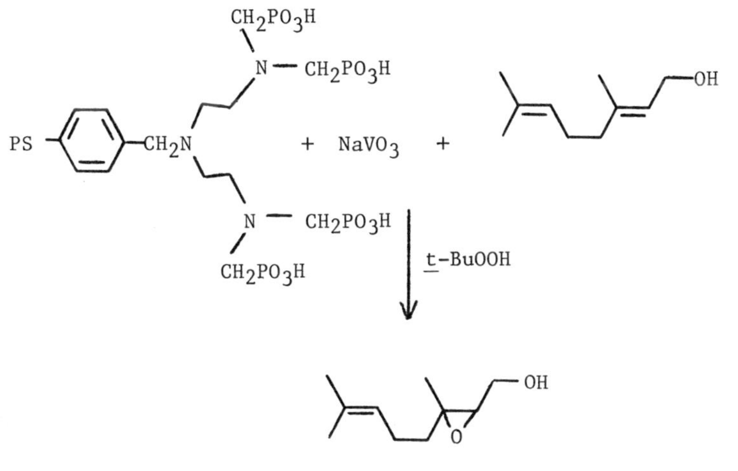

$$(3)$$

Carbonyl and thiocarbonyl ligands. The treatment of chloromethyl-ated polystyrene with acetylacetonate affords a polymer-bound acetylacetone group which has been coordinated to an oxo-vanadium group (14). Bhaduri reports that this catalytic system decomposes tert-butylhydroperoxide and oxidizes sulfides, sulfoxides, and cyclohexene (to cyclohexene oxide). A dithiocarbamate ligand can be attached to chloromethylated polystyrene at the nitrogen through reaction with ethylamine and carbon disulfide. An oxomolybdenum group can be coordinated to the ligand and, in a similar fashion, catalyzes the oxidation of sulfoxides and cyclohexene with tert-butyl hydroperoxide (15). In both instances, recycle of the polymer results in diminished reactivity due to partial loss of the metallic species.

Polyglutamates. In a study of the stereochemical consequences of optically active polymers, a tetrapyridyliron(III) species was anchored to either D or L polyglutamate (16,17). In the oxidation

of ascorbic acid with H_2O_2, Barteri and Pispisa found the catalytic
effect of the polymeric iron species depended on pH and ion
concentration. A catalytic effect was observed only under those
conditions where an α-helix is formed by the polymer. The catalytic
effect is stereospecific: the D-glutamate polymer catalyzes the
oxidation of L-ascorbate to a much higher degree.

Phthalocyanines. Gebler (18) has reported the attachment of a
variety of metal phthalocyanines to both 8% and 20% divinylbenzene
polystyrene copolymer beads. The attachment of the phthalocyanine
unit was either by a sulfonamide or a sulfone linkage. Nickel,
vanadyl, cobalt, iron and manganese complexes were formed in this
way. Since solution aggregation accounts for a diminution of the
catalytic activity, it was anticipated that polymer immobilization
would increase reactivity. Such an effect was not observed and
little advantage over the homogeneous catalysts could be observed in
the oxidation of cyclohexene. Oxidations of thiols by immobilized
phthalocyanines have been reported (19-20) by both Schutten and
Brouwer.

Porphyrins. Rollmann (21) has described the preparation of a
variety of metalloporphyrins on a macroreticular polystyrene copolymer with either divinylbenzene or ethylene glycol dimethacrylate.
Polymer attachment could be made through an amine ester or ketone
linkage. Cobalt(III) derivatives catalyze the oxidation of thiols
in air. The polymers did show a marked tendency to age with
resultant loss of activity.
    In the case of a manganese(III) tetraphenylporphyrin catalyst,
immobilization on a polymeric isocyanide derived from L-tyrosine affords a catalyst which mimics a mono-oxygenase (22). Using either a
single oxygen donor (PhIO or hypochlorite) or oxygen and a reducing
agent the epoxidation of cyclohexene proceeds catalytically. The
polymeric species is reported by van der Made to be threefold more
active than the monomeric species.

Chromium(VI) oxidants

Of all the metallic oxidants which have been subjected to polymer
attachment, the chromium(VI) reagents have found the most general
application to a variety of organic substrates. Since the use of
such homogeneous reagents has been widespread, it is not surprising
that such polymeric reagents have seen rapid development. While
chromium reagents are not particularly expensive, their toxicity and
associated environmental problems make easy recovery and separation
from reaction products a practical necessity.
    Chromic acid has been bound to commercial anion exchange resins
as the CrO_4H^- form by Cainelli et al. (23). This polymeric species
was shown to be quite effective in the oxidation of primary and
secondary alcohols to aldehydes and ketones in high yield.
Overoxidation of aldehydes did not occur.
    In an effort to improve the recycling of the polymer as well as
its efficiency (a large excess of the anion exchange reagent was
needed) improved polymeric reagents were derived, by Fréchet and
coworkers. A 2% crosslinked copolymer of 4-vinylpyridine and
divinylbenzene was treated with chromic anhydride and HCl to afford

poly[vinyl(pyridinium chlorochromate)] (24). While this resin was
rather easily recycled, the reaction still required a rather large
excess of the oxidizing agent. Much of this reduced reactivity was
attributed to the nature of the polymer and the inaccessibility of a
majority of the sites. A slight change in polymerization
conditions, as well as a switch to the less acidic dichromate
species, improved the reagent considerably (25). Only a slight
excess of reagent is required for complete conversion to the ketone
or aldehyde. Best conditions were the wet reagent with a nonpolar
solvent at about 70°C. The spent reagent can be washed with acid to
remove chromium salts then used again. Table I shows the results of
the reaction of polyvinylpyridinium dichromate with a variety of
alcohols.

Table I.   Oxidation of alcohols with polyvinylpyridinium
dichromate [a] (PVPDC) (25).

alcohol	% conversion (time)	
	molar ratio [b] 1.1 : 1	molar ratio [b] 1.7 : 1
benzyl alcohol	75 (15 min) 96 (2 h) > 99 (18 h)	> 99 (1 h)
1-phenylethanol	89 (5 h) > 99 (24 h)	
cinnamyl alcohol	76 (30 min) 98 (4 h)	
cyclopentanol	69 (24 h)	93 (24 h)
cyclohexanol	47 (24 h) 66 (68 h)	76 (24 h) 93 (68 h)
3-pentanol [c]	76 (68 h)	97 (68 h)
1-butanol [c]	81 (68 h)	> 99 (68 h)
1-hexanol [c]		85 (68 h)

[a] Reaction with 1.9 g PVPDC (wet) in 10 ml cyclohexane at 70°C.
[b] Molar ratio Cr(VI) alcohol.
[c] Reaction in sealed tube with hexane as solvent.

Brunelet aand Gelbard report that by adding styrene in the copolymerization, the resulting polymer also exhibits increased swellability and favorable reaction characteristics (26).

Peracids

The multiple uses of peracids in the oxidation of organic compounds make them a valuable asset to the synthetic chemist. In general, an acid is the byproduct of such an oxidation. Polymer immobilization of such species, generally through covalent bonding, allows easy recovery and recycle of the spent reagent. Such polymeric reagents are generally quite stable and react readily, although usually not as rapidly as the homogeneous species. The potential for selective reactions is high.

Perbenzoic acids. The peroxybenzoic acids, due to their relative stability, have been the subject of extensive investigation. Such reagents can be anchored to crosslinked polystyrene beads by a variety of methods. Fréchet and Haque (27) formed the peracid by reaction of polymer-bound benzoic acid with 70% H_2O_2 in the presence of an acid catalyst. In a similar scheme, the polymer bound acid chloride can be treated with sodium peroxide/70% H_2O_2 (Equation 4). Harrison and Hodge (28) used 85% H_2O_2 and higher loads to obtain a level of oxidant as high as 4 mmol/g. Another alternative route to

$$PS \underset{}{\overset{}{\bigcirc}} -CO_2H \; + \; H_2O_2 \longrightarrow PS \underset{}{\overset{}{\bigcirc}} -CO_3H \longleftarrow PS \underset{}{\overset{}{\bigcirc}} -COCl \quad (4)$$

the peracid involves ozonolysis of the aldehyde function (27). The epoxidation of olefins proceeds readily at 40°C with yields generally higher with decreased level of crosslinking. Reactivity parallels the homogeneous reagent and yields are quite good for di- and tri-substituted olefins. Monosubstituted olefins react sluggishly. Acid-catalyzed rearrangements were observed, but only with highly reactive epoxides (28). Reaction workup was accomplished by filtration and the recovered polymer could be recycled to the peracid essentially identical to the original material.

The oxidation of sulfides has also been examined using these polymeric peracids (29). With one equivalent of the peracid, a mixture of sulfoxide and sulfone was normally obtained, as is observed with the homogeneous reagent. The oxidation of penicillins and de-acetoxycephalosporins with the polymeric reagents give sulfoxides in high yield. Upon switching to a macroporous resin, the sulfide oxidation could be adapted to a continuous flow system at 40°C. High yields were obtained within 30 minutes residence time.

Peroxyarsonic acids. While the perbenzoic acid species described above were extremely attractive alternatives to the homogeneous species, the need to recycle polymer still required a reagent preparation step. The concept of using the polymer as a catalyst was an

obvious extention of the concept. What was required to accomplish
this objective was a functional group which could be reoxidized much
more rapidly than the carboxylic acid.

Jacobson, Mares, and Zambri report that the lithiation of 1%
crosslinked polystyrene followed by treatment with triethoxyarsine
and with hydrogen peroxide affords (Equation 5) the polymer-bound
arsonic acid (30).

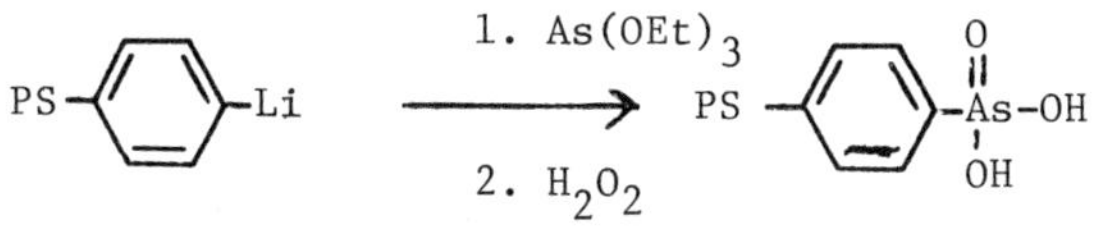

$$PS-\langle\rangle-Li \quad \xrightarrow[\text{2. } H_2O_2]{\text{1. As(OEt)}_3} \quad PS-\langle\rangle-\overset{\overset{O}{\|}}{\underset{OH}{As}}-OH \qquad (5)$$

The peracid species is rapidly generated in situ by the addition of
either 90% or 30% aqueous H_2O_2. The arsonic acid species is an effective catalyst for Baeyer-Villiger oxidations of ketones at 60°-
90°C in dioxane. Roughly, reactivity follows that for other peroxy-
acids (Table II). The presence of protic solvents retards the reaction rate. In addition, some hydrolysis of the lactones has been
observed, especially for water-soluble lactones at high conversions.

These Baeyer-Villiger reactions can also be carried out in a
triphase system of aqueous peroxide, a water immiscible solvent and
the polymer. Such an approach seemed to somewhat retard ester
hydrolysis. All of the above reactions occur in analogy to, but
somewhat slower than, reaction with an equivalent amount of
p-toluenearsonic acid.

The peracid system can also be applied to the oxidation of olefins (31). In dioxane, 90% H_2O_2 is necessary in order to prevent
epoxide hydrolysis to the diol. Operation in a triphase mode allows
the use of 30% H_2O_2. Substitution of the double bond increases the
epoxidation rate, as is expected, though primary alkenes still
react. Allylic alcohols such as geraniol and 2-cyclohexen-1-ol are
also highly reactive. The reaction workup consists of filtration
and separation of liquid layers. The recovered polymer can be used
directly in another reaction. Through five recycles with about 1
mol % catalyst no change in reactivity is seen.

Peroxyseleninic acids. Treatment of a mercurated 2%-crosslinked
polystyrene with selenium dioxide affords a yellow polymer (Equation
6) which has been assigned the structure of polymer-bound
phenylseleninic acid (32).

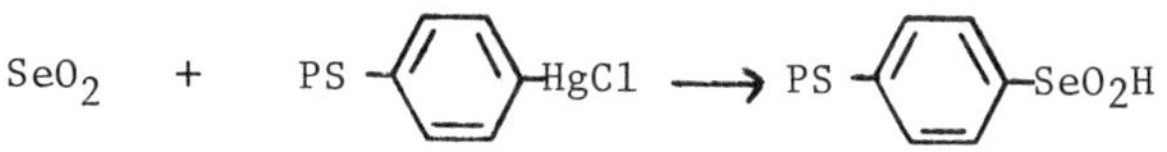

$$SeO_2 \quad + \quad PS-\langle\rangle-HgCl \quad \longrightarrow \quad PS-\langle\rangle-SeO_2H \qquad (6)$$

Table II. Oxidation of Ketones by 90% Hydrogen Peroxide
Catalyzed by Arsonated Polystyrene (50% load) (30)

substrate	time, h	ratio ketone: H_2O_2	%conv[b]	yield,[c] % ester	yield,[c] % acid	yield based on H_2O_2[d]
cyclobutanone	0.5	5	98	100	0	100
cyclopentanone	8	1	39	99	1	62
	3.5	5	59	89	11	89
	8	5	93	85	15	79
2-methylcyclopentanone	11	1	58	92	8	70
	30	1	84	76	24	70
cyclohexanone	7.5	1	40	86	14	73
	3	5	47	100	0	99
	7	5	79	80	20	80
2-methylcyclohexanone	5	5	78	100	0	100
2-phenylcyclohexanone	15	1	85	100	0	90
	4	5	87	100	0	88
cycloheptanone	23	5	31	29	71	11
2-allylcyclohexanone	9	1	65	70	30	52
estrone-3-methyl ether	24	1	60	100	0	71
pinacolone	25	5	83	100	0	89
acetophenone	29	5	11	100	0	14
	46	1	18	39	61	10
5-nonanone	32	5	no rxn			
2-cyclohexenone	23	5	no rxn			
methyl 2-cyclohexanone-carboxylate	26	1	no rxn			
2-chlorocyclohexanone	25	1	adipic acid			

[a] The oxidations were run at 80°C in dioxane with a molar ratio of
H_2O_2:As equal to 30. [b]Conversion to the Baeyer-Villiger products is
based on H_2O_2. [c]ester represents also lactone and acid denotes
hydroxycarboxylic acids formed by hydrolysis of lactones. The yields
are based on the ketone consumed. [d]Yield of lactone or ester based
on the consumed hydrogen peroxide.
Reproduced from reference 30. Copyright 1979 American Chemical
Society.

Treatment of the selenium(IV) polymer with 30% H_2O_2 results in rapid
loss of the color, presumably due to formation of the peracid. In a
triphase system at room temperature 1.5 mol % of the polymer and
1.5 - 1.8 equivalents of H_2O_2 were effective for oxidation of both
olefins and ketones. Unlike the arsonated system, olefins opened
readily to the trans diols; only in the case of tetramethylethylene
could the epoxide be isolated. Such a result may be due to partial
oxidation to the selenonic acid, a more powerful acid, or to a more
aqueous microenvironment. The Baeyer-Villiger reactions of ketones
could be carried out in a similar fashion; in a triphase system some
ester hydrolysis could be observed. Water soluble ketones and
aromatic ketones were both unreactive. The polymer can be recycled
readily but becomes unstable if elevated temperatures are used in
the reaction. Use of the catalyst under reaction conditions for
150h led to loss of less than 5% of the initial selenium content.
    Direct comparison of the arsenic and selenium reagents is some-
what complicated by differences in reaction protocol. It is evident
from the available data that the selenium species is somewhat more
reactive inasmuch as room temperature procedures are used. However,
the arsenic species seems less prone to hydrolysis, even under the
more vigorous conditions utilized.

Other selenium oxidations. Polymer-bound diphenylselenoxide has
been used by Heitz and coworkers to carry out the oxidation reaction
of β-methylnaphthalene to the aldehyde (33, 34). Polymer-bound
phenylseleninic acid has been shown to be an effective catalyst for
the oxidation of activated alcohols (32). Using t-butylhydro-
peroxide as a reoxidant and operating in a biphasic system of
catalyst and refluxing CCl₄, a wide variety of benzylic and cinnamyl
alcohols can be oxidized to the aldehyde or ketone in excellent
yield (Table III). The oxidizing species in this system is probably
the seleninic acid, as indicated by the retention of the yellow
color in the polymer.
    The above catalytic system is also effective at the oxidation
of aromatic systems. The conversion of hydroquinone to benzoquinone
occurs readily, although phenol itself does not react. The conversion of 1,5-dihydroxynaphthalene to juglone (Equation 7) proceeds in
70% isolated yield (32).

Other epoxidizing agents. In efforts to epoxidize naphthoquinone
systems, the treatment of Amberlyst 15 with hydrogen peroxide affords a reagent which has been shown by Jefford and Bernardinelli
to afford epoxyquinones in good yield. The reagent is discussed as
a peracid (35).

Oxygen-transfer polymers

In those systems where an element exceeds its normal valency by
means of a bond to oxygen, the transfer of that oxygen to another
acceptor is readily accomplished. Other than some transition metal
species mentioned above the amine oxides and sulfoxides appear most
prone to such reactivity. The advantage in using such reagents lies
in the mildness and selectivity of such reactions. The reagents of
this type will generally be utilized in a stoichiometric sense in
order to avoid exposure of the substrate to the more powerful
oxidant.

Table III. Oxidation of Alcohols with t-BuOOH/polymer-bound
seleninic acid [a] ($3\overline{2}$)

alcohol	time, h	products	% yield (isolated)
p-NO$_2$-benzyl alcohol	5	p-NO$_2$-benzaldehyde	100
3,4,5-(MeO)$_3$-benzyl alcohol	22	3,4,5-(MeO)$_3$-benzaldehyde	94
benzyl alcohol	24	benzaldehyde	94
benzhydrol	24	benzophenone	100
benzoin	24	benzil	98
xanthen-9-ol	48	xanthone	100
1-phenyl-1,5-pentanediol	48	5-hydroxy-1-phenyl-1-pentanone	69
cinnamyl alcohol	63	cinnamaldehyde	100
3,4,5-trimethoxycinnamyl alcohol	70	3,4,5-trimethoxycinnam-aldehyde	70
cis-1,4-butenediol		no reaction	
cyclopentanol		no reaction	

[a] Conducted in refluxing CCl$_4$, with 1.5 mol % polymer-bound phenylseleninic acid and 1.5 equiv. of t-BuOOH.

Reproduced with permission from reference 32. Copyright 1983 American
Chemical Society.

(7)

Amine oxides. Polymer-bound trimethylamine has been formed by the
addition of dimethylamine to chloromethylated polystyrene (1%
crosslinked) in work by Fréchet and coworkers. Oxidation with 30%
hydrogen peroxide affords the amine oxide (Equation 8). This
polymer efficiently transforms primary halides (preferably not
chlorides) and tosylates into the corresponding aldehyde. The
conversion of secondary halides into ketones is somewhat complicated

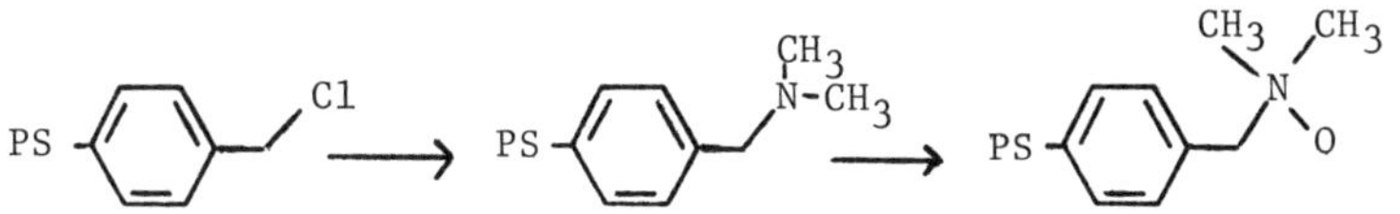

$$\text{(8)}$$

by side reactions, principally elimination. The putative
intermediate in the reaction is the alkoxyammonium ion:

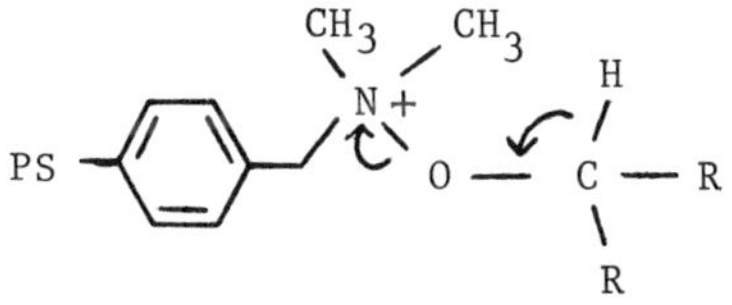

Reaction yields are generally high except where elimination affords
a conjugated product. The amine can be recovered and reoxidized.
At least five recycles proceed without complication (36).

Sulfoxides. Sulfoxides such as DMSO have long been used as
oxidizing agents. Release of the oxygen affords the offensively
malodorous sulfide. A polymer-bound sulfide has been formed by the
reaction of ethyl mercaptan with chloromethylated polystyrene as
described by Davies and Good. Oxidation with MCPBA gives the
sulfoxide which converts benzyl chloride into the aldehyde (Equation
9). The yield is 55% (37).The intermediate in this reaction is best

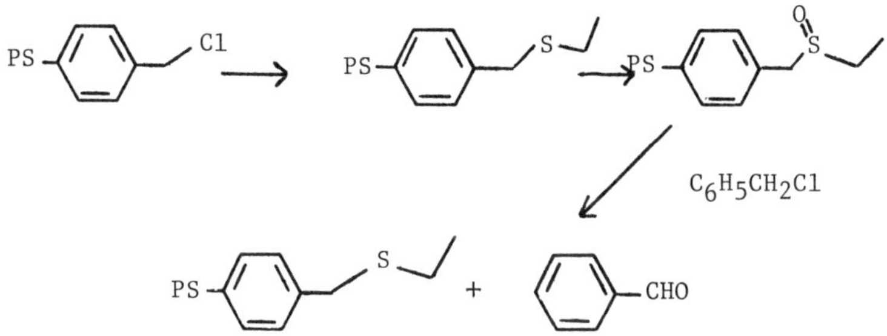

$$C_6H_5CH_2Cl \qquad \text{(9)}$$

formulated, once again, as the alkoxysulfonium ion:

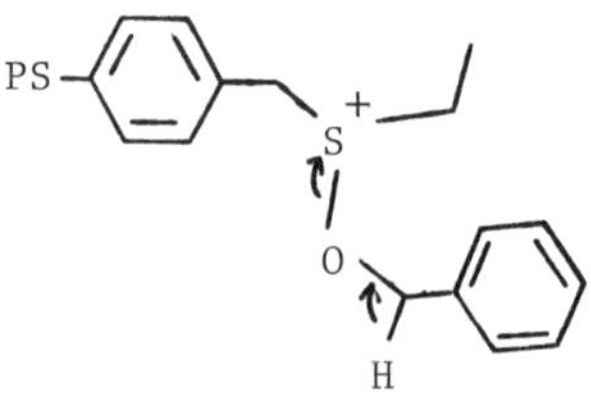

This intermediate can also be implicated in an earlier reagent described by Crosby et al. for the selective oxidation of diols (38). Treatment of polymer-bound thioanisole with chlorine affords the dichloride which can selectively carry out the oxidation by a similar intermediate (Equation 10).

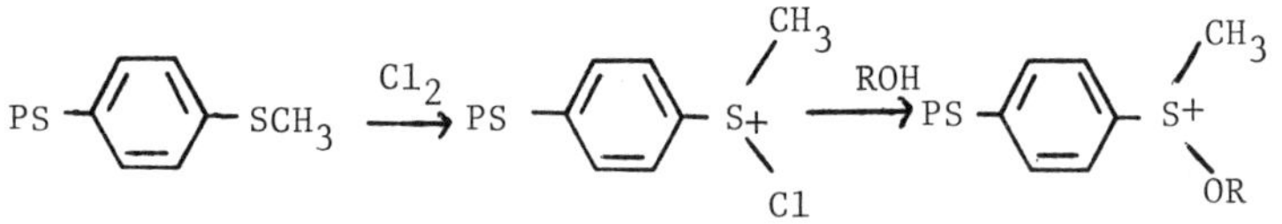

$$(10)$$

The use of a sulfoxide in order to oxidize an alcohol requires the conversion of that alcohol into a leaving group. The use of a polymer-immobilized carbodiimide can permit this Moffat-type oxidation to proceed readily (39). In this work by Weinshenker and Shen, the oxidant is not polymer-bound.

Singlet oxygen. Since the subject of polymer-bound photosensitizers is considered separately in this volume, the photoactivation of oxygen in this fashion will not be pursued, except to note that photosensitizers for such reactions are available commercially.
      A singlet oxygen transfer reagent has been reported (40) by Saito and coworkers. In a multistep sequence, 1,4-dimethyl-6-vinyl-naphthalene and 1,2,4-tri-methyl-6-vinylnaphthalene were prepared. Radical polymerization provides the polymeric methyl substituted naphthalenes which react readily with singlet oxygen (generated at 0°C using methylene blue). Up to 90% of the naphthalene units could be converted in this way to the endoperoxide. The reaction can be monitored by UV spectroscopy.

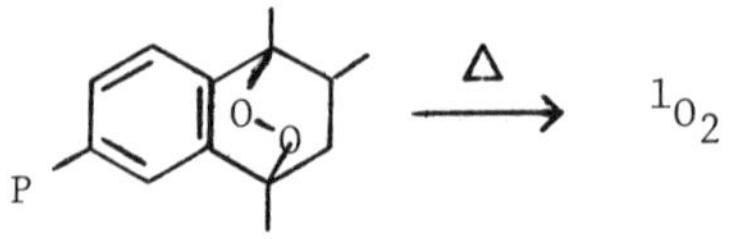

Upon heating to 30-40°C the singlet oxygen is released to the solution (usually CH_2Cl_2). Reactions are typical of singlet oxygen generated in other ways. The recovered polymer can be used over again
with no special treatment and provides a storable source of small
amounts of singlet oxygen (if kept at low temperature). The exact
amount of singlet oxygen, as a stoichiometric reagent, is also easy
to control.

Electron exchange polymers

A reagent which readily exists in two oxidation states provides an
excellent opportunity for its coupling to another redox system. Such
reagents can often act as either oxidizing or reducing agents,
depending on the particular starting oxidation state. The analogy
to the chemistry of the NAD^+/NADH couple in biological systems is
appropriate. Since the redox is so facile, an immobilized species
can easily function as a catalyst, with either air oxidation or
electrolytic oxidation or reduction providing the recycle mechanism.
The benzoquinone-hydroquinone system has been the most intensely
studied one, but others are also of utility.

Benzoquinone-hydroquinone polymers. The attachment of such a redox
system to a polymeric surface can be accomplished by either the
polymerization of a vinylquinone monomer or the attachment of the
quinone functionality to a preformed polymer. Since quinones often
inhibit polymerization, the former strategy requires that the system
be protected during polymerization.
    The work of Manecke has been reviewed in this regard (41).
Through a variety of methods a series of polymers could be formed
which afford a wide range of redox potentials. Examples of such
polymeric systems are presented in Table IV.
    Although such systems should be applicable to a wide variety of
oxidations, one of the most often applied reactions is in dehydro-
genations. The aromatization of dihydroaromatics, the oxidation of
NADH, and the conversion of cycloheptatriene to the tropylium ion
have all been reported, as well as the Strecker oxidation of amino
acids (45,46).
    The capacity of redox polymer systems such as these seems to be
improved by the introduction of charged groups in conjunction with
the quinone. Hydrophilic quinones can be formed by the reaction of
chloromethylated polystyrene with both hydroquinone and trimethylamine as described by Kun (47) or by sulfonation of the hydroquinone
polymer as reported by Prokop and Setinek (47,48).
    In more recent developments, a quinone ester has been bound to
chloromethylated polystyrene and the entire system used or oxidations in a continuous flow mode. The redox capacity can be monitored by the color of the reactor column and such reactions as the

Table IV.   Redox Polymers

Via Polymerization.

Monomer	Redox Capacity	Reference
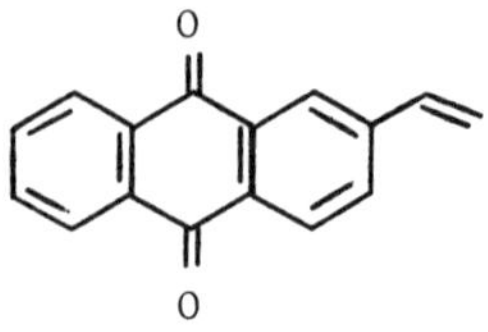	4.76 meq/g	42
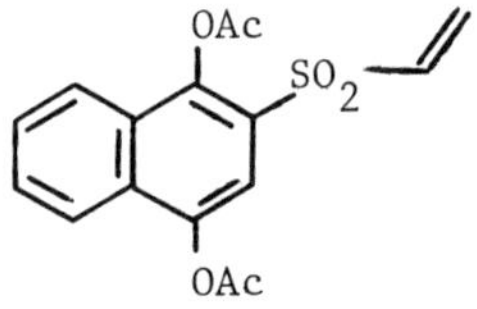	3.2 meq/g	43
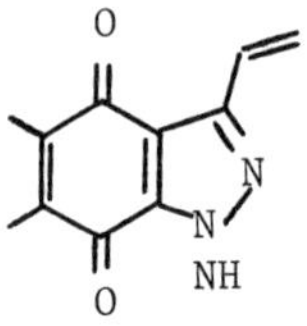	5.0 meq/g	44

Via Attachment.

Ligand	Polymer	Redox Capacity	Reference
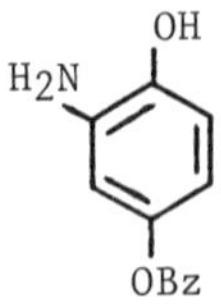	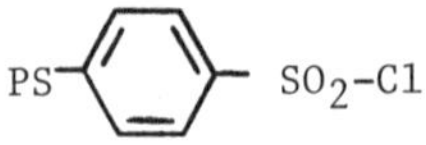	4.7 meq/g	41

preparation of quinones have been accomplished. Conversion of this reagent to the higher potential dicyanoquinone was not successful (49).

Miller has developed a redox electrode based on the above principles (50). Reaction of acryloyl or methacryloyl chloride with dopamine affords the polymeric coating for a glassy carbon electrode. Cyclic voltammetry and oxidations of iron ions have been examined as well as the oxidation of NADH.

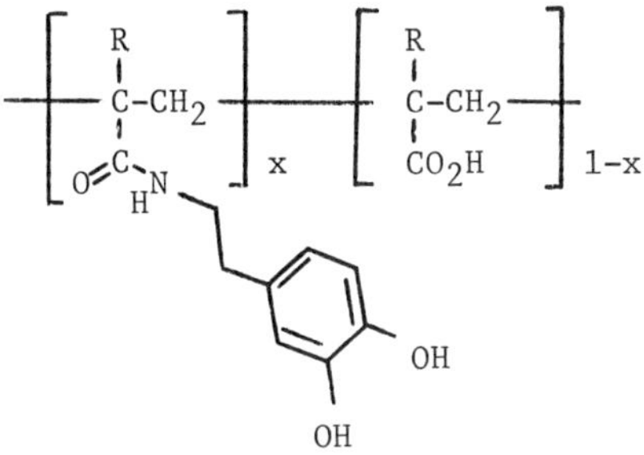

Flavin-based polymers. The treatment of chloromethylated polystyrene with a flavin and various amines affords a flavin redox polymer. Such polymers can be used as models for coenzymes and have

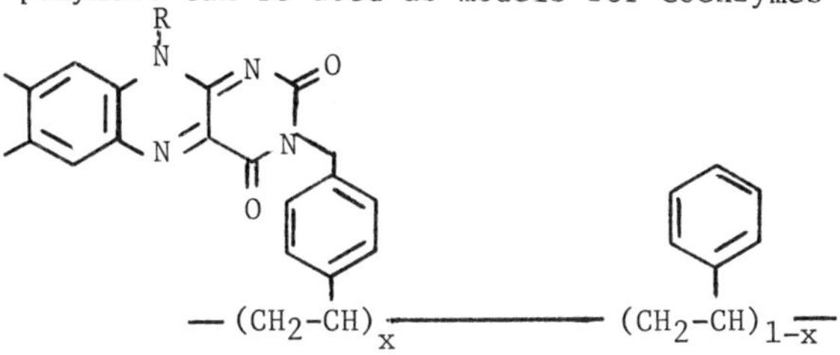

been found by Kunitake et al. to oxidize NADH (51). Similar reagents have been used in a catalytic sense to effect the air oxidation of alcohols. In such a way Yoneda and coworkers converted benzyl alcohol to benzaldehyde (52) and cyclopentanol to cyclopentanone (53).

Viologen-based polymers. Treatment of polyvinyl pyridine with 1,1'-bis(3-bromopropyl)-4,4'-bipyridinium dibromide affords a crosslinked viologen dication (Equation 11). A similar system has been placed on a polystyrene matrix be Endo and coworkers. The viologen system is a useful oxidant and has been used as a catalyst in several reductions. Thus, solid zinc or $Na_2S_2O_4$ could be used to convert azobenzene to hydrazobenzene (54). This redox reaction is of interest, since it can take place with the viologen as a membrane between oxidizing and reducing cells (55).

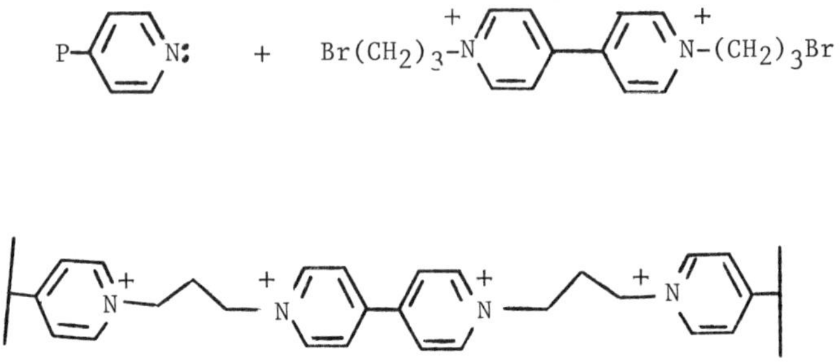

$$P-\text{(pyridine)N:} \quad + \quad Br(CH_2)_3-N\text{(bipyridinium)}N-(CH_2)_3Br$$

(11)

$$\left[\text{(pyridinium)N}^+ - \text{(CH}_2)_n - N^+ \text{(bipyridinium)} N^+ - \text{(CH}_2)_n - N^+\text{(pyridinium)} \right]$$

Halogenating agents

The use of sources of positive halogen to bring about oxidations is well known. While the molecular halogen itself is often
sufficient to bring about the desired reaction, a less nucleophilic
counterion is often quite useful. If this counterion becomes bound
to a polymer, then reaction workup is substantially simplified.
Below are discussed the major polymer-bound halogenating agents.
Those involving hypervalent iodine are discussed later.

Haloamides. In light of the widespread use of reagents such as NBS
and NCS to accomplish the above objectives it is not surprising that
the halogenation of polymeric amides provides a straightforward analogous system.

The most direct analogy to these reagents are N-bromopolymale-
imide and N-chloropolymaleimide (as the copolymers with divinylbenzene). The bromine analog (56) has been shown by Patchornik et al.

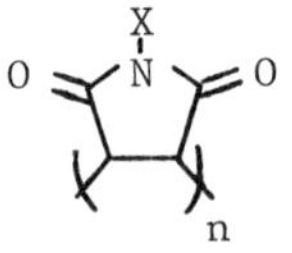

$$O\underset{}{\overset{X}{=}}N\overset{}{=}O$$

$$]_n$$

X=Cl, Br

to be an effective reagent for benzylic bromination. Since the
polymer is an amide group, it is interesting to note that the
products are similar to those of NBS in a polar solvent. The
chlorine analog (57) readily promotes aromatic chlorinations, in
work reported by Yaroslavski and Katchalski.

A more direct haloamide is obtained upon chlorination of nylons. These species have been used for a variety of oxidation reactions, and can be regenerated by the action of hypochlorous acid
as shown by Kaczmar et al. and Schuttenberg and Schulz (58-60). The
oxidation of alcohols to the carbonyl compounds occurs readily, but
primary alcohols are subject to overoxidation.

Pyridine Complexes. The addition of a halogen to pyridine affords a molecular complex. The bromine complex is capable of the halogenation of olefins as reported by Lloyd and Durocher and Zupan et al. (61-63). Anti stereoselectivity is observed and a nucleophilic solvent can replace the second halogen, thus implicating a bromonium ion. A similar complex with polyvinyl pyridine-N-oxide acts in a similar fashion. The polymer-bound pyridinium hydrobromide perbromide has also been formed by Fréchet, Farrall and Nuyens. It reacts with olefins in the same fashion but has also been shown to halogenate ketones (64).

Ion-exchange polymers. The above described perbromide reagent can be obtained somewhat more easily by the treatment of a commercial ion exchange resin (Amberlyst A-26, bromide form) with molecular bromine in CCl_4 (65) according to the method of Bongini et al. (Equation 12). Aside from the bromination of olefins,

$$PS \underset{Br^-}{\underbrace{}} CH_2 - \overset{+}{N}(CH_3)_3 \ + \ Br_2 \rightarrow PS \underset{Br_3^-}{\underbrace{}} CH_2 - \overset{+}{N}(CH_3)_3 \qquad (12)$$

ketones and alkynes can also be brominated (65,66) as described by Cacchi and Cagliotti.

Electrochemically generated species. The treatment of crosslinked polyvinylpyridinium hydrobromide with an electric current in acetonitrile gives what has been proposed to be a hypobromite (Equation 13). Kawabata and coworkers have examined this system and a variety of other electrochemically generated species.

$$P \underset{}{\underbrace{}} NH^+ \ Br^- \ \longrightarrow \ P \underset{}{\underbrace{}} NH^+ \ {}^-OBr \qquad (13)$$

Whatever its nature this electrochemically treated polymer has been shown to oxidize secondary alcohols to ketones in high yield (67). Since the reagent can be recycled by a continually applied current, the reaction proceeds catalytically. No supporting electrolyte is needed. Electrical efficiency is high. Primary alcohols oxidize more slowly to yield the carboxylic acid. The resin can be used many times without loss of reactivity.

Similar polymeric electrolysis conditions are able, in the presence of water, to effect the conversion of olefins into epoxides (68). In this case, choice of an anion exchange resin was crucial (Amberlite IRA-900 (Br)). The electrochemical efficiency is substantially lower (due to the competing electrolysis of water) but yields are high. Unsaturated carbonyl compounds do not react.

In a further extension of this work, a mixture of polyvinyl-pyridinium hydrobromide and hydrosulfate catalyzed the electrochem-

ical oxidation of alkyl sidechains of aromatic compounds (69). A
9:1 mixture of acetonitrile and acetic acid was found to give the
best yields (Table V).

Hypervalent iodine polymers

The use of reagents with iodine in a positive oxidation state
provides a rich source of oxidation reagents. The high atomic
weight of iodine makes the use of an insoluble polymer appropriate
for ease of separation of reaction products. The use of covalently
bonded hypervalent iodine species as well as ion-exchange type reagents are both appropriate given the widely varied nature of the
oxidizing species.

Ion-exchange polymers. The affinity of the iodate and periodate
ions for ion exchange polymers is extremely high. For this reason,
along with the relative insolubility of metal periodates in organic
solvents, the resins make attractive oxidants.

Both the iodate and periodate ions have been exchanged onto
commercial ion exchange resins (Amberlyst A-26 and IRA-904) by
Harrison and Hodge. Most of the known reactions of these ions can
be carried out with the resins (70). The oxidations of sulfides and
phosphines to the oxides occur in high yield. Hydroquinones can be
oxidized to quinones and the cleavage of 1,2-diols has also been
observed. It is noteworthy that the reactions can be carried out in
organic solvents in high yield.

In an interesting extension of the cleavage of 1,2-diols,
Bessodes and Antonakis treated nucleosides with a mixture of two
different resins containing periodate and borohydride ions (71). In
this fashion, the intermediate dialdehyde is not isolated but
directly reduced (Equation 14). In this adaptation of a "Wolf-and-

$$(14)$$

Lamb" reaction it is noteworthy that the two antagonistic reagents
do not react.

A chlorination reagent can be prepared by the treatment of the
hydroiodide of poly(styrene-co-vinylpyridine) with chlorine.
Alternatively, the methylpyridinium iodide polymer can be used.
These tetrachloroiodate polymers can be used as chlorinating
reagents (72). Acetophenone, dimedone, and indanedione are all
chlorinated in high yield.

Iodostyrene polymers. Iodinated crosslinked polystyrene can be
easily prepared in a variety of ways. Hypervalent iodine species
are made by oxidation of the iodine species.

Treatment of the polymer with either chlorine or fluorine gives

Table V.   Electrolytic Side-Chain Oxidation of Alkylbenzenes
Using Polymeric Electron Carrier [a] (69)

substrate	electricity (F/mol)	product	yield, [b] %
$C_6H_5CH_2C_6H_5$	6.0	$C_6H_5COC_6H_5$	78 (81)
$C_6H_5CH_2CH_3$	4.0	$C_6H_5COCH_3$	(72)
$C_6H_5CH_2CH_2CH_2CH_3$	6.0	$C_6H_5COCH_2CH_2CH_3$	(74)
$p\text{-}CH_3OC_6H_4CH_2CH_3$	4.0 [c]	$p\text{-}CH_3OC_6H_4COCH_3$	54 (73)
$p\text{-}CH_3COOC_6H_4CH_2CH_3$	6.0	$p\text{-}CH_3COOC_6H_4COCH_3$	60 (76)
$C_6H_5CH_3$	4.0	C_6H_5CHO	(13)
$p\text{-}CH_3OC_6H_4CH_3$	6.0	$p\text{-}CH_3OC_6H_4CHO$	(15)

[a] The reactions were normally carried out with alkylbenzene (5.0 mmol), water (50 mmol), and a mixture of PVP·HBr (0.50 g) and PVP·H_2SO_4 (0.50 g) in a mixture of acetonitrile (2.7 mL) and acetic acid (0.3 mL) unless otherwise stated.   [b] Yields are given for isolated pure material. Yields in parentheses are GLC determined.   [c] Carried out in DMF.

the derived iodo(dihalide) species. These species both react as
halogenating agents (73). The polymers react with both alkenes and

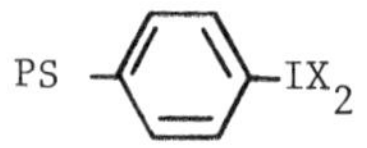

$$PS \left\langle \bigcirc \right\rangle IX_2$$

alkynes with the products formed arising from what is formulated by
Sket and Zupan as a polar reaction, unlike the radical species
postulated in the homogeneous case. The (diacetoxy)iodopolystyrene
species has been used by Hallensleben in the oxidation of amines
(74).
    The dichloroiodopolystyrene has been oxidized with peracetic
acid to the iodyl species. This polymer has been used to effect

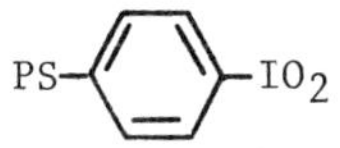

$$PS \left\langle \bigcirc \right\rangle IO_2$$

the oxidation of anthracene to anthraquinone (75).

Conclusion

    The advantages inherent in the use of polymeric reagents have
only begun to be exploited. Many of the best established and widely
used reagents are adaptations of monomeric species in which the
polymer acts as a convenience. In the future, design of the polymer
as an integral component of the reagent should allow new, highly
efficient and selective reagents to be formed.
    Acknowledgment is made to S. J. Burks and T. A. Stevenson for
helpful assistance and discussions in preparation of this work.

Literature Cited

1. Akelah, A. Synthesis, 1981, 413.
2. Akelah, A.; Sherrington, D. C. Chem. Rev., 1981, 81, 557.
3. "Polymer-Supported Reactions in Organic Synthesis"; Hodge, P.;
    Sherrington, D. C., Ed.; Wiley:New York, 1980.
4. Mathur, N. K.; Narang, C. K.; Williams, R. E. "Polymers as Aids
    in Organic Chemistry"; Academic:New York, 1980.
5. Hodge, P., Chem. Br. 1978, 14, 237.
6. McKillop, A.; Young, D. W. Synthesis, 1979, 401.
7. McKillop, A.; Young, D. W. Synthesis, 1979, 481.
8. Challa, G.; Schouten, A. J.; Brinke, G. T.; Meinders, H. C. In
    "Modification of Polymers"; Carraher, C. E. Jr.; Tsuda, M.,
    Ed.; ACS Symposium Series No. 121, American Chemical Society:
    Washington, D.C., 1980; pp. 7-24.
9. Chanda, M.; O'Driscoll, K. F.; Rempel, G. L. Appl. Catal.,
    1984, 9, 291.

10. Chanda, M.; O'Driscoll, K. F.; Rempel, G. L. J. Mol. Catal., 1981, 11, 9.
11. Drago, R. S.; Gaul, J. H. J. Chem. Soc. Chem. Comm., 1979, 746.
12. Drago, R. S.; Gaul, J. H.; Zombeck, A.; Smith, D. K. J. Am. Chem. Soc., 1980, 102, 1033.
13. Yokoyama, T.; Nishizawa, M.; Kimura, T.; Suzuki, M. Chem. Lett., 1983, 1703.
14. Bhaduri, S.; Ghosh, A.; Khwaja, H. J. Chem. Soc. Dalton Trans, 1981, 447.
15. Bhaduri, S.; Khwaja, H. J. Chem. Soc. Dalton Trans, 1983, 415.
16. Barteri, M.; Pispisa, B. J. Chem. Soc. Faraday Trans. I, 1982, 78, 2073.
17. Barteri, M.; Pispisa, B. J. Chem. Soc. Faraday Trans. I, 1982, 78, 2085.
18. Gebler, M. J. Inorg. Nucl. Chem., 1981, 43, 2759.
19. Schutten, J. H.; Beelen, T. P. M. J. Mol. Catal., 1980, 10, 85.
20. Brouwer, W. M.; Piet, P.; German, A. L. J. Mol. Catal., 1984, 22, 297.
21. Rollmann, L. D. J. Am. Chem. Soc., 1975, 97, 2132.
22. vanderMade, A. W.; Smeets, J. W. H.; Nolte, R. J. M.; Drenth, W. J. Chem. Soc., Chem. Comm., 1983, 1204.
23. Cainelli, G.; Cardillo, G.; Orena, M.; Sandri, S. J. Am. Chem. Soc., 1976, 98, 6737.
24. Fréchet, J. M. J.; Warnock, J.; Farrall, M. J. J. Org. Chem., 1978, 43, 2618.
25. Fréchet, J. M. J.; Darling, P.; Farrall, M. J. J. Org. Chem., 1981, 46, 1728.
26. Brunelet, T.; Gelbard, G. Nouv. J. Chim., 1983, 7, 483.
27. Fréchet, J. M. J.; Haque, K. E. Macromolecules, 1975, 8, 130.
28. Harrison, C. R.; Hodge, P. J. Chem. Soc., Perkin Trans. I, 1976, 605.
29. Harrison, C. R.; Hodge, P. J. Chem. Soc., Perkin Trans. I, 1976, 2252.
30. Jacobson, S. E.; Mares, F.; Zambri, P. M. J. Am. Chem. Soc., 1979, 101, 6938.
31. Jacobson, S. E.; Mares, F.; Zambri, P. M. J. Am. Chem. Soc., 1979, 101, 6946.
32. Taylor, R. T.; Flood, L. A. J. Org. Chem., 1983, 48, 5160.
33. Michels, R.; Kato, M.; Heitz, W. Makromol. Chem., 1976, 177, 2311.
34. Kato, M.; Michels, R.; Heitz, W. J. Polym. Sci., Polym. Lett. Ed., 1976, 14, 413.
35. Jefford, C. W.; Bernardinelli, G. Tetrahedron Lett., 1985, 26, 615.
36. Fréchet, J.M.J.; Farrall, M. J.; Darling, G. React. Polym., Ion Exch. Sorbents, 1982, 1, 27.
37. Davies, J. A.; Good, A. Makromol. Chem. Rapid Commun., 1983, 4, 777.
38. Crosby, G. A.; Weinshenker, N. M.; Uh, H.S. J. Am. Chem. Soc., 1975, 97, 2232.
39. Weinshenker, N. M.; Shen, C. M. Tetrahedron Lett., 1972, 3285.
40. Saito, I.; Nagata, R.; Matsuura, T. Tetrahedron Lett., 1981, 22, 4231.
41. Manecke, G. Pure Appl. Chem., 1974, 38, 181.

42. Manecke, G. Angew. Chem., 1962, 74, 903.
43. Manecke, G.; Ruhl, C. S.; Wehr, G. Makromol. Chem., 1972, 154, 121.
44. Manecke, G.; Ramlow, G. J. Polymer. Sci. Pt.C, 1969, 22, 957.
45. Manecke, G.; Bahr, C.; Reich, C. Angew. Chem., 1959, 71, 646.
46. Manecke, G.; Kossmehl, G.; Gawlik, R.; Hartwich, G. Angew. Makromol. Chem., 1969, 6, 89.
47. Kun, K. A. J. Polymer Sci., Part A, 1966, 4, 847.
48. Prokop, Z.; Setinek, K. Coll. Czech. Chem. Comm., 1981, 46, 1237.
49. Perry, G. J.; Sutherland, M. D. Tetrahedron, 1982, 38, 1471.
50. Fukui, M.; Kitani, A.; Degrand, C.; Miller, L. J. Am. Chem. Soc., 1982, 104, 28.
51. Shinkai, S.; Yamada, S.; Kunitake, T. Macromolecules, 1978, 11, 65.
52. Yoneda, F.; Sakuma, Y. Heterocycles, 1978, 9, 1763.
53. Yoneda, F.; Yamato, H.; Nagamatsu, T.; Egaw, H. J. Polym. Sci., Polym. Lett. Ed., 1982, 20, 667.
54. Saotome, Y.; Endo, T.; Okawara, M. Macromolecules, 1983, 16, 881.
55. Ageishi, K.; Endo, T.; Okawara, M. Macromolecules, 1983, 16, 884.
56. Yaroslavsky, C.; Patchornik, A.; Katchalski, E. Tetrahedron Lett., 1970, 3629.
57. Yaroslavsky, C.; Katchalski, E. Tetrahedron Lett., 1972, 5173.
58. Schuttenberg, H.; Klump, G.; Kaczmar, U.; Turner, S. R.; Schulz, R. C. J. Macromol. Sci., Chem., 1973, 7, 1085.
59. Kaczmar, B. C. Angew. Chem., Int. Ed. Engl., 1973, 12, 430.
60. Schuttenberg, H.; Schulz, R. C. Angew. Makromol. Chem., 1971, 18, 175.
61. Lloyd, W. G.; Durocher, T. E. J. Appl. Polym. Sci., 1963, 7, 2025.
62. Zupan, M.; Sket, B.; Johar, Y. J. Macrom. Sci., Chem., 1982, 17, 759.
63. Johar, Y.; Zupan, M.; Sket, B. J. Chem. Soc., Perkin Trans. I, 1982, 2059.
64. Fréchet, J.M.J.; Farrall, J. M.; Nuyens, L. J. J. Macromol. Sci., Chem. A, 1977, 11, 507.
65. Bongini, C.; Cainelli, G.; Contento, M.; Manescalchi, F. Synthesis, 1980, 143.
66. Cacchi, S.; Caglioti, L. Synthesis, 1979, 64.
67. Yoshida, J.; Nakai, R.; Kawabata, N. J. Org. Chem., 1980, 45, 5269.
68. Yoshida, J.; Hashimoto, J.; Kawabata, N. J. Org. Chem., 1982, 47, 3575.
69. Yoshida, J.; Ogura, K.; Kawabata, N. J. Org. Chem., 1984, 49, 3419.
70. Harrison, C. R.; Hodge, P. J. Chem. Soc. Perkin Trans I, 1982, 509.
71. Bessodes, M.; Antonakis, K. Tetrahedron Lett., 1985, 26, 1305.
72. Sket, B.; Zupan, M. Tetrahedron, 1984, 40, 2865.
73. Sket, B.; Zupan, M.; Zupet, P. Tetrahedron, 1984, 40, 1603.
74. Hallensleben, M. L. Angew. Makromol. Chem., 1972, 27, 223.
75. Taylor, R. T.; Stevenson, T. A. unpublished data.

RECEIVED August 19, 1985

Wittig Reactions on Polymer Supports

Warren T. Ford

Department of Chemistry, Oklahoma State University, Stillwater, OK 74078

Polymer-supported Wittig reagents give excellent yields of alkenes using a wide variety of bases and solvents. The polymer-supported reagents are preferred to soluble Wittig reagents when separation of triphenylphosphine oxide from the product alkene is difficult. The most useful support for laboratory scale syntheses is 1-2% cross-linked p-polystyryltriphenylphosphine prepared by bromination of polystyrene and replacement of the bromine by lithium diphenylphosphide. More highly cross-linked supports may be needed for large scale reactions because of greater ease of filtration and suitability for use in flow reactors. Stabilized carbanions from phosphonates can be prepared in anion exchange resins and give high yield modified Wittig reactions by a method readily adapted to continuous processes.

The Wittig reaction is the most general method for regiospecific introduction of carbon-carbon double bonds in organic synthesis (Scheme 1). The starting materials are usually readily available alkyl halides and aldehydes or ketones. Alkylation of triphenylphosphine gives a phosphonium salt that can be used directly, but often is recrystallized. Treatment of the phosphonium salt with a strong base converts it to a phosphorane, also known as an ylide from its zwitterionic resonance structure. The phosphorane cycloadds [2+2] to the carbonyl compound to form an intermediate oxaphosphetane, which decomposes to the alkene and triphenylphosphine oxide. Reviews of the synthetic scope and the mechanism of the Wittig reaction are available (1-5).

Scheme 1

$$Ph_3P + R^1R^2CHX \longrightarrow Ph_3PCHR^1R^2\ X^-$$

$$Ph_3PCHR^1R^2\ X^- + base \longrightarrow Ph_3P{=}CR^1R^2 + base{-}HX$$

$$Ph_3P{=}CR^1R^2 + R^3R^4CO \longrightarrow Ph_3PO + R^1R^2C{=}CR^3R^4$$

0097–6156/86/0308–0155$08.75/0

Although not all Wittig reactions proceed by exactly the same mechanism, an oxaphosphetane intermediate is common to all of the mechanisms (Scheme 2). The overall rates of the reactions depend upon the rates of formation of the oxaphosphetane and of its conversion to alkene and phosphine oxide. The stereochemistry of alkene formation depends on the rates of formation and conversion to product of the stereoisomeric oxaphosphetanes, and upon equilibration of the oxaphosphetane with starting materials and with zwitterionic intermediates in some cases (3-5).

Scheme 2

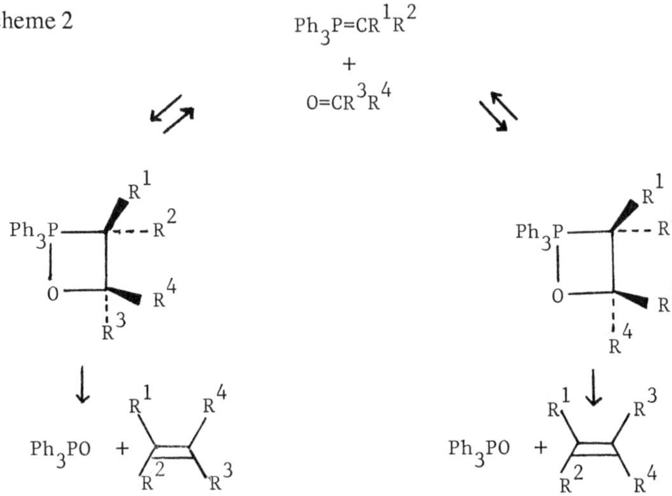

$$Ph_3P=CR^1R^2$$

$$+$$

$$O=CR^3R^4$$

In spite of its wide use, there are still three major problems with the Wittig reaction. 1) The stereochemistry often cannot be controlled. 2) Ketones and hindered aldehydes fail to react with phosphoranes that are hindered or are stabilized by strongly electron withdrawing substituents. 3) The by-product triphenylphosphine oxide can be difficult to separate from the product alkene. Often the alkene and the triphenylphosphine oxide cannot be separated by extraction, distillation, or crystallization, and column chromatography is required.

Polymer-supported Wittig reagents overcome the problem of separation of the triphenylphosphine oxide from the alkene (6-11). With polystyryldiphenylphosphine in place of triphenylphosphine, the by-product triarylphosphine oxide is bound to an insoluble polymer and removed from a solution of the alkene by filtration. Such a reagent could be used in a column in a continuous process in place of the conventional batch processes of small scale organic synthesis. The polymer-bound phosphine oxide can be reduced to the starting phosphine and recycled. The cost of polymer-supported Wittig reagents is much higher than the cost of conventional soluble Wittig reagents. Polymer-supported Wittig reagents do not solve the remaining problems of stereochemical control and of low reactivity of ketones, hindered aldehydes, and stabilized ylides. The polymer may exert stereochemical effects, but there are too few results so far to allow prediction of the courses of new reactions.

This chapter is a comprehensive review of alkene syntheses with polymer-supported Wittig reagents to produce soluble and polymer-bound alkenes. It does not include polymerizations of difunctional phosphoranes with difunctional carbonyl compounds, polymer-bound ylides of elements other than phosphorus, or other synthetic and catalytic uses of polymer-bound phosphines.

Reactions of Polymer-Bound Phosphonium Ions

Phosphonium Ions from *p*-Polystyryldiphenylphosphine. These reagents are polymer-bound analogs of the classic Wittig reagents prepared from alkyl halides and triphenylphosphine. The results are in Table I. In some cases yields in the original papers were corrected for recovered starting materials. Table I reports yields on the basis of the limiting reagent, either polymer-bound phosphonium ion or carbonyl compound, and it reports, when known, the amounts of starting materials that could have reacted but failed to react. Most of the reactions have employed polystyrenes cross-linked with 0.5-2% divinylbenzene. The data available do not show consistently better results with any one level of cross-linking. Degrees of functionalization (DF, the mol fraction of functionalized repeat units) of the phosphine polymers from 0.06 to 0.81 have been used, and high yield syntheses have been found throughout the range. The major factors that affect the results of the Wittig reactions are the method of synthesis of the polymer support and the solvent and base employed to generate the ylide. The methods used for the reactions are in the footnotes of Table I.

Reactions of (*p*-Polystyrylmethyl)triphenylphosphonium Ions. In these reactions the polymer-bound phosphorane reacts with an aldehyde or ketone to produce an alkene substituent on the polymer and leave the by-product triphenylphosphine oxide in solution, where it can be washed away from the polymer. The results are in Table II. Many of the same methods of phosphorane generation used to produce soluble alkenes were used. The yields in Table II are less accurate than the yields of micromolecular alkenes in Table I because of the difficulty of analysis of insoluble polymers. The yields have been based on weight changes of the polymer, analysis of the polymer for an element introduced during the reaction, the yield of triphenylphosphine oxide, or the amount of bromine consumed by the modified polymer.

Polymer Functionalization

p-Polystyryldiphenylphosphine. Copolymers of styrene and *p*-styryldiphenylphosphine (**1**) (*28*) in a 3/1 mole ratio cross-linked with 2 wt % divinylbenzene were used for some of the first polymer-supported Wittig reagents (*12,20*). Modest yields of alkenes and frequently unreacted aldehyde or ketone were recovered, as shown in the entries in Table I where "monomer" is the source of phosphine. Copolymer reactivity ratios for styrene (M_1) and *p*-styryldiphenylphosphine (M_2, $r_1 = 0.46$ and $r_2 = 1.11$ (*29*) or $r_1 = 0.52$ and $r_2 = 1.43$ (*30*)), indicate that the copolymer initially formed from the 75/25 monomer mixture has a 64/36 composition. The mixture of *m*- and *p*-divinylbenzene likewise is incorporated rapidly into copolymer (*31*) with 2 wt % in the monomer mixture giving 4 wt % in the initially formed polymer. Consequently the *p*-styryldiphenylphosphine is incorporated preferentially into the more highly cross-linked parts of the heterogeneous polymer matrix. Although most of these sites can be alkylated to phosphonium ions, the more sterically demanding Wittig reaction fails at some of the least accessible sites.

$$\text{H}_2\text{C=CH} - \langle \bigcirc \rangle - \text{PPh}_2 \qquad\qquad \mathbf{1}$$

The more effective methods of preparation of polymer-bound triarylphosphine for Wittig reactions are 1) lithiation of the polymer followed by reaction with chlorodiphenylphosphine, and 2) reaction of brominated polystyrene with lithium or sodium diphenylphosphide as shown in Scheme 3.

Table I. Soluble Alkenes from Soluble Carbonyl Compounds and Phosphonium Ions Bound to 0.5-2.0% Cross-linked Polystyrenes

alkyl halide	source of phosphine	DF	carbonyl compound	method[a]	mol ratio P+/C=O	product	% yield isolated (gc)	% recovered C=O	lit.
MeI	monomer	0.25	$n\text{-}C_5H_{11}CHO$	A	1.0	$n\text{-}C_5H_{11}CH\text{=}CH_2$	(56)	(3)	12
MeI	$LiPPh_2$	0.35	$(CH_2)_5 C\text{=}O$	A	1.05	$(CH_2)_5 C\text{=}CH_2$	(99)	0	13
MeI	monomer	0.25	$(CH_2)_5 C\text{=}O$	A	1.0	$(CH_2)_5 C\text{=}CH_2$	(63)	(30)	12
MeI	monomer	0.35	$PhCHO$	A	1.0	$PhCH\text{=}CH_2$	(50)	(39)	12
MeI	$LiPPh_2$	0.70	$p\text{-}ClC_6H_4CHO$	B	1.5	$p\text{-}ClC_6H_4CH\text{=}CH_2$	0	0	14
MeI	monomer	0.25	$PhCOMe$	A	1.0	$Ph(Me)C\text{=}CH_2$	(14)	(82)	12
MeI	$LiPPh_2$	0.35	$PhCH\text{=}CHCHO$	A	1.05	$PhCH\text{=}CHCH\text{=}CH_2$	(95)	0	13
MeI	monomer	0.25	Ph_2CO	A	1.0	$Ph_2C\text{=}CH_2$	(72)	(23)	12
MeI	$LiPPh_2$	0.35	Ph_2CO	A	1.09	$Ph_2C\text{=}CH_2$	(94)	0	13
MeI	$LiPPh_2$	0.35	$(n\text{-}C_9H_{19})_2C\text{=}O$	A	1.5	$(n\text{-}C_9H_{19})_2C\text{=}CH_2$	(96)	(0.4)	13

MeI	LiPPh$_2$	0.35	cholest-4-en-3-one	A	1.5	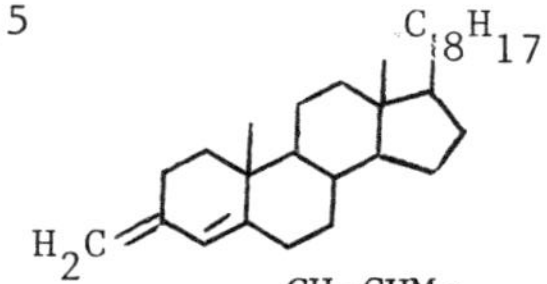	91	0	13
EtBr	NaPPh$_2$	b	9-formylanthracene	C	c	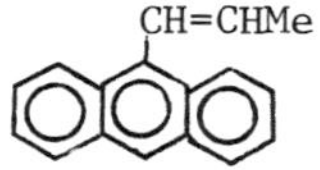	66		15
EtI	monomer	0.25	(CH$_2$)$_5$CO	A	1.0	(CH$_2$)$_5$C=CHMe	(50)	(44)	12
MeOCH$_2$Cl	LiPPh$_2$	0.81	(CH$_2$)$_4$CO	D	1.0	(CH$_2$)$_4$C=CHOMe	89		16
MeOCH$_2$Cl	LiPPh$_2$	0.81	(CH$_2$)$_5$CO	D	1.0	(CH$_2$)$_5$C=CHOMe	95		16
MeOCH$_2$Cl	LiPPh$_2$	0.81	PhCHO	D	1.0	PhCH=CHOMe	85		16
MeOCH$_2$Cl	LiPPh$_2$	0.81	Ph$_2$CO	D	1.0	Ph$_2$C=CHOMe	92		16
MeSCH$_2$Cl	LiPPh$_2$	0.60	PhCHO	E	1.0	PhCH=CHSMe	86		16
MeSCH$_2$Cl	LiPPh$_2$	0.60	PhCOMe	E	1.0	PhMeC=CHSMe	81		16
MeSCH$_2$Cl	LiPPh$_2$	0.60	Ph$_2$CO	E	1.0	Ph$_2$C=CHSMe	79		16
MeO$_2$CCH$_2$Br	LiPPh$_2$	0.19	p-C$_6$H$_4$(CHO)$_2$	F	1.0	p-OHCC$_6$H$_4$CH=CHCO$_2$Me	63	32	17

Continued on next page

Table I (continued)

MeO_2CCH_2Br	$LiPPh_2$	0.24	$p\text{-}C_6H_4(CHO)_2$	G	1.0	$\underline{p}\text{-}OHCC_6H_4CH=CHCO_2Me$	80	18	17
$CH_2=CHCH_2Br$	$LiPPh_2$	0.70	$\underline{p}\text{-}ClC_6H_4CHO$	H	1.5	$\underline{p}\text{-}ClC_6H_4CH=CHCH=CH_2$	78		14
$CH_2=CHCH_2Br$	$LiPPh_2$	0.70	9-formylanthracene	B	1.5	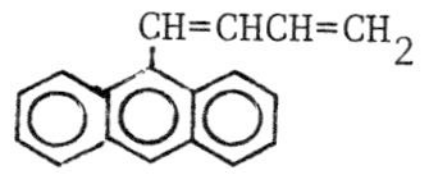 [anthracene]$CH=CHCH=CH_2$	trace		14
$\underline{i}\text{-}PrBr$	$NaPPh_2$	$\underline{b}$	9-formylanthracene	C	$\underline{c}$	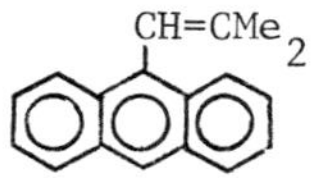 [anthracene]$CH=CMe_2$	10		15
$\underline{n}\text{-}BuBr$	$NaPPh_2$	0.43	PhCOMe	C	1.3	$PhMeC=CHPr$	97		15
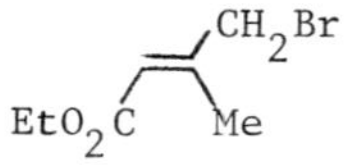 EtO_2C [structure with CH_2Br, Me]	$LiPPh_2$	0.35	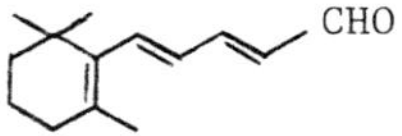 [structure with CHO]	I	1.08	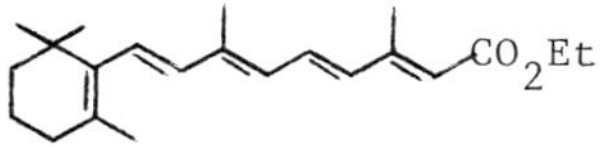 [structure with CO_2Et]	70		18
$\underline{n}\text{-}C_6H_{13}Br$	$LiPPh_2$	0.70	$\underline{p}\text{-}ClC_6H_4CHO$	Q	$\underline{d}$	$\underline{p}\text{-}ClC_6H_4CH=CHC_5H_{11}$	26		19
$PhCH_2Cl$	$LiPPh_2$	0.81	HCO_2Me	J	1.0	$MeOCH=CHPh$	93		16

Halide	Base	Equiv	Carbonyl		Ratio	Product	Yield		Ref
$PhCH_2Cl$	$LiPPh_2$	0.70	$(CH_2)_5CO$	B	1.5	$(CH_2)_5C{=}CHPh$	0		14
$PhCH_2Cl$	monomer	0.25	$PhCHO$	K	1.0	$PhCH{=}CHPh$	40		20
$PhCH_2Cl$	monomer	0.25	$PhCHO$	L	1.0	$PhCH{=}CHPh$	60		20
$PhCH_2Br$	monomer	0.25	$PhCHO$	A	1.0	$PhCH{=}CHPh$	(35)	(51)	12
$PhCH_2Br$	$LiPPh$	0.35	$PhCHO$	M	0.97	$PhCH{=}CHPh$	(93)		13
$PhCH_2Cl$	$LiPPh_2$	0.70	$PhCHO$	B	1.5	$PhCH{=}CHPh$	92		14
$PhCH_2Br$	$LiPPh_2$	0.17	$\underline{p}{-}BrC_6H_4CHO$	Q	1.2	$\underline{p}{-}BrC_6H_4CH{=}CHPh$	71		21
$PhCH_2Cl$	$LiPPh_2$	0.70	$\underline{n}{-}C_7H_{15}CHO$	B	1.5	$\underline{n}{-}C_7H_{15}CH{=}CHPh$	(93)		14
$PhCH_2Br$	$LiPPh_2$	0.12	$\underline{p}{-}C_6H_4(CHO)_2$	N	1.0	$\underline{p}{-}OHCC_6H_4CH{=}CHPh$	83	0	17
$PhCH_2Br$	$LiPPh_2$	0.12	$\underline{p}{-}C_6H_4(CHO)_2$	F	1.0	$\underline{p}{-}OHCC_6H_4CH{=}CHPh$	54	36	17
$PhCH_2Br$	$LiPPh_2$	0.12	$\underline{m}{-}C_6H_4(CHO)_2$	F	1.0	$\underline{m}{-}OHCC_6H_4CH{=}CHPh$	67		17
$PhCH_2Cl$	$LiPPh_2$	0.06	$\underline{p}{-}C_6H_4(CHO)_2$	F	1.0	$\underline{p}{-}OHCC_6H_4CH{=}CHPh$	49	22	17
$PhCH_2Cl$	$LiPPh_2$	0.35	$PhCH{=}C(Me)CHO$	M	1.0	$PhCH{=}C(Me)CH{=}CHPh$	(89)	(6)	13

Continued on next page

Table I (continued)

Reagent			Substrate	Method	Ratio	Product	Yield	Ref
PhCH$_2$Cl	LiPPh$_2$	0.70	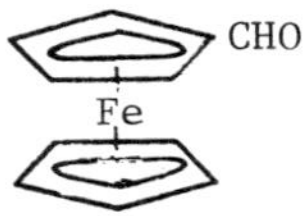 (ferrocenyl)CHO	B	1.5	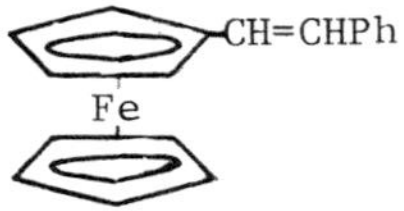 (ferrocenyl)CH=CHPh	62	19
PhCH$_2$Cl	LiPPh$_2$	0.70	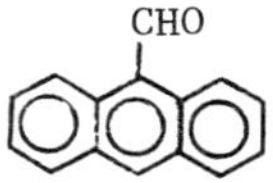 (anthracenyl)CHO	Q	c	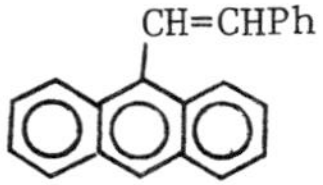 (anthracenyl)CH=CHPh	94	19
PhCH$_2$Br	LiPPh$_2$	0.17	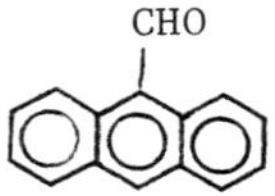 (anthracenyl)CHO	Q	1.1	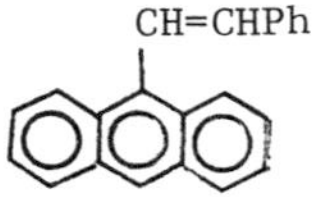 (anthracenyl)CH=CHPh	57	21
p-Me$_3$CC$_6$H$_4$CH$_2$Cl	LiPPh$_2$	0.70	CH$_2$O	B	1.5	p-Me$_3$CC$_6$H$_4$CH=CH$_2$	95	14
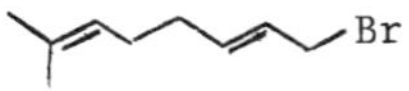 (geranyl)Br	NaPPh$_2$	b	p-ClC$_6$H$_4$CHO	C	c	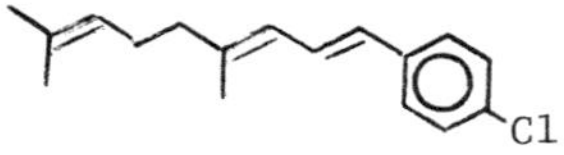 p-ClC$_6$H$_4$CH=... Cl	48	15
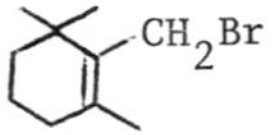 (cyclohexenyl)CH$_2$Br	LiPPh$_2$	0.35	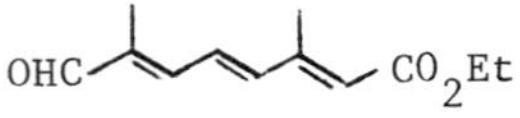 OHC...CO$_2$Et	I	0.67	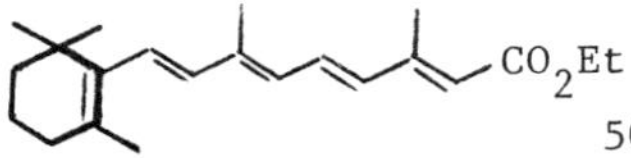 ...CO$_2$Et	50	18

Substrate	Base		Carbonyl	Method		Product	Yield	Ref.
naphthyl-CH$_2$Br	NaPPh$_2$	b	PhCO(CH$_2$)$_4$Br	C	c	cyclopentene–Ph	32	15
naphthyl-CH$_2$Br	LiPPh$_2$	0.70	CH$_2$O	B	1.5	naphthyl–CH=CH$_2$	67	14
naphthyl-CH$_2$Br	LiPPh$_2$	0.70	naphthyl-CHO	B	1.5	naphthyl–HC=CH–naphthyl	65	14
fluorenyl-Br	NaPPh$_2$	b	PhCHO	C	c	fluorenylidene=HCPh	86	15
anthracenyl-CH$_2$Cl	LiPPh$_2$	0.70	PhCHO	B	1.5	anthracenyl–CH=CHPh	20	22

Footnotes on next page

Table I. (Footnotes)

<u>a</u> A. Add a solution of the sodium salt of DMSO in DMSO to polymer in tetrahydrofuran (THF).

B. Add 50% aqueous NaOH and a phase transfer catalyst, either cetyltrimethylammonium bromide or tetra-*n*-butylammonium iodide, to polymer in dichloromethane.

C. Add *n*-butyllithium in hexane to polymer in dioxane.

D. Add phenyllithium in ether to polymer in ether.

E. Add phenyllithium (solvent not specified) to polymer in THF.

F. Add excess ethylene oxide to preformed phosphonium ion polymer in benzene.

G. Add excess ethylene oxide to the phosphine polymer in a solution of methyl bromoacetate in benzene.

H. Add 50% aqueous NaOH to polymer in dichloromethane without any phase transfer catalyst.

I. Add sodium ethoxide in ethanol to polymer in ethanol.

J. Add sodium methoxide in methanol to polymer in methanol.

K. Add potassium *t*--butoxide to polymer in THF.

L. Add sodium hydride to polymer in THF.

M. Add sodium methoxide in methanol to polymer in THF.

N. Add *n*-butyllithium in ether to polymer in THF.

O. Add 50% aqueous NaOH to polymer in aqueous 37% formaldehyde.

P. Add *n*-butyllithium in hexane to polymer in benzene.

Q. Add solid potassium carbonate and dicyclohexyl-18-crown-6 ether to polymer in THF and reflux.

R. Add solid potassium carbonate to polymer in THF and reflux.

S. Add aqueous sodium hydroxide to polymer in methanolic formaldehyde.

T. Add methyl bromoacetate and ethylene oxide to the phosphine polymer in THF.

<u>b</u> Not reported, but probably 0.3-0.5.

<u>c</u> Not reported, but probably <u>ca</u>. 0.4.

<u>d</u> Not reported specifically, but in the range 1.0-1.3.

Table II. Polymer-Bound Alkenes from Polymer-Bound Benzyl
Phosphonium Ions and Soluble Carbonyl Compounds [a]

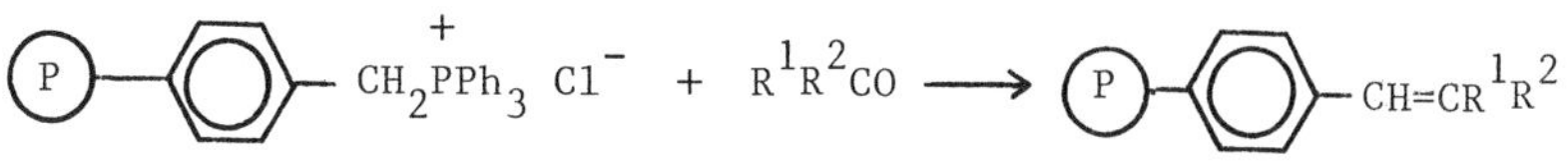

$$\text{P}-\bigcirc-CH_2\overset{+}{P}Ph_3\ Cl^- + R^1R^2CO \longrightarrow \text{P}-\bigcirc-CH=CR^1R^2$$

R^1R^2CO	method[b]	% yield	lit.
CH_2O	B	13	22
CH_2O	O	79	23
$ClCH_2CHO$	P	50	24
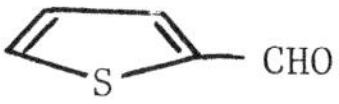 thiophene—CHO	B	83	25
PhCHO	H	76	19
$p\text{-}ClC_6H_4CHO$	B	97	22
$p\text{-}BrC_6H_4CHO$	Q	79	19
$p\text{-}O_2NC_6H_4CHO$	B	49	25
$p\text{-}BrC_6H_4COMe$	Q	17	22
 alkenyl CHO	B	41	25
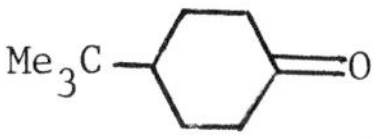 Me_3C—cyclohexanone=O	B	0	25
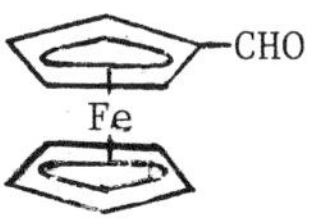 ferrocene—CHO	B	62	25

Continued on next page

Table II (continued)

Structure	R/B	Yield	Value
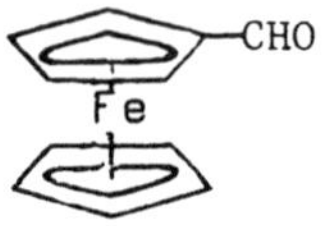 ferrocene-CHO	R	55	19
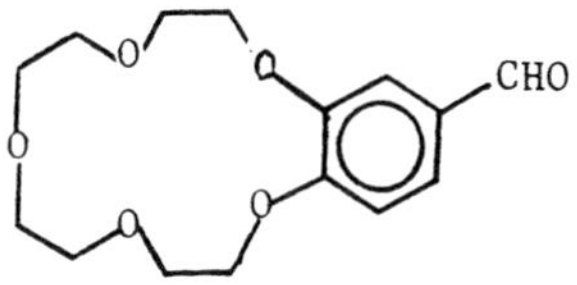 benzo-crown-CHO	B	$\underline{ca.}$ 50	22
$\underline{p}$-Ph$_2$PC$_6$H$_4$CHO	B	80	25
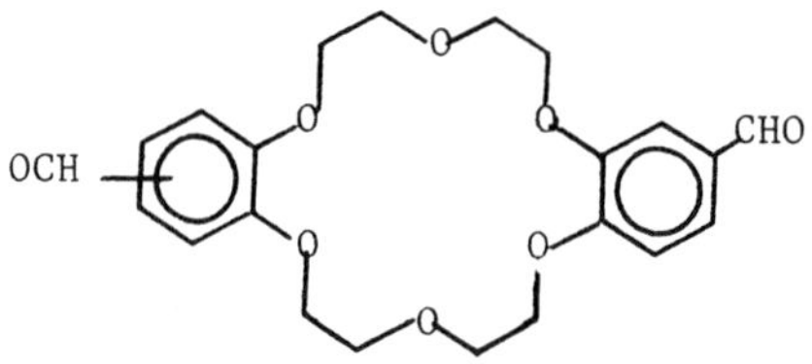 dibenzo-crown OCH₃/CHO	B	$\underline{ca.}$ 25	22
	R	$\underline{ca.}$ 40	22

$\underline{a}$ All reactions were carried out on DF 0.18-0.55, 1-2% cross-linked polymers with 1.0-1.5 molar equiv of carbonyl compound, except for 17-50 molar equiv of formaldehyde.

$\underline{b}$ See Table I.

Scheme 3

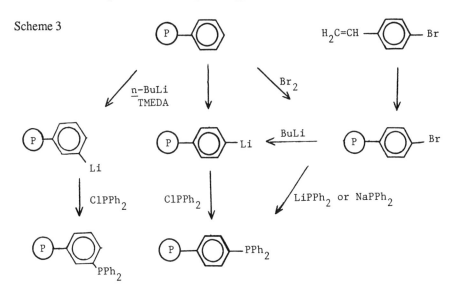

Polystyrene may be lithiated directly with *n*-butyllithium and tetramethylethylene-diamine (TMEDA) in cyclohexane (32,33), or brominated polystyrene may be metala-ted by metal-halogen exchange with *n*-butyllithium in benzene or toluene (15,32). The *n*-BuLi/TMEDA method gives a 2/1 *m/p* isomeric mixture of methylstyrene repeat units by C-13 NMR spectral analysis after quenching with methyl iodide (34). At 65 °C a DF of only 0.27 was attained, but use of *t*-butyllithium enabled a DF of 0.5 (34). Treatment of the lithiated polymers with chlorodiphenylphosphine gives triarylphos-phine groups, but the phosphine isomer distribution has not been analyzed. At de-grees of functionalization of <0.16, the active sites in a 1.8% cross-linked gel polysty-rene are found preferentially near the bead surface, according to electron microprobe analysis (33). The BuLi/TMEDA route to polystyryldiphenylphosphine has not been used for Wittig reagents. It offers a potential advantage of having most of the func-tional groups near the particle surface, but the reactivity of the meta isomer is un-known.

Brominated polystyrenes may be prepared by copolymerization of styrene (M_1) and *p*-bromostyrene (M_2, $r_1 = 0.67$ and $r_2 = 1.10$ (35)), which gives slightly greater incorporation of the functional monomer during the early stages of copolymerization. This method was reported as a route to Wittig reagents, but no specific results were given (12). An easier method is Lewis acid-catalyzed bromination of cross-linked polystyrene with bromine in carbon tetrachloride and thallic acetate, thallic chloride, ferric chloride, or boron trifluoride as catalyst (32,36). Quantitative incorporation of bromine at the para position can be attained, allowing easy control of the degree of functionalization. Iron powder also is an effective catalyst, but the residual iron must be removed from the reaction mixture with a magnet (37). Lithiation of the bromin-ated polymers with *n*-butyllithium in benzene or toluene proceeds quantitatively, and trapping with chlorodiphenylphosphine proceeds in 80-84% functional yield on 1-2% cross-linked polystyrenes (13,32). Yields are lower on more highly cross-linked polystyrenes (13). Alternatively brominated polystyrene may be treated with lithium diphenylphosphide or sodium diphenylphosphide (15,38). On 2% cross-linked poly-mers functional yields from lithium diphenylphosphide are 90-100%, but yields are lower on more highly cross-linked polymers, presumably because the bulky diphenyl-phosphide cannot form the required P-C bond (13).

There are differences of opinion as to the DF required for a useful polymer-sup-ported reagent. Some workers contend "the higher the better". It is mol/L of reaction mixture, not mmol/g of dry polymer, that is important. That is the quantity required for design of a manufacturing process. Two factors control mol/L: the dry reagent capacity in mmol/g, and the swelling ratio under reaction conditions. When the func-tional repeat unit has a much larger formula weight than the other copolymer repeat units, as is the case with the phosphines and phosphonium ions in Wittig reagents, a wide range of DF leads to high capacity in mmol/g. For a styryldiphenylphosphine-styrene copolymer, DF = 1.0 corresponds to 3.47 mmol/g, and DF = 0.30 corres-ponds to 1.88 mmol/g. For Wittig reactions on 2% cross-linked polystyrenes, all capacities within that range can give high yield reactions. For Wittig reactions on more highly cross-linked polystyrenes, lesser DF may be desirable for reasons explained in the section on polymer cross-linking. The highest yields in Wittig reactions are normally obtained with low cross-linking and high swelling ratio. Those conditions are fine for laboratory scale reactions but not for large scale processes, in which reactor loading and physical stability and ease of filtration of the polymer also are important.

Alkyl(polystyryl)(diphenyl)phosphonium Ions. Many methods have been used suc-cessfully for the alkylation of polystyryldiphenylphosphine. A swelling solvent is the only critical requirement. Benzyl chloride and benzyl bromide have been used without solvent. Benzene, toluene, chlorobenzene, THF, N,N-dimethylformamide (DMF), and dimethylsulfoxide (DMSO) have been used with other alkyl halides. The labile ß-cyclogeranylphosphonium ion in Scheme 4 was prepared by conversion of the poly-meric phosphine to its hydrobromide with HBr in benzene, removal of the excess HBr by washing with benzene and ether, and reaction of the hydrobromide with ß-cyclo-geraniol (18).

Scheme 4

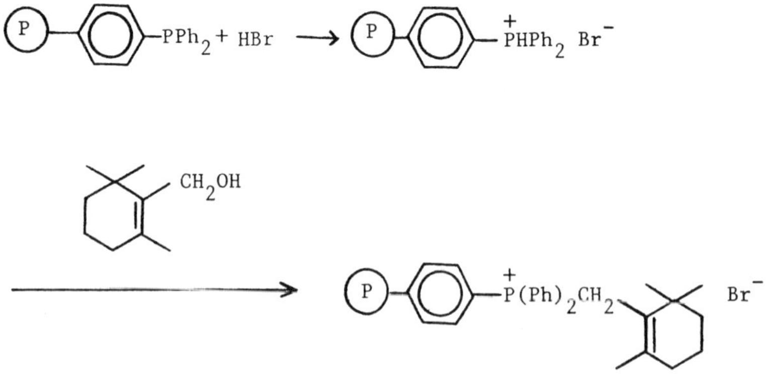

(Polystyrylmethyl)triphenylphosphonium Ions. The phosphonium ions used for for-mation of alkenes on cross-linked polystyrenes by the Wittig reaction have been pre-pared by reaction of triphenylphosphine with chloromethylated polystyrene in chloro-benzene at reflux (25). More highly nucleophilic phosphines such as tri-n-butylphos-phine require only 80-100 °C to form phosphonium salts with chloromethylated polystyrene. Copolymers of chloromethylstyrene (commercially available as a 70/30 *m/p* mixture) could be used in place of the chloromethylated polymers. Linear phosphonium ion polymers have been prepared by polymerization of (*p*-styrylmethyl) triphenylphosphonium chloride (2) with and without styrene (19,22,27).

$$H_2C=CH-\langle\bigcirc\rangle-CH_2\overset{+}{P}Ph_3Cl^-$$ **2**

<u>Analysis of Phosphonium Ion Polymers</u>. The equivalent weight of a reagent must be known for synthetic use. That is not necessarily easy with polymer-bound reagents. Polymer-bound phosphonium ions, however, can be analyzed well. The halide counterions at the phosphonium sites can be determined titrimetrically after they have been displaced from a small sample of the reagent by another anion such as nitrate (13). The solvent swollen reagent can be analyzed qualitatively by C-13 and P-31 NMR, and P-31 NMR can even be used quantitatively (although with less accuracy than the titrimetric analysis for halide) by peak area comparison with an internal standard (39). Elemental analyses for phosphorus and halide should be used periodically to confirm the results of analyses performed in the chemist's own laboratory.

<u>Wittig Reaction Conditions</u>

Solvents ranging from ether to methanol have been used for the Wittig reaction, even though neither ether nor methanol is a good solvent for polystyrene. Swelling of the polymer usually is needed for the base and the carbonyl compound to penetrate to all of the phosphonium ion sites. Swelling generally requires that solvation of the polymer be exothermic or no more than slightly endothermic. Polar solvents such as methanol, ethanol, DMF, and DMSO solvate well the ionic phosphonium halide sites but are not good solvents for polystyrene. Nonpolar solvents such as benzene, toluene, and THF solvate polystyrene but not the phosphonium halide sites. Chloroform and dichloromethane solvate both. In water, methanol, and DMSO the phosphonium ions have coulombic attraction to the conjugate bases of the solvent. In solvents of low polarity, strong bases such as organolithium reagents must penetrate the polymer as electrically neutral species.

Phase transfer catalyzed reactions in which ylides are formed from allylic and benzylic phosphonium ions on cross-linked polystyrenes in heterogeneous mixtures, such as aqueous NaOH and dichloromethane or solid potassium carbonate and THF, are particularly easy to perform. Ketones fail to react under phase transfer catalysis conditions. A phase transfer catalyst is not needed with soluble phosphonium ion polymers. The cations of the successful catalysts, cetyltrimethylammonium bromide and tetra-*n*-butylammonium iodide, are excluded from the cross-linked phosphonium ion polymers by electrostatic repulsion. Their catalytic action must involve transfer of hydroxide ion to the polymer surface rather than transport of the anionic base into the polymer. Dicyclohexyl-18-crown-6 ether was used as the catalyst for ylide formation with solid potassium carbonate in refluxing THF. Potassium carbonate is insoluble in THF. Earlier work on other solid-solid-liquid phase transfer catalyzed reactions indicated that a trace of water in the THF is necessary (40), so the active base for ylide formation is likely hydrated, even though no water is included deliberately in the reaction mixture.

A side reaction has been identified in the phase transfer catalyzed Wittig reactions. Hydroxide ion attacks positively charged phosphorus to give a phosphine oxide and the methylaromatic hydrocarbon from cleavage of the benzylic carbon-phosphorus bond (Equation 1). The phosphonium ion derived from 9-chloromethylanthracene

gave 80% 9-methylanthracene in addition to the desired alkene (22). The less hindered phosphonium ion from 4-t-butylbenzyl chloride gave 16% of 4-t-butyltoluene (14).
With poly(p-styrylmethyl)triphenylphosphonium chloride) (Table II) the by-product is
a p-methylstyrene repeat unit (Equation 2).

$$(1)$$

$$(2)$$

Polymer Cross-linking

Although most Wittig reactions on polymer supports have employed 0.5-2% cross-
linked polystyrenes, a number of phase transfer catalyzed reactions have been carried
out with soluble polymers, both for synthesis of soluble alkenes and for introduction
of alkene functional groups to the polymer. The soluble polymer can be isolated from
a reaction mixture after precipitation into ether. Results are in Table III. Particularly
noteworthy are the reactions with formaldehyde that produce vinyl groups on the
polymer, for the vinyl groups can be used to prepare graft copolymers. The last
example in Table III shows how the soluble phosphonium ion polymer can be cross-
linked by Wittig reaction with a dialdehyde. Similarly soluble aldehyde-functionalized
polystyrenes have been cross-linked by Wittig reactions with a bisphosphonium salt
(Equation 3) (22,27). These cross-linked polymers are insoluble but still swellable.

$$(3)$$

More important for industrial applications of polymer-supported Wittig reagents are
polystyrenes cross-linked with more than 2% divinylbenzene. The swollen 1-2%
cross-linked polymers may be too gelatinous for filtration on a large scale or for use in
a flow reactor. Higher degrees of cross-linking give more rigid, more easily filtered
polymers. However, more cross-linking may result in slower reactions and in totally
unreactive sites within the polymer matrix due to diffusional limitations. Table IV
compares 2% and 8% cross-linked gel polystyrenes and a 20% cross-linked macroporous polystyrene as supports for Wittig reactions. Respectable but somewhat lower
yields were obtained with the more highly cross-linked supports even for the reactions

Table III.

Wittig Reactions of Linear Poly(styrylphosphonium halide)s

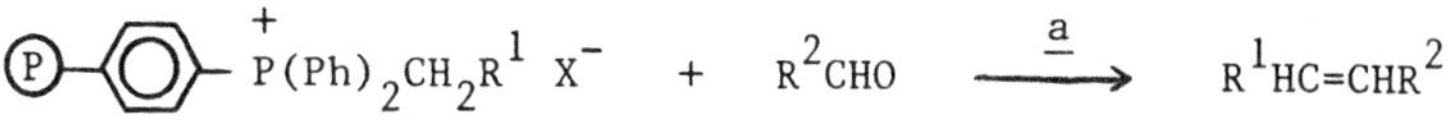

$$\text{(P)}-\text{⬡}-\overset{+}{\text{P(Ph)}}_2\text{CH}_2\text{R}^1 \text{ X}^- + \text{R}^2\text{CHO} \xrightarrow{a} \text{R}^1\text{HC=CHR}^2$$

R^1CH_2X	R^2CHO	% yield	lit.
CH_2=CHCH$_2$Br	p-ClC$_6$H$_4$CHO	78	19
PhCH$_2$Cl	PhCH=CHCHO	75	22
PhCH$_2$Cl	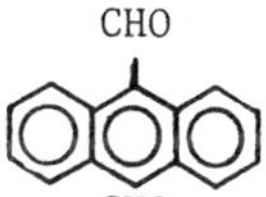	92	19
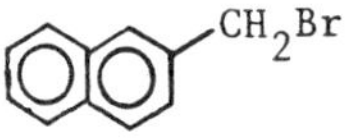	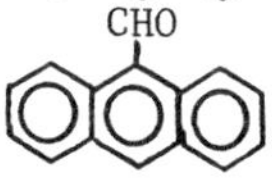	100	22

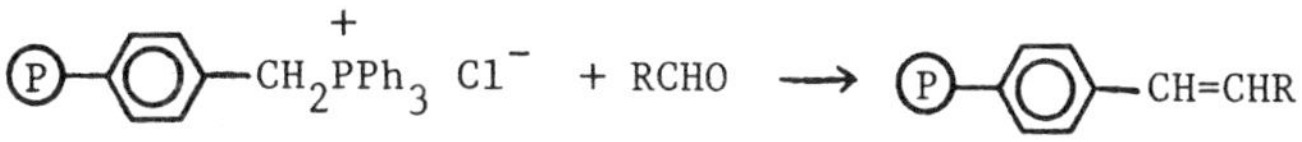

$$\text{(P)}-\text{⬡}-\overset{+}{\text{CH}_2\text{PPh}_3} \text{ Cl}^- + \text{RCHO} \longrightarrow \text{(P)}-\text{⬡}-\text{CH=CHR}$$

DF	RCHO	method[c]	% yield	lit.
0.29	CH$_2$O	S	b	23
0.42	CH$_2$O	S	86	26
0.29	PhCHO	H	55	19
0.29	p-ClC$_6$H$_4$CHO	H	40	22
0.29	p-C$_6$H$_4$(CHO)$_2$	H	100 (insoluble polymer)	27

[a] All reactions used method H with DF 0.55 phosphine.

[b] Not reported.

[c] See Table I.

of large ketones such as 10-nonadecanone and cholest-4-en-3-one. The more highly cross-linked polymers had lower degrees of functionalization because the reactions of brominated polystyrene with lithium diphenylphosphide, or with butyllithium followed by chlorodiphenylphosphine, proceeded in lower functional yield. The subsequent alkylations of the phosphines and the Wittig reactions gave good yields. The phosphine-forming reactions in the preparations of the Wittig reagents are restricted more by the polymer matrix than are the Wittig reactions. Sites in the polymer matrix that are accessible to lithium diphenylphosphide are also accessible to the carbanion from DMSO and to large ketones such as 10-nonadecanone and cholest-4-en-3-one. If the polymeric phosphines had been prepared by copolymerization of p-styryldiphenylphosphine, the Wittig reaction yields from such large ketones would probably have been unreportably low.

Examination of Table IV reveals that yields were generally better with the 20% cross-linked macroporous support than with the 8% cross-linked gel support. The permanent pores of the macroporous polymer provide a way for reagents to be transported most of the way from external solution to the reactive sites through solvent-filled pores rather than through polymer matrix. Although the diffusion path within the 20% cross-linked matrix is more tortuous, its average length is only a few nanometers, whereas in the 8% cross-linked gel the average length of the path from the particle surface to the reactive site is tens of micrometers. The major difficulty encountered with the macroporous resins is that in organic solvents they are often more friable than gel resins. Powdered resins clog filters and cannot be recycled easily. This problem might be solved by a thorough investigation of how the mode of synthesis of macroporous polystyrenes relates to their fragility in organic solvents. Such investigations have given macroporous ion exchange resins of high physical stability.

Stereochemistry

Table V compiles the Z/E product isomer ratios of reactions from Table I for which such information has been reported. With 0.5-2% cross-linked polystyrene supports, the isomer ratios are similar to those obtained from Wittig reactions in solution. The exceptions appear to be the reactions on more highly cross-linked supports shown in Table IV. There is a clear trend toward greater E selectivity as the degree of cross-linking of the polymer increases. This probably should be explained as an environmental effect, but comparison with solvent effects on stereochemistry of Wittig reactions in the literature reveals no tendency for aromatic solvents, structurally similar to polystyrene, to increase E selectivity.

One particularly puzzling result is the difference in Z/E stilbenes obtained from 2% cross-linked, DF 0.18 polystyryldiphenylphosphines prepared by lithiation/phosphination and by copolymerization of p-styryldiphenylphosphine and styrene (41).

Wittig reactions of unstabilized phosphoranes performed in solution under salt-free conditions give high Z selectivity (1-5). After formation of the phosphorane with an organolithium base, separation of the lithium halide from the reaction mixture is particularly easy when the phosphorane is bound to insoluble polystyrene, for only thorough washing of the polymer is required (42). Deliberate addition of lithium salts to the reaction mixture gives more of the E alkene, due either to reversible formation of the oxaphosphetane or to lithium ion catalyzed equilibration of the oxaphosphetanes with zwitterionic intermediates. These methods for control of Wittig reaction stereochemistry are effective for reactions of unstabilized ylides from n-alkyl halides, but not for reactions of more stabilized ylides from allyl and benzyl halides. Table VI shows the results of application of lithium free and lithium ion promoted Wittig reaction conditions on polymer supports. In some cases stereoselectivity was as high as

Table IV. Effects of Polymer Cross-linking on Yield and Stereochemistry

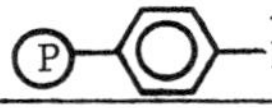

$$\underset{\text{(P)}}{\overset{+}{\bigcirc}}\!-\!\text{P(Ph)}_2\text{CHR}^1\text{R}^2\text{X}^-\ \xrightarrow{\ \text{a}\ }\ \text{R}^3\text{R}^4\text{CO}\ \longrightarrow\ \text{R}^1\text{R}^2\text{C=CR}^3\text{R}^4$$

R^1R^2CHX	R^3R^4CO	method	% DVB	% yield[b]	$Z/\underline{E}$	lit.
MeI	PhCH=CHCHO	A	2	95		13
			8	52		
			20	83		
MeI	Ph_2CO	A	2	94		13
			8	61		
			20	74		
MeI	$(n\text{-}\underline{C}_9H_{19})_2CO$	A	2	96		13
			20	62		
MeI	Cholest-4-en-3-one	A	2	91		13
			20	87		
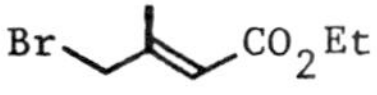 Br~~~CO₂Et	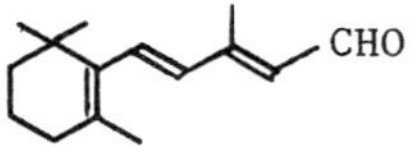 [β-ionylideneacetaldehyde, CHO]	I	2	70	46/54	18
			20	65	26/74	
$PhCH_2Br$	PhCHO	M	2	93	57/43	13
			8	73	48/52	
			20	80	28/72	
$PhCH_2Br$	PhCH=CHCHO	M	2	89	40/60	13
			8	77	35/65	
			20	72	17/83	

[a] DF = 0.32–0.38, 0.28–0.32, and 0.10–0.17 for 2, 8, and 20% cross-linked polymers.

[b] GC yields for all reactions except those of cholest-4-en-3-one and β-ionylideneacetaldehyde, from which yields are isolated.

Table V. Stereochemistry of Alkene Formation

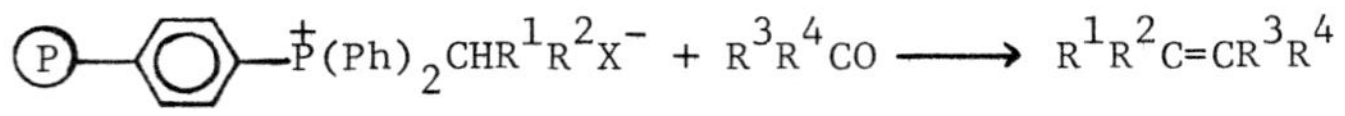

$$\text{P}-\!\!\bigcirc\!\!-\overset{+}{\text{P}}(\text{Ph})_2\text{CHR}^1\text{R}^2\text{X}^- + \text{R}^3\text{R}^4\text{CO} \longrightarrow \text{R}^1\text{R}^2\text{C}=\text{CR}^3\text{R}^4$$

$\text{R}^1\text{R}^2\text{CHX}$	$\text{R}^3\text{R}^4\text{CO}$	method[a]	$\underline{Z/E}$	lit.
EtI	PhCHO	A	84/16	12
EtI	PhCHO	C	70/30	15
$\text{MeO}_2\text{CCH}_2\text{Br}$	PhCHO	T	9/91	41
$\text{MeO}_2\text{CCH}_2\text{Br}$	$\underline{p}\text{-C}_6\text{H}_4(\text{CHO})_2$	F,G	0/100	17
$\underline{n}$-BuBr	PhCHO	L	85/15	41
$\underline{n}$-BuBr	PhCOMe	C	59/41	15
$\underline{n}\text{-C}_6\text{H}_{13}\text{Br}$	EtCHO	C	60/40	15
PhCH_2Br	MeCHO	C	60/40	15
PhCH_2Br	PhCHO	A	42/58	12
PhCH_2Br	PhCHO	L	60/40[b]	41

$PhCH_2Br$	PhCHO	L	20/80[c]	41
$PhCH_2Cl$	$p\text{-}ClC_6H_4CHO$	B	56/44	14
$PhCH_2Cl$	$p\text{-}MeC_6H_4CHO$	B	43/57	14
$PhCH_2Cl$	$p\text{-}C_6H_4(CHO)_2$	F	30/70	17
$PhCH_2Br$	$p\text{-}C_6H_4(CHO)_2$	N	79/21	17
$PhCH_2Cl$	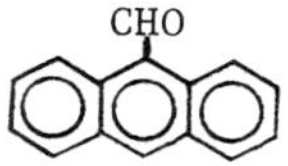	B	0/100	14

[a] See footnotes of Table I. [b] Polystyryldiphenylphosphine prepared by 1) bromine, Tl(OAc)$_3$;
2) n-BuLi, PhMe; 3) Ph$_2$PCl, PhMe.

[c] Polystyryldiphenylphosphine prepared from p-styryldiphenylphosphine.

Table VI. Stereochemistry of Wittig Reactions under Z-Selective and E-Selective Conditions (42) [a]

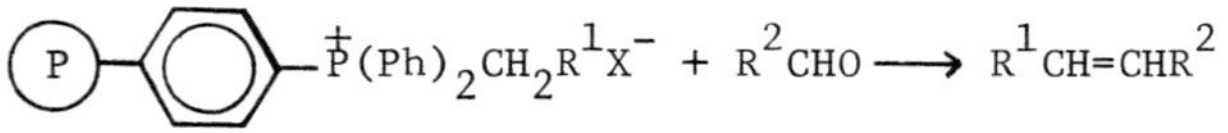

$$\text{P}\!-\!\!\bigcirc\!\!-\!\overset{+}{\text{P}}(Ph)_2CH_2R^1X^- + R^2CHO \longrightarrow R^1CH=CHR^2$$

R^1CH_2X	R^2CHO	Z/E [b]	
		Z-selective	E-selective
EtI	PhCHO	92/8	3/97
		(96/4)	(3/97)
n-BuBr	PhCHO	97.5/2.5	14/86
n-C_6H_{13}Br	EtCHO	75/25	30/70
		(96/4)	(1/99)
PhCH$_2$Br	MeCHO	60/40	60/40
		(44/56)	

[a] Adapted from ref. 42 with permission. Copyright 1973, VCH Verlagsgesellschaft mbH.
[b] Isomer ratios from solution reactions in parentheses.

attained in solution Wittig reactions. Why others did not show similarly high selectivity is not known.

Recycling of the Polymer

Two methods have been used to reduce polymeric phosphine oxides to phosphines. A refluxing mixture of trichlorosilane and benzene effects direct reduction to the phosphine (13,43-45). Several different procedures have been reported, using as little as 1.1 and as much as 6 molar equivalents of trichlorosilane, with triethylamine or N,N-dimethylaniline to trap the HCl liberated, or no base at all. The by-product siloxane oligomers must be removed from the resin by thorough washing with anhydrous solvent, so use of only a slight excess of trichlorosilane is recommended.

Triphenylphosphine oxide from the commercial production of vitamin A by a Wittig reaction is recycled by reaction with phosgene to give the phosphine dichloride and reduction of the dichloride to triphenylphosphine with elemental phosphorus (46). Polymeric phosphine oxides have been converted to phosphine dichlorides with oxalyl chloride and then reduced to phosphine with diisobutylaluminum hydride (47).

The extent of conversion from phosphine oxide to phosphine in cross-linked polystyrenes can be determined by P-31 NMR spectral analysis (13,39,45). This method also has been used to prove that solvent-swollen polystyryldiphenylphosphine resins in air are oxidized to the phosphine oxide (48).

Reuse of the phosphine resins for Wittig reactions has been reported without details several times. In the only detailed results 92% conversion of phosphine oxide to phosphine with trichlorosilane was attained with a 2% cross-linked polymer, and repeated synthesis of stilbene by quaternization with benzyl bromide and Wittig reaction with benzaldehyde gave 97% gc yields in both the second and the third cycles based on the amount of phosphonium bromide used (13). Identical recycling of a 20% cross-linked macroporous polymer gave 75% gc yield of stilbene compared with 80% from the first use (13).

Reactions of Polymer-Bound Aldehydes with Wittig Reagents

High yields have been obtained in reactions of singly polymer-protected aromatic dialdehydes with ylides from allylic and benzylic triphenylphosphonium ions as shown in Scheme 5 and Table VII (49-52). The monoaldehydes are readily recovered from the polymer by acidic hydrolysis. Yields were much higher with the six-membered cyclic acetals 4 and 5 than with the five-membered acetal protecting group obtained from 3. The diol-functionalized resin 3 could be regenerated and reused (49), but diol resins 4 and 5 could not be regenerated satisfactorily (50). The method has been extended to synthesis of unsymmetrical carotinoids from dial 6 and Wittig reagents 7 (51). This monoprotection method failed with aliphatic dialdehydes and with diketones. Another drawback is the low degree of functionalization of the diol polymers. Typically only DF = 0.04 was attained from starting chloromethylated 2% cross-linked polystyrene with DF = 0.19 or 0.26. However, much higher functional conversions have been reported for a thioglycerol bound to polymer as diol 8 (53). Attempts to monoprotect terephthalaldehyde with 8 and with a related diol were not particularly successful (53,54). It is not known if as low a DF as 0.04 is necessary to achieve monoprotection of aromatic dialdehydes. Successful Wittig reactions of only one aldehyde function of isophthalaldehyde and terephthalaldehyde were achieved with polymer-bound Wittig reagents with DF = 0.12 (17). The higher DF should be preferred.

Scheme 5

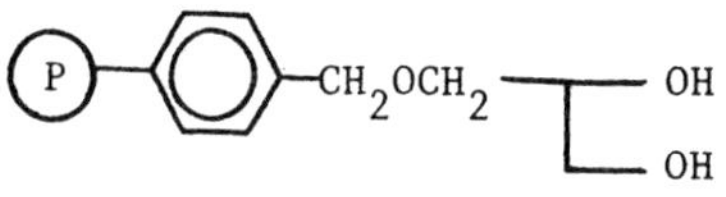

3, 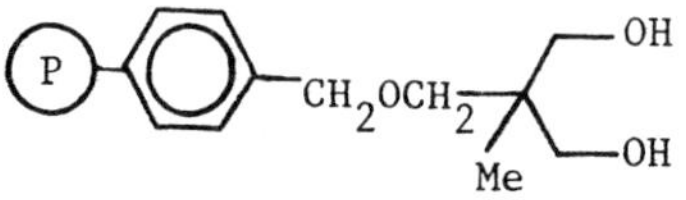 4, or 5 + OHC—⟨ ⟩—CHO   ⟶

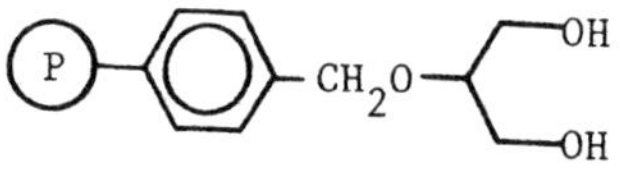

+
$Ph_3PCH_2(CH=CH)_nPh\ X^-$
⟶
$n = 0, 1$
$MeONa,\ DMF$

$\xrightarrow{H^+}$ OHC—⟨ ⟩—$(CH=CH)_{n+1}Ph$ + 3, 4, or 5

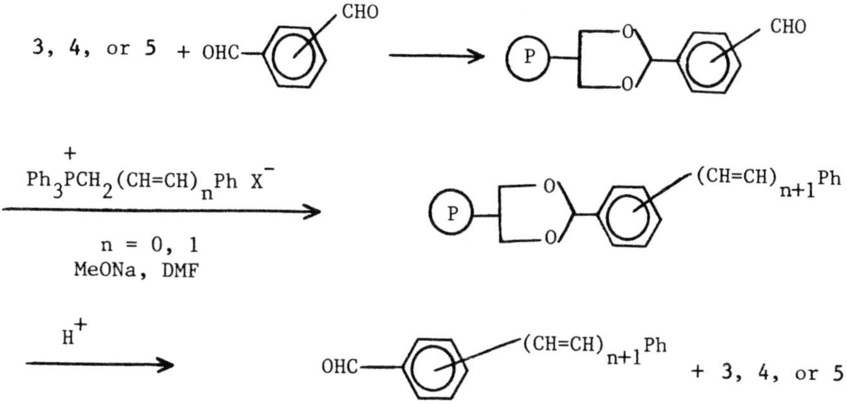

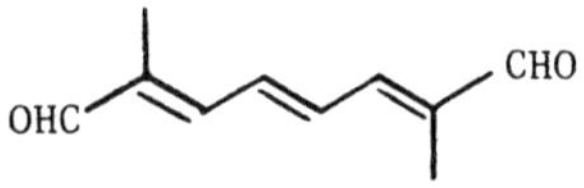

6

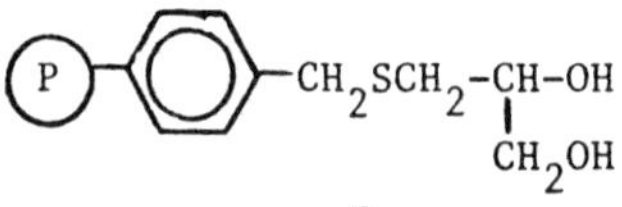

8

+
$RPPh_3\ Br^-$ R =
7

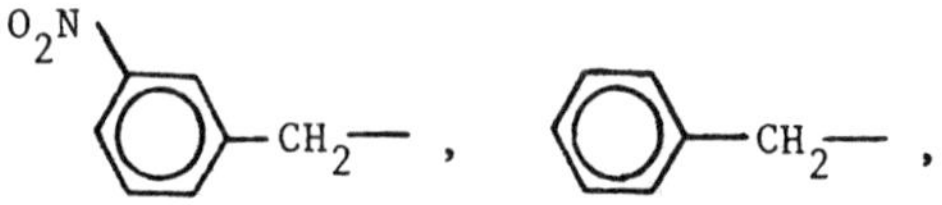

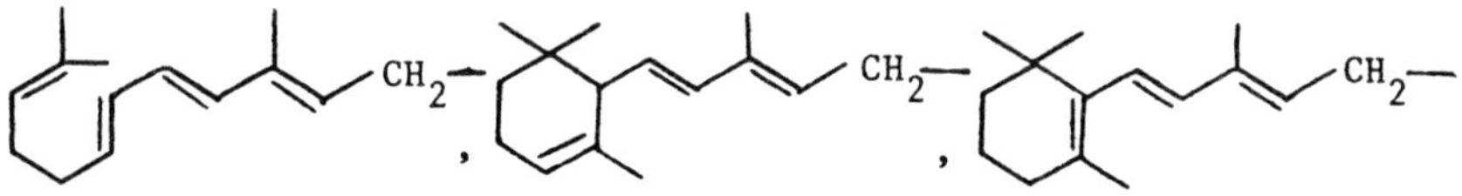

Table VII. Wittig Reactions of Polymer-Protected Dialdehydes

polymer diol	DF	product	% yield	ref.
3	0.038	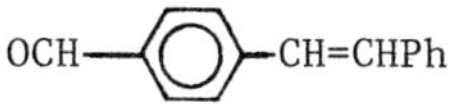 OCH—⟨O⟩—CH=CHPh	76	49
3	0.040	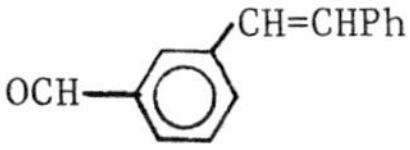 OCH—⟨O⟩ (CH=CHPh)	50	49
5	0.044	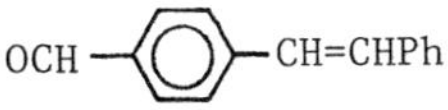 OCH—⟨O⟩—CH=CHPh	89	50
5	0.016	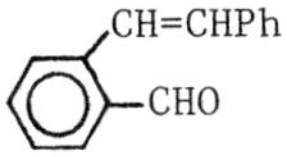 ⟨O⟩ (CH=CHPh)—CHO	100	50
4	0.016	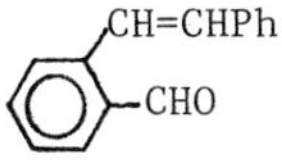 ⟨O⟩ (CH=CHPh)—CHO	100	50

Phosphorus Ester Analogs of Wittig Reagents

Diethylphosphonoacetonitrile and methyl diethylphosphonoacetate can be converted to carbanions by hydroxide ion in a strongly basic anion exchange resin as shown in Scheme 6 (55). The polymer-bound nitrile-stabilized carbanion gives alkenes in 75-97% isolated yields from aldehydes and ketones as shown in Table VIII. The polymer-bound ester-stabilized carbanion reacts with aldehydes but not with ketones. The experimental procedure is simple: Amberlyst A-26 macroporous resin in hydroxide form is shaken with a methanolic solution of the phosphonic ester and washed to produce phosphonate carbanion reagents with dry capacities of 3.5 mmol/g. The wet resin is stirred with a THF solution of the aldehyde or ketone at room temperature. Filtration of the resin and distillation of the filtrate give the prodect alkene. Alternatively the reagent may be prepared by percolation of a solution of the phosphonic ester through a column of the hydroxide resin in ether. A clever extension of this method involves generation of the aldehyde or ketone by hydrolysis of the dioxolan derivative with the sulfonic acid resin Amberlyst 16 present in the same mixture with the phosphonate carbanion resin as shown in Scheme 7 (55).

Polymer-bound α,β-unsaturated nitriles, esters, and ketones could be prepared by modified Wittig reactions of known polymer-bound species such as the β-keto-phosphonate **9** (56).

$$\text{P} - \text{C}_6\text{H}_4 - \text{CH}_2\underset{\underset{\text{COMe}}{|}}{\overset{\overset{\text{P(O)(OEt)}_2}{|}}{\text{CH}}} \qquad \qquad \textbf{9}$$

The polymer-bound phosphinate **10** has been prepared by successive treatments of 2% cross-linked brominated polystyrene with *n*-butyllithium, diethylchlorophosphite, and ethyl bromoacetate. It was used to form alkenes from aldehydes and ketones, but the yields were not consistently high, and the reagent was contaminated with an unidentified polymer-bound phosphorus species (57).

$$\text{C}_6\text{H}_4 - \underset{\underset{\text{OEt}}{|}}{\overset{\overset{\text{O}}{\|}}{\text{P}}}\text{CH}_2\text{CO}_2\text{Et} \qquad \qquad \textbf{10}$$

Conclusions

A wide variety of polymer-bound Wittig reagents give excellent yields of alkenes under many different solvent/base conditions. The polymer-bound reagents are preferred to soluble Wittig reagents when separation of triphenylphosphine oxide from the product alkene is difficult and when functional conversion of just one group of an aromatic dialdehyde is needed. Commercial 1-2% cross-linked polystyryldiphenylphosphine is suitable for laboratory scale procedures provided the phosphine has been introduced after polymerization and not by copolymerization of *p*-styryldiphenylphosphine. The most effective method of polymer functionalization for Wittig reagents is reaction of brominated polystyrene with lithium diphenylphosphide.

Scheme 6

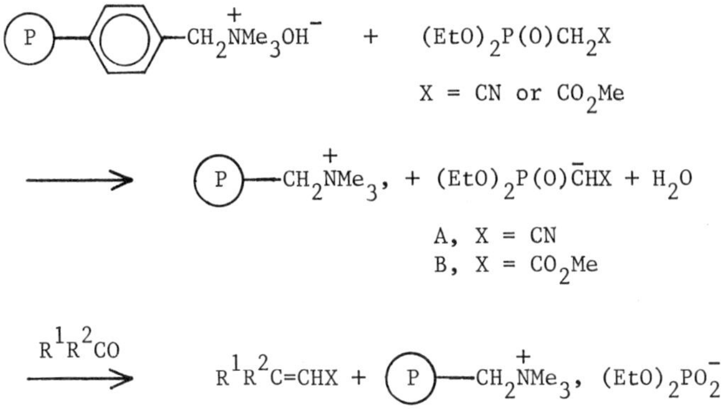

Scheme 7

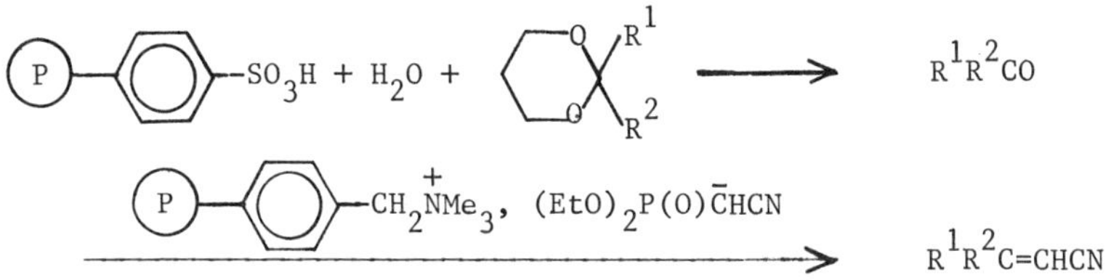

Table VIII. Modified Wittig Reactions with Phosphonate Carbanions
Bound to an Anion Exchange Resin $(\underline{55})^{\underline{a}}$

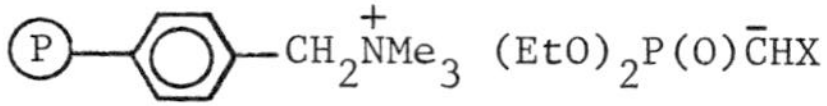

$$\text{(P)}-\!\!\bigcirc\!\!-CH_2NMe_3^+ \; (EtO)_2P(O)\bar{C}HX$$

X	aldehyde or ketone	product	% yield	Z/E
CN	p-ClC$_6$H$_4$CHO	p-ClC$_6$H$_4$CH=CHCN	83	0/100
CN	n-C$_6$H$_{13}$CHO	n-C$_6$H$_{13}$CH=CHCN	75	25/75
CN	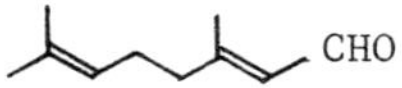 CHO	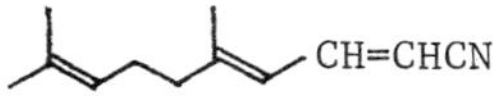 CH=CHCN	97	50/50
CN	n-C$_9$H$_{19}$COMe	n-C$_9$H$_{11}$C(Me)=CHCN	95	33/67
CN	PhCOMe	PhMeC=CHCN	90	0/100
CN	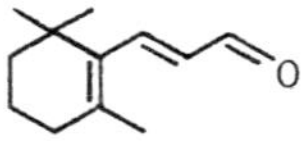	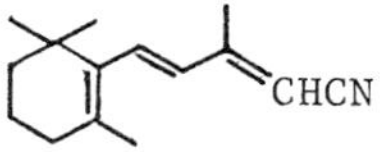 CHCN	93	33/67

Table VIII.   (continued)

CO_2Me	p-ClC$_6$H$_4$CHO	p-ClC$_6$H$_4$CH=CHCO$_2$Me	97	0/100
CO_2Me	n-C$_8$H$_{17}$CHO (structure)	n-C$_8$H$_{17}$CH=CHCO$_2$Me	64	0/100
CO_2Me	(structure, CHO)	(structure, CO_2Me)	95	0/100

$\underline{a}$ Adapted from ref. 55 with permission. Copyright 1980, The Royal Society of Chemistry.

More highly cross-linked polystyrene supports may be needed for large scale reactions because of ease of filtration of the polymer and ease of use in a flow reactor. Wittig reactions proceed in high yield even with 20% cross-linked macroporous polystyrene supports and with large ketones. The highly cross-linked supports show a greater preference for formation of the E-alkene, whereas 2% cross-linked supports give stereochemistry similar to that of analogous Wittig reactions in solution.

Carbanions from diethyl phosphonoacetonitrile and methyl diethylphosphonoacetate on anion exchange resins give high yield modified Wittig reactions by a technique readily adapted to continuous processes.

Acknowledgment

The research cited from this laboratory has been supported by the U. S. Army Research Office.

Literature Cited

1. Maercker, A. Org. React. 1965, *14*, 270.
2. Gosney, I.; Rowley, A. G. In "Organophosphorous Reagents in Organic Synthesis"; Cadogan J. I. G. Ed.; Academic Press: New York, 1979; Chap. 2.
3. Bestmann, H. J. Pure Appl. Chem. 1980, *52*, 771.
4. Schlosser, M. Topics in Stereochemistry 1970, *5*, 1.
5. Vedejs, E.; Meier, G. P.; Snoble, K. A. J. J. Am. Chem. Soc. 1981, *103*, 2823.
6. Akelah, A.; Sherrington, D. C. Chem. Rev. 1981, *81*, 557.
7. Frechet, J. M. J. Tetrahedron 1981, *37*, 663.
8. Akelah, A. Synthesis 1981, 413.
9. Hodge, P. In: "Polymer-Supported Reactions in Organic Synthesis"; Hodge, P., Sherrington, D. C., Eds.; Wiley: New York, 1980; pp. 139-146.
10. Mathur, N. K.; Narang, C. K.; Williams, R. E. "Polymers as Aids in Organic Chemistry"; Academic Press: New York, 1980; pp. 189-191.
11. Akelah, A.; Sherrington, D. C. Polymer 1983, *24*, 1369.
12. McKinley, S. V.; Rakshys, J. W., Jr. J. Chem. Soc., Chem. Commun. 1972, 134.
13. Bernard, M.; Ford, W. T. J. Org. Chem. 1983, *48*, 326.
14. Clarke, S. D.; Harrison, C. R.; Hodge, P. Tetrahedron Lett. 1980, *21*, 1375.
15. Heitz, W.; Michels, R. Angew. Chem., Int. Ed. Engl. 1972, *11*, 298.
16. Akelah, A. Eur. Polym. J. 1982, *18*, 559.
17. Castells, J.; Font, J.; Virgili, A. J. Chem. Soc., Perkin Trans. 1 1979, 1.
18. Bernard, M.; Ford, W. T.; Nelson, E. C. J. Org. Chem. 1983, *48*, 3164.
19. Hodge, P.; Hunt, B. J.; Khoshdel, E.; Waterhouse, J. Nouv. J. Chim. 1982, *6*, 617.
20. Camps, F.; Castells, J.; Font, J.; Vela, F. Tetrahedron Lett. 1971, 1715.
21. Hodge, P.; Khoshdel, E.; Waterhouse, J. Makromol. Chem. 1984, *185*, 489.
22. Harrison, C. R.; Hodge, P.; Hunt, B. J.; Khoshdel, E.; Waterhouse, J. In: "Crown Ethers and Phase Transfer Catalysis in Polymer Science"; Mathias, L. J., Carraher, C. J., Jr., Eds.; Plenum: New York, 1984; p. 35.
23. Frechet, J. M. J.; Eichler, E. Polym. Bull. 1982, 7, 345.
24. Frechet, J. M. J.; Schuerch, C. J. Am. Chem. Soc. 1971, *93*, 492.
25. Hodge, P.; Waterhouse, J. Polymer 1981, *22*, 1153.
26. Farrall, M. J.; Alexis, M.; Trecarten, M. Polymer 1983, *24*, 114.

27. Hodge, P.; Hunt, B. J.; Waterhouse, J.; Wightman, A. Polymer Commun. 1983, *24*, 70.
28. Rabinowitz, R.; Marcus, R. J. Org. Chem. 1961, *26*, 4157.
29. Braun, D.; Daimon, H.; Becker, G. Makromol. Chem. 1963, *62*, 183.
30. Rabinowitz, R.; Marcus, R.; Pellon, J. J. Polym. Sci. 1964, *A2*, 1241.
31. Frick, C. D.; Rudin, A.; Wiley, R. H. J. Macromol. Sci.-Chem. 1981, *A16*, 1275.
32. Farrall, M. J.; Frechet, J. M. J. J. Org. Chem. 1976, *41*, 3877.
33. Grubbs, R. H.; Su, S.-C. H. J. Organomet. Chem. 1976, *122*, 151.
34. Farrall, M. J.; Frechet, J. M. J. Macromolecules 1979, *12*, 426.
35. Braun, D.; Czerwinski, W.; Disselhoff, G.; Tudos, F.; Kelen, T.; Turcsanyi, B. Angew. Makromol. Chem. 1984, *125*, 161.
36. Akelah, A.; El-Borai, M. Polymer 1980, *21*, 255.
37. Pittman, C. U., Jr.; Wilemon, G. M. J. Org. Chem. 1981, *46*, 1901.
38. Relles, H. M.; Schluenz, R. W. J. Am. Chem. Soc. 1974, *96*, 6469.
39. Ford, W. T.; Mohanraj, S.; Periyasamy, M. Brit. Polym. J. 1984, *16*, 179.
40. MacKenzie, W. M.; Sherrington, D. C. Polymer 1980, *21*, 791.
41. Camps, F.; Castells, J.; Vela, F. Ann. Quim. 1974, *70*, 374.
42. Heitz, W.; Michels, R. Liebig's Ann. Chem. 1973, 227.
43. Michels, R.; Heitz, W. Makromol. Chem. 1975, *176*, 245.
44. Regen, S. L.; Lee, D. P. J. Org. Chem. 1975, *40*, 1669.
45. Appel, R.; Willms, L. Chem. Ber. 1981, *114*, 858.
46. Pommer, H.; Nurrenbach, A. Pure Appl. Chem. 1975, *43*, 527.
47. Kobayashi, S.; Suzuki, M.; Saegusa, T. Polym. Bull. 1981, *4*, 315.
48. Grubbs, R. H.; Su, S.-C. H. In: "Enzymic and Non-Enzymic Catalysis"; Dunnill, P., Wiseman, A., Blakebrough, N., Eds.; Wiley/Halstead: Chichester, 1980; p. 223.
49. Leznoff, C. C.; Wong, J. Y. Can. J. Chem. 1973, *51*, 3756.
50. Leznoff, C. C.; Greenberg, S. Can. J. Chem. 1976, *54*, 3824.
51. Leznoff, C. C.; Sywanyk, W. J. Org. Chem. 1977, *42*, 3203.
52. Leznoff, C. C. Acc. Chem. Res. 1978, *11*, 327.
53. Hodge, P.; Waterhouse, J. J. Chem. Soc., Perkin Trans. 1 1983, 2319.
54. Frechet, J. M. J.; Bald, E.; Svec, F. Reactive Polym. 1982, *1*, 21.
55. Cainelli, G.; Contento, M.; Manescalchi, F.; Regnoli, R. J. Chem. Soc., Perkin Trans. 1 1980, 2516.
56. Sturtz, G.; Clement, J. C. Polym. Bull. 1983, *9*, 125.
57. Qureshi, A. E.; Ford, W. T. Brit. Polym. J. 1984, *16*, 231.

RECEIVED November 14, 1985

9

Molecular Recognition in Polymers Prepared by Imprinting with Templates

Günter Wulff

Institute of Organic Chemistry II, University of Düsseldorf, Universitätsstr. 1, D-4000 Düsseldorf, Federal Republic of Germany

This is a review on attempts to produce specific binding sites in synthetic polymers analogous to those of biological receptors or natural enzymes. For this goal an imprinting procedure was used with the aid of templates in crosslinked polymers. Suitable polymerizable binding groups were bound to a template molecule. This was copolymerized into highly crosslinked polymers. After the removal of the templates the remaining cavities possessed a shape and an arrangement of their functional groups corresponding that of the template.

Applications of these polymers for optical resolution of racemates, for selective metal ion binding, and for stereoselective reactions are reviewed. The optimization of the polymer structure is described and the origin of the memory effect discussed. Similar imprints can be obtained on surfaces of silica.

In recent years the construction of molecular size cavities containing a defined arrangement of functional groups has attracted much attention. These arrangements could be used for selective binding for example in chromatography or could be used as receptor models. Furthermore it is possible to study the cooperativity of functional groups and anchimeric effects. Especially interesting would be their application as specific catalysts working in a fashion similar to natural enzymes. Therefore many investigations on this route towards the construction of enzyme models have been made.

During the last few years, remarkable progress has been made in the design of low-molecular-weight molecules based on molecular recognition. As binding sites, the

0097-6156/86/0308-0186$12.25/0
© 1986 American Chemical Society

cavities of cyclodextrines (1-3), crown ethers (4-6),
cryptates (4-6), cyclophanes (7-9), and similar ring
systems have been used. Intensive effort has been made
to introduce functionalities into these cavities in a
defined manner (1-3), or to introduce these functionalities during the preparation of the cavity (4-9). Other
cavities used for this purpose are the channels of certain
crystals, like desoxycholic acid (10) or, in the case of
very small cavities, those of certain zeolites (11).

Of special interest in this respect would be the use
of polymeric substances for preparing specific binding
sites and catalytically active compounds as receptor- and
enzyme-models. The use of polymeric substances would have
some advantages since enzymes are polymers as well. Many
of their unique features are directly connected to their
polymeric nature. This is especially true for the high
cooperativity of the functional groups and the dynamic
effects such as the induced fit, the allosteric effect,
and the steric strain as shown by the enzymes. But in
spite of these advantages of polymers, progress with
polymeric enzyme models has been slow [for reviews see
Refs. (12-17)]. A major problem is achieving proper
stereochemistry at the active site.

It is common practice (12-17) to introduce catalytically active groups and binding groups by the copolymerization of monomers containing these groups into a
polymer. By this method one obtains a polymer with
randomly distributed groups (see Figure 1, A), if the
monomer reactivities are not greatly different or highly
alternating. Another possibility is the grafting of side
chains already containing the desired arrangement of
functional groups onto a polymer (see Figure 1, B). A
third possibility is the polymerization of monomers
already containing the desired functional groups in the
backbone. In this case the groups are localized, as in
some hormone receptors, in the main chain one after
another (see Figure 1, C). In this case, called
"continuate words" arrangement (18), only two dimensional
information can be transferred.

In contrast to these arrangements, the functional
groups responsible for the specificity of natural enzymes
are located at quite distant points along the peptide
chain and are brought into spatial relationship as a
result of the specific folding of the chain. Here, both
the functional group sequence on the chain and the
peptide's tertiary structure, i.e. its topochemistry, are
decisive (see Figure 1, D). This type of arrangement has
been called the "discontinuate words" by R. Schwyzer (18).
By the specific folding cavities of defined shape are
formed simultaneously in the protein structure. It was
our aim to prepare this type of arrangement in synthetic
polymers (enzyme-analogue built polymers) as a prere-
quisite for enzyme modelling.

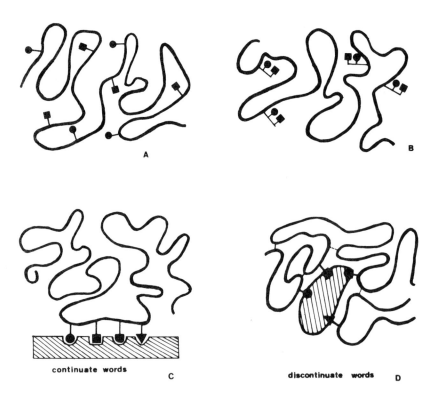

Figure 1.   Possible arrangements of functional groups
in synthetic and natural polymers.

<u>Molecular Imprinting - a New Approach</u>

To prepare specific cavities with functional groups in a
"discontinuate words" arrangement several years ago we
introduced a new approach (19-23). The functional groups
to be introduced were bound to a suitable template
molecule in the form of polymerizable vinyl derivatives.
This monomer then was copolymerized under such conditions
that highly crosslinked polymers were formed having chains
in a fixed arrangement. After removal of the template,
free cavities were formed (see Figure 2). They possessed
a shape and an arrangement of functional groups corresponding to that of the template. The functional groups
in this polymer are located at quite different points of
the polymer chain and they are held in a spatial relationship by crosslinking.

This approach to prepare a cavity contrasts with the
use of a low molecular weight moiety carrying the desired
stereochemical information (4-9). Our method resembles
to some extent that of Dickey (24,25), who demonstrated
that silica gel, prepared from sodium silicate and acetic
acid in the presence of methyl orange, had a specific
affinity for methyl orange in the presence of ethyl,
n-propyl and n-butyl isomers. See "Imprinting in Modified
Silicas" for more details. In contrast to Dickey's work,
using our method not only are imprints of the substances
obtained, but it is possible at the same time to introduce
functional groups into the cavities in a specific arrangement. This increases specificity and broadens the applicability of the method.

One example illustrating the principle is described
in somewhat greater detail. Monomer 1 was used to some
extent for the optimization of the method. The template
is phenyl-α-D-mannopyranoside, to which two molecules of
4-vinylphenylboronic acid are bound by ester linkages to
two hydroxyl groups each. Compared to Figure 2 , this is
a simpler case with only two identical functional groups
being involved. Boronic acid was chosen as a binding
group because it undergoes easily reversible interactions
with diol groups. Since there is a chiral template, the
ability to create the steric arrangement in the cavity
can be tested (after removal of the template) by the
ability of the polymer to resolve a racemic mixture of
template molecules.

The monomer 1 was copolymerized by free radical
initiation in the presence of an inert solvent with a
large amount of a bifunctional crosslinking agent. Under
these conditions, macroporous polymers were obtained which
possessed a permanent pore structure, a high inner surface
area, and good accessibility. Additionally, low polymer
swellability would imply limited mobility of the polymer
chains. From a polymer of this type, 40 to 90% of
template molecules can be split off by treatment with
water or alcohol (see Figure 3). If this polymer is

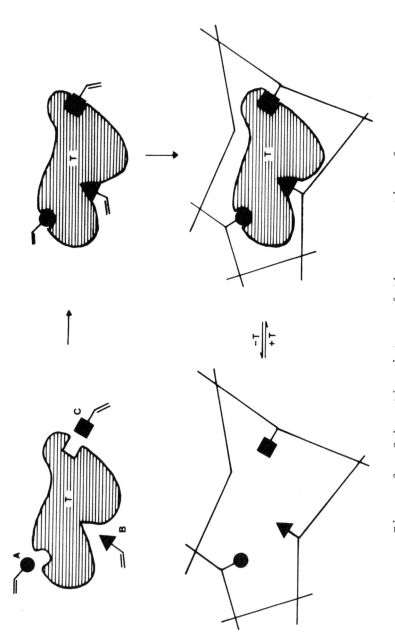

Figure 2.   Schematic picture of the preparation of microcavities with functional groups in a "discontinuate word" arrangement.

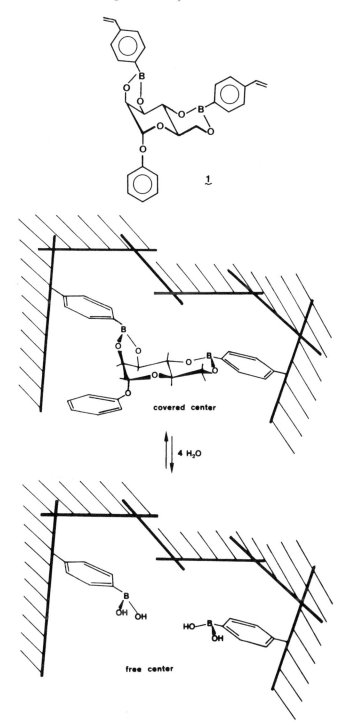

Figure 3.   Removal and uptake of the template.

treated in a batch procedure with a racemate of the
template, the enantiomer that has been used for the
preparation of the polymer should preferably be incorporated.
        If the enantiomer specificity is expressed by the
separation factor α, [the ratio of the distribution
coefficients between solution and polymer of L- and D-form]
values of α in this and similar cases range from 1.20 to
3.66 depending upon the equilibration conditions and
polymer structure. The highest α-value obtained to date
was 3.66. In this case, the simple batch procedure gives
an enrichment of 13% of the L-form in the filtrate and
40% of the D-form on the polymer (23).
        The high specificity obtained for optical resolution
shows that it was in fact possible to copy the shape of
the template and the arrangement of the functional groups
into the polymer by this imprinting method. It should be
noted that this method of resolution of racemates is a
novel one. In this case no interactions of the racemate
with molecular parts carrying asymmetric carbon centers
take place, but the enantiomers are bound within a chiral
environment. The stability of these inclusion complexes
is apparently different for the two enantiomers. This type
of optical resolution is somewhat analogous to urea
inclusion compounds (26) or to silica imprinted by chiral
compounds (27-29). Certain aspects of our method have
been reviewed by our group (30-34) and by others (35-39).

Polymer Optimization

It is apparent that the quality of the imprinted polymers
is crucially dependent on the polymer structure. After
splitting off the templates the remaining cavities have
to restore their shape. Therefore the polymer structure
has to be optimized for this purpose (22,23,40-42).
        In nearly all of our investigations we used macro-
porous (macroreticular) polymers. For details on the
structure of macroporous polymers see Refs. (43,44).
Macroporous polymers consist of primary particles (so-
called nuclei) grown together during polymerization.
In our case these particles possess diameters of 100-200 nm
and between them a system of permanent pores (macro pores)
is formed which have diameters of 20-50 nm (see Figure 4)
It can be assumed that the diffusion within macropores is
completely unhindered. At higher magnification (right
side of Figure 4) the micro-cavities are shown. The
black particles represent the templates which are on the
order of 1-2 nm and can be situated near the inner surface
or inside the denser nuclei. Our polymers were optimized
along several lines.

Rigidity. The polymers should possess high rigidity to
preserve the shape of their cavities after splitting off
the template. This can be done by high crosslinking. It

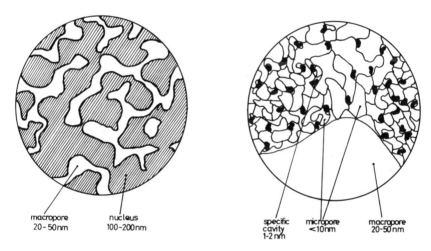

Figure 4.   Schematic diagrams of a macroporous polymer with chiral microcavities at two magnifications.

was shown that below 10% crosslinking (see Figure 5) the
specificity disappears (23,41), at a higher percentage
the specificity increases. Figure 5 (41) shows the remarkable behavior of ethylene glycol dimethacrylate as the
crosslinking agent. Up to 50% crosslinking, an increase
to an α-value of 1.50 is observed. From 50% to 66.7% a
dramatic increase from 1.50 to 3.04 occurs, which means
a fourfold increase in selectivity in this range. Further
increase to 95% crosslinking yields a specificity of
α = 3.66. Butanediol dimethacrylate and especially
p-divinylbenzene as crosslinking agents show a much lower
increase.

     Polymers crosslinked with divinylbenzene or ethylene
glycol dimethacrylate retained their specificity for a
long time. Even under high pressure in an HPLC column
the activity remained for months. Above 60-70°C the
specificity begins to decrease gradually (40). The same
is true if oxygen is not avoided in the eluent, since then
a slow C-B-bond breaking occurs and the capacity and the
specificity of the column decreases. Several batch equi-
librations with the same polymer showed nearly the same
specificity.

Flexibility. In contrast to the requirement for rigidity,
the polymers should also possess some flexibility in their
whole arrangement. This is necessary to allow a fast
binding and splitting of the substances within the cavities. Cavities of accurate shape but without any flexibility will show kinetic hindrance to reversible binding.
     Table I (23,41) shows the influence of different
crosslinking agents on the properties of the polymers.
An indication of flexibility is the swellability of the
polymers. o, m and p-divinylbenzenes give only low
specificity; swellability and splitting percentage are
rather low. These crosslinking agents produced highly
rigid cavities, and the most specific cavities could not
be freed from the template. Glycol dimethacrylate, by
contrast produces cavities which are flexible enough for
fast binding and splitting. Higher flexibility of the
crosslinking agent as in di-(DEGDM in P7) and triethylene
glycol dimethacrylates (TEGDM in P8) produces less
selective cavities. The use of the new optically active
crosslinking agents 2 and 3 (41) does not give an additional increase in selectivity.
     In order to use the stabilizing effect of a rather
rigid crosslinked polymer in combination with a more
flexible one we prepared interpenetrating networks (IPN)
(see Table II). The best polymer in this row (P14) consisted of a mother polymer from 1 and glycol dimethacrylate with a relatively high inner pore volume. This
unmodified polymer, after the removal of the template,
showed a separation factor α = 2.68. This polymer was
swollen with technical divinylbenzene and solvent and
again polymerized. The second polymer was expected to

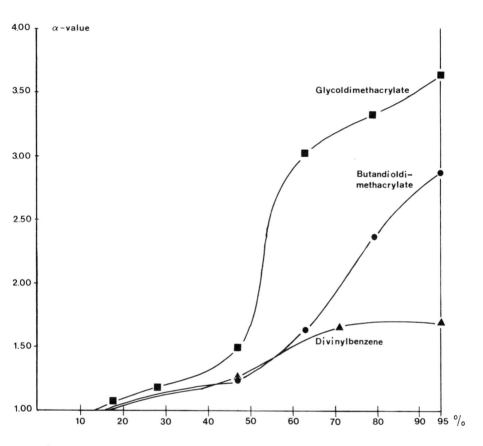

Figure 5. Dependence of the specificity of the polymers on the kind and the amount of crosslinking.
Polymers imprinted with 1 were prepared with varying
amount of the crosslinking agents ethylene glycol
dimethacrylate, butanediol dimethacrylate, and p-divinylbenzene. After splitting off the templates the
separation factor α for the optical resolution of
phenyl-α-D,L-mannoside in the batch procedure was
determined.

Table I. Dependence of the Properties of the Polymers on the Kind of Crosslinking Agent (41)

Polymer	Crosslinking Agent	Comonomer	Percentage of Crosslinking [%]	Inner Surface Area [m²/g]	Splitting Percentage [%]	Swellability	Separation Factor α
P 1	o-DVB	Styrene	71	348	49.2	1.45	1.64
P 2	m-DVB	Styrene	71	577	42.4	1.26	1.65
P 3	p-DVB	Styrene	71	661	42.2	1.36	1.67
P 4	Techn. DVB	Ethyl	95	825	41.3	1.27	1.70
		styrene	50	402	39.1	1.33	1.42
P 5	EGDM	MMA	63	169	89.5	1.87	3.04
P 6	EGDM	–	95	383	81.7	1.69	3.66
P 7	DEGDM	–	95	30	90.0	2.42	1.24
P 8	TEGDM	–	95	n.d.	n.d.	2.71	1.41
P 9	BDDM	–	95	128	92.3	2.27	2.88
P 10	2	–	95	214	81.4	2.04	2.05
P 11	3	–	95	344	77.6	2.28	2.68
P 12	4	–	95	339	86.5	1.96	1.78

EGDM = ethylene glycol dimethacrylate; TEGDM = triethylene glycol dimethacrylate;
BDDM = butanediol dimethacrylate; DVB = divinylbenzene; MMA = methyl methacrylate.
In all cases 5% l, the crosslinking agent, and in some cases additional monomer were copolymerized radically in presence of the inert solvent acetonitrile/benzene 1:1 (1 g monomeric mixture/1 ml solvent mixture). Swellability: ratio of the specific gel bed volume to the bulk volume.
Splitting percentage: percentage of templates that could be split off.

2

3

4

Table II. Chiral Cavities in Interpenetrating Networks (41)

	Composition of		Inner Surface Area [m²/g]	Splitting Percentage [%]	Swellability	Separation Factor α
First Polymer	First Polymer	Second Polymer				
P 13	5% 1̃ 52% DVB 43% ethyl styrene 1.5 ml/g solvent	75% EGDM 25% MMA 0.67 ml/g solvent	402	39.0	1.33	1.42
P 14	5% 1̃ 95% EGDM 1.9 ml/g solvent	54% DVB 46% ethyl styrene 0.67 ml/g solvent	441	43.6	1.39	2.81
P 15	100% EGDM 2.0 ml/g solvent	3.5% 1̃ 52% DVB 43% ethyl styrene 0.67 ml/g solvent	437	44.4	1.37	1.23

The first polymer (the mother polymer) was prepared as usual, but with a higher amount of inert solvent. After preparation and drying, 40% of the polymer weight of another monomeric mixture containing azobisisobutyronitrile dissolved in 0.67 ml/g acetonitrile/benzene was sucked in and equilibrated at room temperature for 1.5 hrs. Afterwards polymerization was performed at 80°C.

stabilize the more flexible initial one without affecting the kinetic exchange. The selectivity was somewhat higher (2.81), but the splitting percentage was considerably lower than the previous value for macroporous polymer P6.

Table III.   Dependence of Polymer Properties on the
              Kind of Solvent Used

Polymer	Solvent	Inner Surface Area (m^2/g)	Splitting Percentage (%)	Separation Factor α
P 6a	Acetonitrile/ benzene	383	81.7	3.66
P 6b	Acetonitrile	290	76.6	2.68
P 6c	Benzene	453	54.6	2.86
P 6d	DMF	208	90.9	2.59
P 6e	DMSO	117	87.6	2.68
P 6f	Ethylacetate	380	75.6	2.74
P 6g	Dioxane	211	79.2	2.65

The polymer was prepared as in case of P 6, but the kind of solvent was varied.

There is a strong influence of inert solvent used during polymerization on the polymer structure (41), but the influence on selectivity is astonishingly low (see Table III). A strong influence on the selectivity is observed if the amount of inert solvent during polymerization is changed from 0.29 up to 1.76 ml/g monomeric mixture, the optimum being around 1.0 (42).

Accessibility. As many cavities as possible in the polymer should be accessible. The accessibility can be measured by the percentage of the templates that can be split off. This depends on the flexibility of the polymer chains, on the inner surface area, and on the pore size distribution within the macroporous polymer. As in all instances of higher crosslinking relatively high inner surface areas (see Tables I, II and III) are present. The porous structure, especially in the swollen state, can best be investigated by obtaining GPC calibration curves with polystyrene standards. The distribution of the pore sizes strongly varies with the conditions of the preparation of the polymer (40).

Mechanical Stability. For application in HPLC and as catalysts the mechanical stability should be high. Especially polymers with ethylene glycol dimethacrylate as a crosslinking agent show good mechanical stability. Further improvement can be achieved by preparing these polymers on a silica support (see "Imprinted Polymer Layers on Silica").

On the Origin of the Memory Effect

The selectivity of the imprinted polymers for the preferred adsorption of their templates can be explained by
crosslinking and stabilizing a favorable conformation of
the polymer chains around the template. In the case of
chiral templates this is an asymmetric crosslinking and
asymmetric cavities are formed. The "right" enantiomer
can preferentially be embedded in these cavities. Another
possible reason for the observed optical resolution of
racemates could be a selective adsorption at the optically
active templates that could not be split off. This can
be ruled out for the following reasons:

i)   Polymers from which the original templates had not
     been split off displayed no optical resolution in
     cases where the racemic templates could not take
     part in exchange equilibria (22,45).

ii) Racemates other than that of the template were either
     not at all or only very poorly separated by the
     polymer (see later this section).

iii) Polymers prepared from 5b, showed in aqueous solution
     in equilibrium with the ˜racemate of the template,
     not borondiester bond formation but rather strong
     hydrophobic interaction. This does not lead to any
     optical resolution (46).

iv) The specificity of the polymers could be destroyed
     completely and irreversibly by heating in dioxane.
     In this case a strong swelling took place. Apparent-
     ly, in this case the glass transition temperature of
     the solvated network was exceeded and the original
     form of the cavities could not be restored after
     deswelling (22,47).

     For these reasons we conclude that optical resolu-
tion of racemates takes place within the cavities formed
by the template. Another possibility might be that
separation takes place at polymer chain segments that
exhibit a preference for one enantiomeric arrangement due
to the chirality of the main chain. Although it has been
demonstrated quite recently (23,48,49) that polyvinyl-
compounds can be optically active due to the chirality of
the main chain, this can be excluded for polymerization
in the presence of 1 (49). Additionally, arguments ii,
iii, and iv would speak against the participation of
chiral chain segments in racemic resolution.
     Belokon and co-workers (50,51) attributed their
template effects (see "Distance Accuracy of Two Functional
Groups") to the existence of cyclopolymerization, since
polymers of low crosslinking also showed a good selectivity. What takes place is an intrachain reaction rather
than an interchain one. Only in one case have we observed
a similar behavior (23,49,52). In our case a cyclopolymerization was proved by copolymerization of 3.4-0-
isopropylidene-D-mannitol 1,2,5,6-bis-0-[(4-vinylphenyl)

boronate] with another comonomer. After the template was
split off, linear optically active copolymers with asymmetric diads of 4-vinylphenyl boronic acid moieties were
obtained. Other monomers, like 1, did not produce optically active polymers in the same procedure.

Another question is to what extent selectivity can be
attributed to the shape of the cavity and to what extent
it is due to the arrangement of the binding groups. Even
one binding group in the cavity can produce selectivity
(46,53-55), but the selectivity is lower than with two
binding groups. In polymers of 1 also the interaction of
the template within the cavity is not always through two
binding groups (56-58). Only the more specific cavities
show full cooperativity of the two binding sites. In this
respect it is important to know that around 92% of the
monomer 1 is connected by the two point attachment during
the polymerization procedure and 8% of the cavities
contain only one boronic acid binding group (22).

This might be one reason for the broad distribution
of specificity among the single cavities as we determined
in one case from the dependence of the specificity of the
resolution of racemates on the covering percentage of the
cavities (59). This lack of homogeneity should arise in
part during the polymerization itself. Whereas the
functional groups to be introduced are fixed during polymerization in a definite steric arrangement by the
template, the surrounding chains adapt with varying
accuracy to the structure of the template. This is due to
the fact that the chain links are about the same size as
the template itself. Thus it is impossible to copy the
template exactly. Furthermore, the different chain links,
including the crosslinks,are distributed somewhat randomly
along the chains. Another possibility for reduced specificity exists if the template molecule is polymerized very
near to the surface, so that only an incomplete complementary imprint can be formed. This is more likely with the
high surface area polymers.

The distribution of specificity obtained during polymerization can be altered after cleavage of the template.
It is possible that during polymerization, stresses in the
chains will have been built up, which are suddenly compen-
sated for after the cleavage of the template. For an
elastomer network the Flory picture shows the polymer to
be formed in the relaxed state. Also in this case cleavage would create strain that could be relieved by swelling or shrinkage, altering the shape of the cavity and
the position of the functional groups.

It is amazing that polymers with high crosslinking
and high selectivity freed from the templates can possess
a swellability of more than 100% in the solvent used for
the equilibration in the batch procedure (see Table I).
On adding the template, the original volume is restored
by deswelling (42). Assuming that the swelling occurs
with solvation of the functional groups in the micro-

cavities, the microcavities apparently restore their
original form on binding the substrate as the reduced
swellability and the substantial specificity indicate.
This could be regarded as an analogy to the "induced fit"
in enzymes.

With respect to the exactness of copying the shape
of a compound the question arises as to whether these
polymers are able to resolve racemates other than that of
the template. Polymers were prepared from different
esters of glyceric acid phenylboronate (5a, 5b, 5c (46)).
Table IV shows the selectivity of the polymers for
resolution of different racemates. The polymers were most
specific for resolution of the racemate of their own
templates. Small differences in structure led to a
decrease in specificity; higher differences resulted in a
complete disappearance of specificity. Similar results
were obtained in other cases (47,60).

Table IV. Separation Factors for Various Racemates

Racemates Tested	Polymer Prepared from			
	Methyl ester 5a	Benzyl ester 5b	Nitrobenzyl ester 5c	Nitrobenzyl ester 5c*
D,L-glyceric acid methyl ester	1.07	1.04	1.03	1.06
D,L-glyceric acid benzyl ester	1.04	1.10	1.05	1.08
D,L-glyceric acid nitro-benzyl ester	–	1.09	1.09	1.22
p-Nitrophenyl α-D,L-mannoside	1.00	1.00	1.00	1.00

*Polymer prepared in the presence of an equimolar amount
of 4-N,N-dimethylamino styrene

Imprinting with Different Kinds of Binding Interaction

With our method of preparing binding sites by an imprinting procedure, the binding groups have a twofold
function. First, during polymerization a strong interaction between template and binding groups should be
present, so that the template molecule can fix the binding
groups at the growing polymer chains in a defined stereochemistry. Second, after splitting off the template, the
binding groups should be able to undergo an easily reversible binding interaction with the template. For the
first purpose binding should be strong, for the second
the activation energy should be low.

Possible interactions are hydrophobic, electrostatic,

charge-transfer, and dipole-dipole as well as covalent.
We have chosen in most cases covalent binding since it is
strictly oriented in space during polymerization. For
a fast and reversible binding reaction, however, the activation energies of covalent bonds are too high in most
cases. They could be lowered by the addition of suitable
catalysts (e.g. piperidine or ammonia in the case of
boronic acids (33,57,61) or toluene sulfonic acid in the
case of Schiff base formation (55)) or by the introduction
of appropriate neighboring groups (33,61-63).

Boronic Acids as Binding Sites. Poly(vinylphenylboronic
acids) are commercially available and can be used in an
alkaline aqueous medium for the purification and chromatography of diol-containing compounds (64-66). In organic
solvents they usually form trigonal borondiester bonds
(see Equation 1 and 2).
     Several template monomers have been polymerized by
using just one boronic acid for binding. Among the
templates are glyceric esters (32,42,46,47,54), propane-
diol (53), and mandelic acid (67). Generally, with one
binding interaction the selectivity of the resulting
polymer is only moderate, however in the case of mandelic
acid prepared by Sarhan (67) from 6, a rather high
separation factor α of 1.59 was observed. With one
boronic acid binding group the influence of the mobility
of the functional groups was investigated (54,68).
Templates with two boronic acid interactions gave much
higher selectivity (22,23,40,41,56-59) as has been shown
under the heading "Optimization of the Polymer Structure".
     In principle, three binding interactions during
polymerization and during equilibration with the racemate
should be better than two. In all cases investigated
until now (21,53,57), with three binding groups (also
different ones) specificity was reduced. At the present
time it is not completely clear whether the correct
cooperativity of three groups cannot be reached in the
cavity, whether the template cannot easily enter a highly
specific cavity with three binding groups, or whether
there are other reasons. In the case of a polymer from
D-mannitol-tris-0-1,2;3,4;5,6-(4-vinylphenylboronate)
(61,53) a selectivity of α = 1.09 for the optical
resolution of D,L-mannitol was observed.

Boronic Acids in Combination with other Interactions.
In the case of monomer 7 (60) a 4-vinylphenyl-boronic acid
is bound to the two hydroxyl groups of L-DOPA methyl ester.
In addition, 5-vinylsalicylaldehyde (69) is linked to the
amino-group through an azomethine-bond. Polymerization
under the usual conditions and   splitting off the
template, L-DOPA methyl ester, yielded a polymer that
showed a separation factor α = 1.98 for the solution of
the racemate of the template. The same polymer did not
resolve the racemate of tyrosine methyl ester or

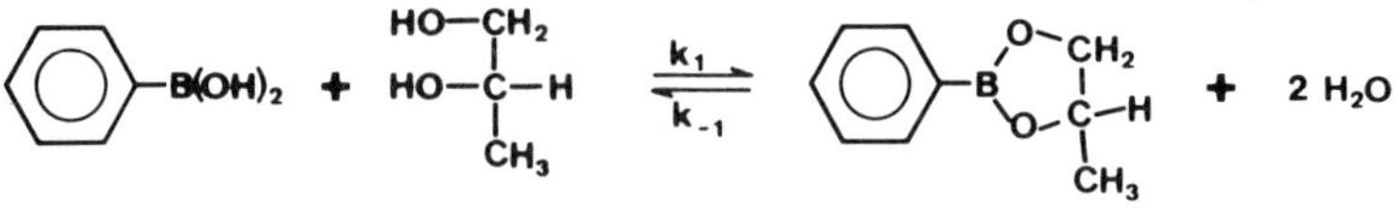

$$(1)$$

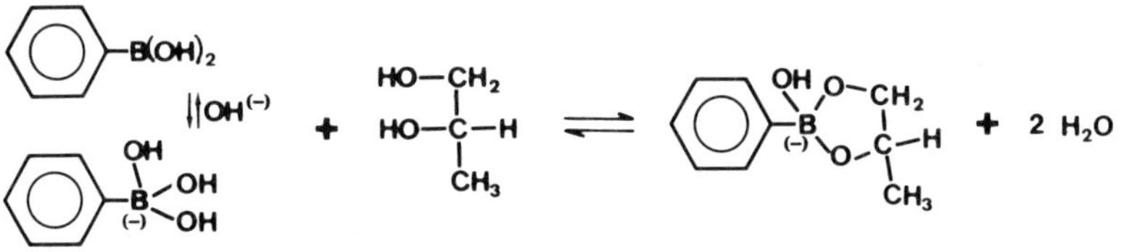

$$(2)$$

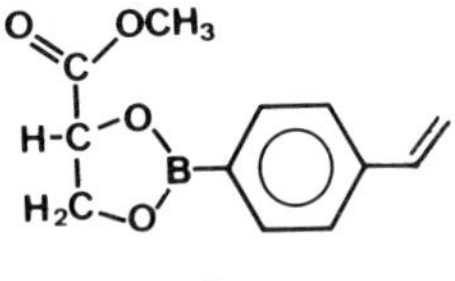

5a

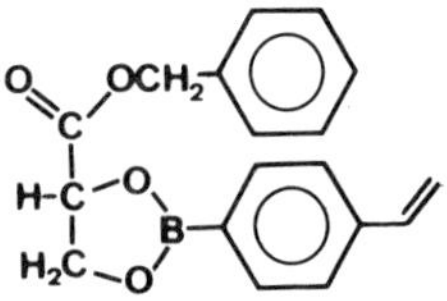

5b

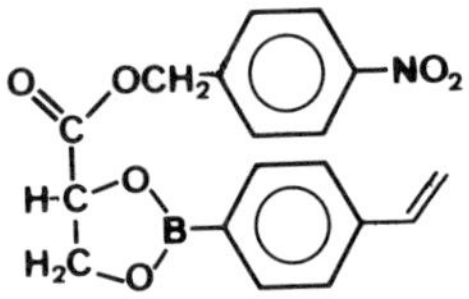

5c

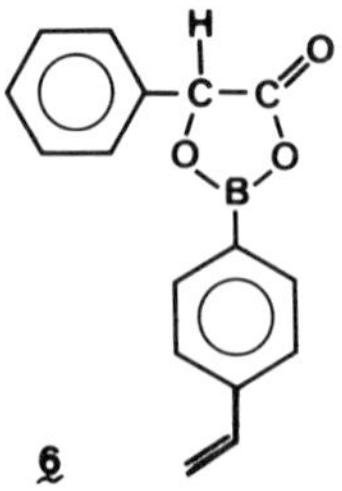

6

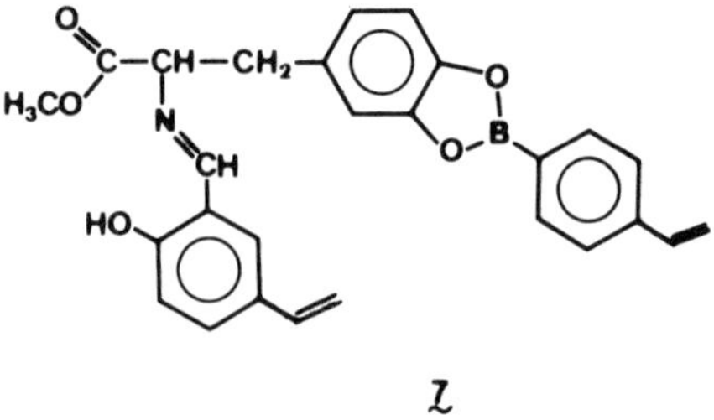

7

phenylalanine methyl ester. Figure 8 (see "Imprinting
in Modified Silicas") shows the combination of two phenyl-
boronic acid interactions with an azomethine bond interaction. In this case the selectivity was not enhanced
by the third binding group (57).

In the first example of our group, in monomer 8 a
combination of a boronic acid interaction with an amide
bond was used (19-21,68). In this case it was assumed
that after splitting off the template D-glyceric acid from
the polymer, the equilibration with the racemate would
yield borondiester bond formation and moreover an electrostatic interaction. The amide bond could only be split
under rather rigorous conditions and the splitting
percentage was rather low (~20%). The selectivity could
be improved to some extent by optimizing the polymer
structure (α = 1.034 → 1.20) (32,42,47). but more specific
polymers showed unexpected behavior (42). The glyceric
acid was bound on equilibration not only by a borondiester
bond but by an amide bond instead of an electrostatic
interaction. Apparently, in the most specific cavities
the amide bond can be formed under rather mild conditions
by a cooperative effect.

It was investigated whether non-covalent interactions
between the template and added comonomers were strong
enough during polymerization to allow the formation of
specific structures within the cavities. In the case of
D-glyceric acid-2,3-O-(4-vinylphenylboronate) an additional weak electrostatic interaction was used by polymerization in the presence of an equimolar amount of 2-vinyl-
pyridine (32). By this procedure the selectivity was
enhanced. On the other hand, a strong additional electrostatic interaction decreased selectivity (32,55). An
additional charge transfer interaction also improved the
selectivity as can be seen from Table IV. Here, an
electron deficient ester moiety (4-nitrobenzyl-) was
polymerized in the presence of N,N-dimethylaminostyrene.

Other Covalent Interactions. Azomethine bond (Schiff
base) formation represents a possibility to use aldehydes
or amines as binding groups (33). Somewhat problematic
is the slow equilibration in the batch procedure or, even
more pronounced, in chromatography. Nevertheless, it was
possible in the case of L-phenylalanine-anilide to use the
Schiff bases of polymerizable aldehydes. The selectivities obtained are in the range α = 1.22 - 1.36 (55). For
a one-point binding this is quite a good result.

In a recent paper an extremely efficient optical
resolution of D,L-N-benzylvaline is reported by Fujii
and coworkers (70). They used the amino acid Schiff base
Cobalt (III) complex (9) as the template monomer. After
splitting off the template L-N-benzylvaline from the
crosslinked polymer it showed an optical resolution for
the racemate of the template with a separation factor α
of 682. An enantiomeric excess of 99.5% of one antipode

8

9

could be obtained, in the batch procedure. In order to
evaluate the template effect, the chiral recognition of
the chiral cobalt complex itself and also the complex
attached to a polymer without a template effect were
investigated. The low molecular weight cobalt complex
showed an α-value of 24.8 and the non-template polymer
complex of 18.5. These results clearly indicate that the
template effect operates most effectively in this system.

   For binding Damen and Neckers (71) used an ester
group to introduce t-boc-L-proline as a template. Rebin-
ding was effected through an ester bond as well. This
binding occurred under kinetic control and yielded a
separation factor of only α = 1.015 - 1.038. Under non-
equilibrium conditions, in our case (55), no enrichment
of enantiomers was observed. This means that this kind
of separation is thermodynamically and not kinetically
controlled.

   Amide bonds between template and binding group, as
already pointed out, can be used to cause an electrostatic
interaction afterwards. This interaction alone yielded a
low specificity of α = 1.04 (55). Amide bonds have also
been used by Hopkins and Williams (72) to introduce
cavities containing amino groups on the surface of micro-
gels. These polymers possess selectivity for the molecular size of bound substrates due to finite cavity size.
The selectivity may be modified by varying the pH which
causes a change in size of the cavities.

Non-covalent Interactions. Mosbach et al. (73) prepared
L-phenylalanine-ethylester-selective polymers using the
ion-pair association of template and carboxyl containing
monomers in the polymerization step. The selectivity for
racemic resolution with α = 1.04 - 1.08 was not very
marked, but these polymers can easily be prepared.

   Electrostatic interactions may also play the main
role in the preparation of a number of dye-selective
sorbents. In these cases there seems not to be a directed
single interaction present, but rather the statistical
superposition of different interactions. Takagishi and
Klotz (74) were the first to prepare this type of polymer.
Polyethyleneimine was crosslinked in the presence of
methyl orange and its homologs and the dye was removed
from the polymer. These polymers exhibited substantially
higher binding affinity for its particular dye. Later,
poly(vinylpyrrolidone) also was treated in a similar
fashion (75,76).

   Another approach was used by Mosbach et al. (77,78).
They performed crosslinking copolymerization in the
presence of certain dyes and added some carboxyl-con-
taining monomers. The prepared polymers showed a marked
selectivity for the template that was used in the preparation of the polymer.

Imprinted Biopolymers

The most specific formation of shape- and site-selective
biopolymers is occurring in nature itself. Highly specific
antibodies are produced in response to the presence of
antigens. Such an antibody forms a specific complex with
only the antigen that evoked its formation.

In chemistry biopolymers have been used to prepare
specific substances by imprinting or template procedures.
Shinkai (<u>79</u>) used water soluble starch and added methylene
blue as template. The mixture was crosslinked in water
with cyanuric chloride. A reference was prepared without
the dye. The apparent binding constant for methylene blue
was twice as high for the imprinted product. A more
careful investigation showed that the main difference
between the two resins was not the binding constant but
the number of the binding sites.

Recently Keyes (<u>80-82</u>) described an interesting
approach to the synthesis of semisynthetic enzymes. A
protein was partially denatured and then brought in
contact with an inhibitor of an enzyme to be modelled.
The protein was then crosslinked in the presence of the
inhibitor. In this way a protein with some of the activity of the model enzyme was obtained. The activity was
quite different from that of the original native protein.

It is assumed that the partially denatured protein
binds the inhibitor whereby a new active center is formed
which is similar to that of the model enzyme. In one of
several examples transformation of a trypsin to a chymo-
trypsin-active protein is described with indole as
inhibitor and glutaraldehyde as crosslinking agent. The
new product showed an increase of chymotrypsin activity
of 400% and a decrease of trypsin activity of 14%.

Bünemann, Müller and co-workers (<u>83</u>) looked for
polymers which can bind distinct sequences of DNA. They
used the target DNA as the template to synthesize the
complementary polymer. To the target DNA a mixture of
base-specific dyes carrying polymerizable acrylic derivatives was bound. It was then possible to copolymerize
these acrylic derivatives while bound to a native DNA in
an aqueous solution. The resulting polymers were separated from the template DNA and assayed for DNA-binding
specificity (<u>84</u>). It was possible with these polymers
to strongly inhibit transcription from that DNA which had
served as template for polymer synthesis.

Distance Accuracy of Two Functional Groups

Imprinting in most cases has been used for the preparation
of chiral cavities with functional groups, and the test
for accuracy of the arrangement has usually been the
ability to resolve the racemate of the template. Thus the
selectivity is a result of the combination of the exact
cavity-shape and the exactness of the arrangement of the

functional groups. Since there is some mobility in the
chains even in a highly crosslinked polymer (85), separations due to the distinct distance of two functional
groups are expected to be less effective.

   To test the distance accuracy of two functional
groups disulfide 10 was copolymerized (86,87) under
comparable conditions as in the foregoing examples. By
subsequent reduction of the disulfide bond with diborane,
polymers with two closely positioned mercapto groups were
prepared (see Figure 6). Under comparable conditions
polymers with randomly distributed mercapto groups were
obtained from (4-vinylbenzyl)-thioacetate. The degree of
cooperativity of each two mercapto groups was determined
quantitatively by oxidation with I_2. The percentage of
reoxidation by I_2 to disulfide depends on the exact
position of each pair of mercapto groups, which in turn
depends on the degree of crosslinking of the polymer, the
concentration of the mercapto groups, the degree of
swelling and the reoxidation temperature. Under the
proper conditions it was possible to obtain 99% reoxidation of polymers from 10. With randomly distributed
mercapto groups it was possible under certain conditions
to obtain complete site separation and no reoxidation
(86,87).

   High cooperativity between two groups was also
obtained by Belokon and co-workers (50,51), who introduced
an aldehyde and an amino group into a crosslinked polymer
through an azomethine bond. Internal azomethine bond
reformation proceeded easily after splitting.

   To get two amino groups in defined distance in cross-
linked polymers, relatively rigid aromatic dialdehydes
were used (see Table V) (88). The diimines 11 and 12 were
polymerized, the dialdehydes were split off and the resulting polymers were equilibrated with a mixture of three
dialdehydes of varying distances between their two
aldehyde groupings. Formation of a diimine should only be
possible with the dialdehyde with the correct group
distance. The selectivity of these polymers is rather
high. Similar experiments with bisketals have been
performed by Shea and Dougherty (89). This represents a
shape selective two point binding. But it is also
possible to attach the two functional groups by a template
in a distinct distance to a flat surface of a solid with
no shape selection (90) (see Figure 10). In the section
"Imprinting in Modified Silicas" it is shown that even
in this case there is a distance specificity of α = 1.74-
1.67.

Template Effects in Metal Ions Sorption

Kabanov, Efendiev, and Orudzhev (91) were the first to use
the imprinting procedure for the preparation of improved
sorbents for metal ions. They used copolymers of vinyl
phosphonic acid and acrylic acid and complexed them with

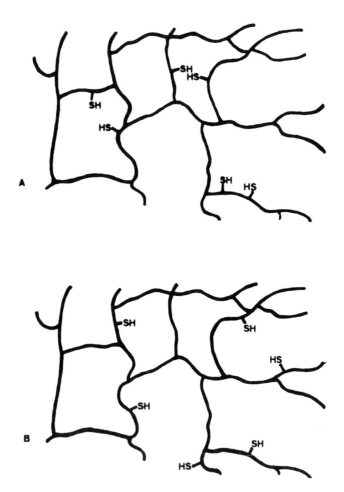

Figure 6. Mercapto groups as nearest neighbors by copolymerization of 10 and subsequent reduction (A) or distributed at random by copolymerization of (4-vinyl-benzyl)-thioacetate and subsequent reduction (B).

Table V. Selectivity of Polymers With Each Two Amino Groups in a Defined Distance

Monomer 11

Monomer 12

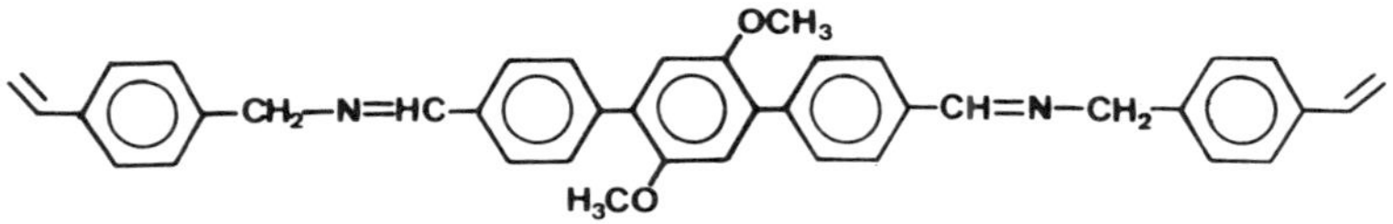

	Splitting Percentage	Distance r of groups (nm)	Apparent Binding Constants of			Selectivity
			B	C	D	
Polymer A–11 from 11	84%	0.72	1.38	0.20	0.20	$\alpha'_{B,C}$ = 4.60 $\alpha'_{B,D}$ = 5.37
Polymer A–12 from 12	91%	1.56	0.42	0.30	0.59	$\alpha'_{D,C}$ = 1.69 $\alpha'_{D,B}$ = 1.81
Polymer A–A from A	~90%	—	0.18	0.12	0.14	—

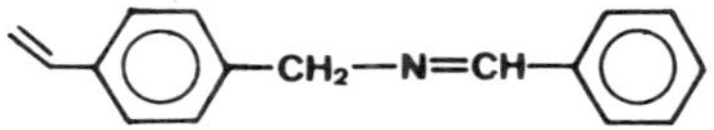

A

B

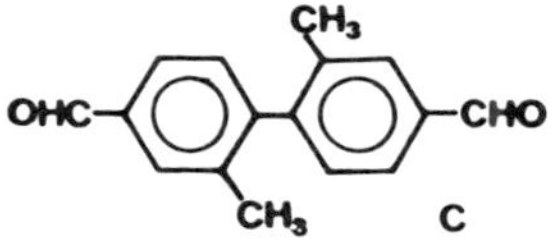

C

D

copper ions. After a prearrangement of the macromolecules
around the metal ion this favorable conformation was
subsequently fixed by means of an intermolecular cross-
linking with butadiene. After removing the copper ions
from the crosslinked sorbent with acid the polymer showed
an increased sorption capacity and an improvement in its
kinetic characteristics in comparison with a copolymer
crosslinked by the same method but without the addition
of a metal ion.

Independently from the Russian scientists the
Japanese group of Nishide and Tsuchida (92,93) reported
the crosslinking of a metal complex between a poly(4-vinyl-
pyridine) ligand (partially quaternized) and a metal ion,
by adding 1,4-dibromobutane to the solution. These authors
reported in addition improved selectivity. The resins
showed a preference for adsorbing the metal ions used as
a template. This effect is shown in Table VI. Most
specific is the copper-resin. The selectivity for copper
is increased by a factor of 4.8, the Zn-resin shows an
increase in selectivity of 3.0, the Cd-resin of 2.1 and

Table VI.   Adsorption of Metal Ions Crosslinked
            Poly(4-vinylpyridine) Resins (93)

Resin	Adsorbed Metal Ions, %			
	Cu^{2+}	Co^{2+}	Zn^{2+}	Cd^{2+}
Cross-linked (Cu)	52	6	8	9
Cross-linked (Co)	16	10	8	9
Cross-linked (Zn)	8	6	11	6
Cross-linked (Cd)	9	4	7	8
Cross-linked without metal ion	15	7	6	6

Resins imprinted by Cu^{2+}, Co^{2+}, Zn^{2+}, and Cd^{2+}.
Equilibration in $CH_3COOH-CH_3COONa$ buffer of PH 5.5.
The uptake percentage of an equimolar mixture of the
four metal ions, is listed above.

the Co-resin of 1.1 for its own metal ion. Later Kabanov
and co-workers (94-96) prepared resins by their method
imprinted by Cu^{2+}, Ni^{2+}, and Co^{2+} and found as well that
the sorbents exhibited higher selectivity with respect to
metals for the ion from which they were prepared.
Tsuchida and co-workers (97) tried similar crosslinking
with branched and linear poly(ethylene imine) and with
poly(vinylimidazole) in the presence of Cu^{2+} or Co^{2+} ion
templates, but no enhancement of selectivity towards the
same metal ions was achieved. The same was true when they
prepared a polymer from a monomeric complex (98). In this
case the metal-1-vinylimidazole complex (from Ni^{2+}, Co^{2+}
or Zn^{2+} as the template) was copolymerized and crosslinked

with 1-vinyl-2-pyrrolidone by x-ray irradiation. This
polymer showed improved sorption characteristics but no
enhancement of selectivity.

Gupta and Neckers (99) used another complexing
monomer. They copolymerized 4-methyl-4-vinyl-2,2'-
bipyridine with divinylbenzene in the presence of metal
ions Ni^{2+}, Co^{2+}, and Cu^{2+}. After removing the metal, the
polymers retained some memory of the original chelating
metal. Braun and Kuchen (100) changed the procedure
somewhat since they prepared and characterized the metal-
monomer complex before polymerization. They used bis[bis
(4-vinylphenyl)dithiophosphinato]nickel(II) and cobalt(II)
and copolymerized these to macroporous polymers. After
removal of the metal ions the nickel polymer showed an
uptake ratio of nickel: cobalt of 1.4 and the cobalt
polymer of cobalt:nickel of 2.4.

On the whole the selectivity enhancements by template
effects for metal ions until now are not great enough for
practical application. In contrast, the improved absorption properties, the capacity and the kinetic improvement
may be practical even now, especially, since these resins
can be prepared by relatively easy procedures.

There is some uncertainity about the origin of the
enhanced selectivity (99). Since the ionic radii of Ni
and Co for example are very similar, the hole-size of the
template should not be all that efficient. Small distance
effects should not cause strong binding differences since
the mobility of the crosslinked polymer inhibits this.
A selectivity enhancement can more easily be explained if
coordination geometry of the metal ions are different.
In cases where this is not so the origin of the selectivity remains to be explained.

Imprinting in Modified Silicas

When polymers with chiral cavities were used as chromatographic support in HPLC columns, near complete separations
of racemates were possible (22,40,101). Figure 7 shows
a chromatographic separation of phenyl-α-D,L-mannoside
with a chromatographic resolution of R_s = 1.2 (101). But
under fast chromatography the peaks showed a strong
tailing indicating a slow mass transfer during the chromatography. One reason for this slow mass transfer might be
a slow diffusion of substances to be separated through
the polymer matrix to reach the cavities. To reduce the
influence of diffusion thin layers of polymers on a solid
support were prepared (56,57,78).

Imprinted Polymer Layers on Silica. Wide pore silicas
(mean pore diameter 500 and 1000 Å) were surface-modified
by 3-(trimethoxysilyl)propyl methacrylate by formation of
siloxane bonds. To such a silica, layers of approximately 50 - 100 Å thickness of the usual monomeric mixture
containing 1 were applied (56,57). The monomeric mixture
was radically polymerized giving silicas with pores coated

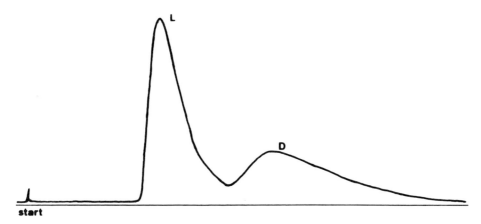

start

Figure 7.   Chromatography of the racemate of phenyl-α-
D-mannoside on a crosslinked polymer prepared with the
template monomer 1.   [Polymer of the type as in P 6,
Table I].   Eluent~acetonitrile with 4% NH_3 (25% in
water) and 5% water.   Flow-rate 0.1 ml/min, room
temperature.

on the surface with the polymer but still allowing free
diffusion through them. The silicas after coating were
still macroporous, the inner surface areas of the products
remained in the same order of magnitude. The templates
could be split off to a high percentage (see Figure 8).
These polymer-modified silicas showed high selectivity
comparable to the best macroporous polymers mentioned
before. In this way it is possible to prepare non-
swellable particles of the desired particle size for HPLC.

The mass-transfer for the separation of enantiomers
was still slow. This will be discussed in the next sub-
section. If prepared as described, the template monomer 1
should have an irregular orientation relative to the   ~
surface of the silica. To give it a more regular orientation a third binding site to the silica was introduced
by an azomethine bond (see Figure 9). Until now the
specificity of polymer layers of this type has been
reduced instead of increased (57). More work is necessary
to optimize this promising method.

Independent of our work, Mosbach et al. (78) prepared
cavities in thin polymer layers on the surface of silica.
They used certain dyes for imprinting. The polymer-
modified silicas were used also for HPLC investigations.

Quite another type of imprinting on the surface of
silica was used by Sagiv (102). Mixed monolayers of
n-octadecyltrichlorosilane and surfactant dyes are
absorbed and chemically bound to glass. The dye molecules
are then removed, leaving holes entrapped within a stable
network of chemisorbed and polymerized silane molecules.
These layers show a preferred adsorption of the dyes used
as templates. Problems are encountered with the kinetics
of the sorption-desorption process which has to proceed
through a channel formed from the dense arrangement of
long alkyl chains in the monolayer.

Chromatography with Imprinted Polymers. Chromatographic
separations of the two enantiomers of phenyl-α-mannoside
on a stationary silica phase with a polymer layer
imprinted with 1 were performed (57). In spite of the
thin layer of polymer the peaks were still rather broad.
This cannot, therefore, be attributed to diffusion
resistance in the polymer matrix, since these very thin
layers should not show substantial diffusion hindrance.
The reasons for this appear to be:
i) The rate of equilibration in the covalent binding of
    the substrate is too slow. Although the binding
    reaction can be considerably enhanced by the addition
    of piperidine or ammonia, the formation of the boronic
    diester bond is strongly sterically demanding. This
    means that a bulky substrate in a small cavity can
    only with difficulty adapt to the transition state
    of this esterification. Other types of binding
    should be somewhat better in this respect.
ii) A two point binding in a cavity seems to be kineti-

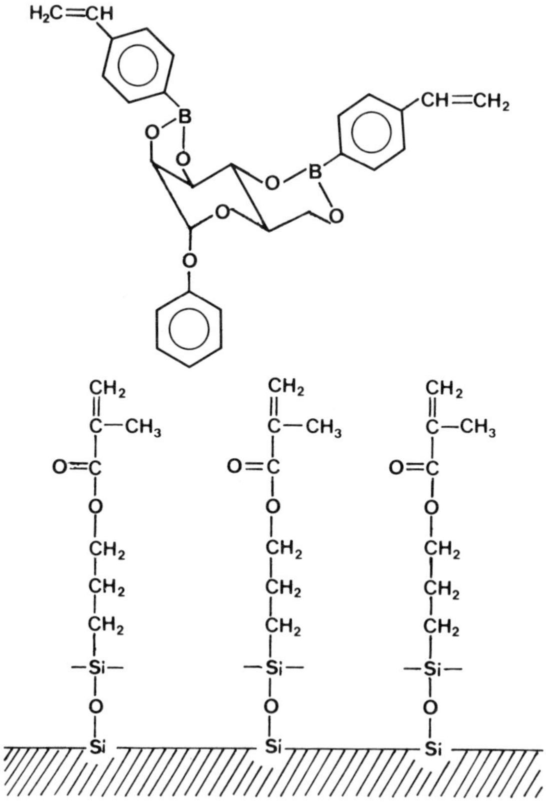

Figure 8. (a) Schematic picture of a polymer coated silica imprinted by 1.

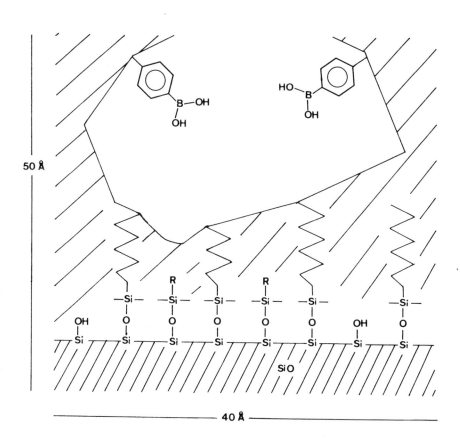

Figure 8. (b) Schematic picture of a polymer coated silica imprinted by 1.

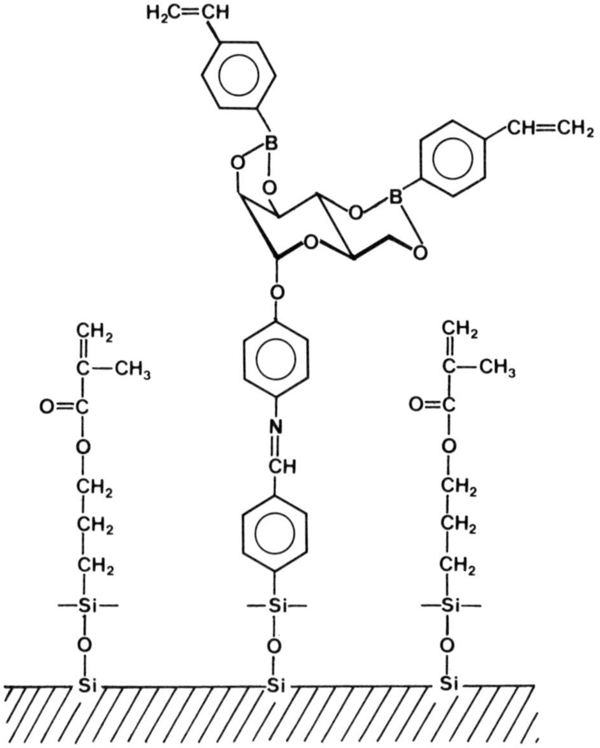

Figure 9. (a) schematic picture of a polymer coated to an aldehyde containing silica and imprinted by an amino group containing template.

b

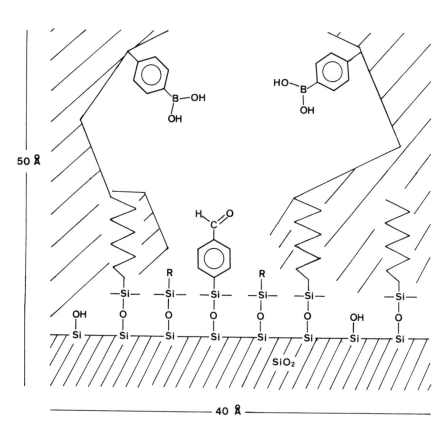

Figure 9. (b) Schematic picture of a polymer coated to an aldehyde containing silica and imprinted by an amino group containing template.

cally restricted. A one-point binding might be
possible for both enantiomers; this is apparently a
much faster process. The transformation from a one-
point binding to a two-point binding, i.e. the
complete embedding into the cavity, is a slow process
(56).
Similar slow sorption-desorption kinetics were observed
with some other enantioselective phases in liquid chromatography (103-105). Also with highly selective crown
ethers or cryptates the release of substrates from their
complexes can be extremely slow (4-6).
    Nevertheless with polymers or polymers coated on
silica, highly selective chromatography can be performed.
It is also possible to use this procedure for preparative
separations (57,101).

Direct Imprinting in Silica. As already pointed out at
the beginning Dickey (24) used the imprinting procedure
to produce selectively adsorbing silicas. Other research
groups applied this technique to various substrates (106,
107), and it was also possible to prepare silicas
imprinted by optically active substances and achieve some
enantiomeric resolution (27-29,108).
    In order to test whether it is possible to attach two
functional groups to the surface of silica at a distinct
distance the monomers 13 and 14 were directly condensed
(90) (see Figure 10). After splitting off the dialdehyde,
two amino groups each remained. In this case their
position at the silica should not be changed by chain
mobility, swelling, or shrinking of a polymer. The
distance can only be altered by conformational changes
of the functional group part. The specificity was tested
as before by equilibrating the mixture of two dialdehydes
(see Table VII). Each silica showed an enhanced
specificity for its template. One-point binding is
possible for both dialdehydes on both silicas, but only
the original template should give a diimine with a
resultant higher binding constant.

Reactions in Imprinted Polymers

One goal for the use of imprinted polymers is the preparation of catalysts and enzyme models. Experiments perform-
ing stereoselective reactions in the cavities of the
imprinted polymers are a first step in this direction.
Shea and co-workers (109,110) were the first to try this
approach. They prepared polymers from the optically
active (-)-trans-1,2-cyclobutanedicarboxylic diester 15
and obtained cavities after splitting off the template,
containing two benzyl alcohols each. To these cavities
fumaric acid could be bound. It was reacted with a
methylene-transfer reagent [(dimethylamino)methyl-phenyl-
oxosulfonium tetrafluoroborate]. The cyclopropanedicarb-

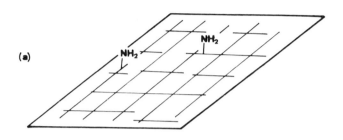

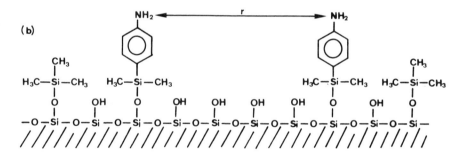

Figure 10. Preparation of two amino groups in a
distinct distance on a flat silica surface
(a) schematic picture (b) chemical structure.

Table VII.   Selectivity of Modified Silicas With Each Two Amino Groups in a Defined Distance

Monomer 13

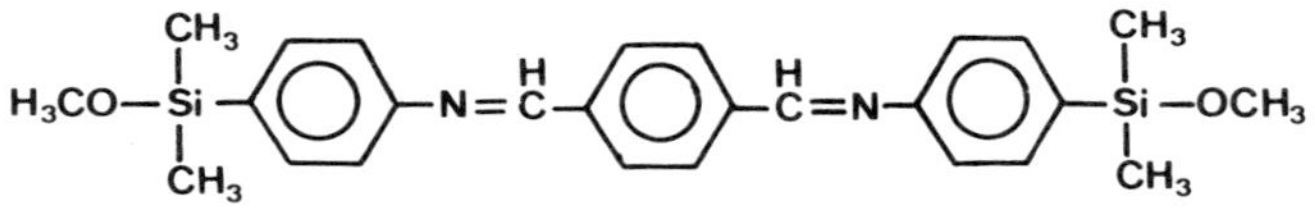

Monomer 14

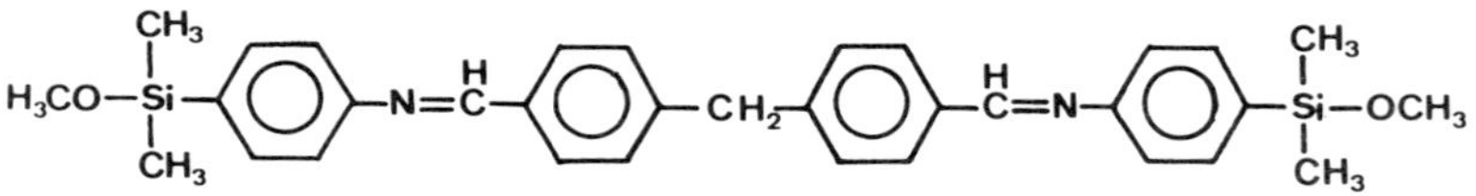

	Splitting Percentage	Distance r of groups (nm)	Apparent Binding Constants of			Selectivity α'
			OHC—⬡—CHO	OHC—⬡—CH₂—⬡—CHO	OHC—⬡—CH₂—⬡—CHO	
Silica modified with 13	>95%	0.72	4.91	2.58		1.74
Silica modified with 14	>95%	1.05	9.07	13.77		1.67
Silica modified with A (at random)	—	—	2.26	2.05		—

A $H_2N\!-\!(CH_2)_3\!-\!Si(OCH_3)_3$

oxylic acid thus prepared was liberated by hydrolysis with
a 34% overall yield and an optical enrichment of 0.05% ee
was observed. This low enantiomeric excess does not give
a clear indication on the applicability of this concept.
    Damen and Neckers (111) were more successful in this
respect. They also used cyclobutane derivatives and
copolymerized the bis(vinylbenzyl)ester of α-truxillic
acid, β-truxinic acid, and δ-truxinic acid. After
removing the template, trans-cinnamic acid was bound to
each pair of benzyl alcohols in a cavity. Irradiation
of the polymers yielded dimerization of the trans-cinnamic
acid. Table VIII shows the composition of the products
from hydrolysis of the polymers.

Table VIII.   Photochemical Reaction of trans-Cinnamate-
               Esters Bound to Imprinted Cavities (111)

Polymer (prepared from)	Composition of the Hydrolysate in Mol%		
	α-Truxillic Acid	β-Truxinic Acid	δ-Truxinic Acid
$P_{\alpha-truxill.}$	100	0	0
$P_{\beta-truxin.}$	47	53	0
$P_{\delta-truxin.}$	47.3	0	52.7

    For the dimerization of trans-cinnamate esters bound
to a random polymer the exclusive formation of α-truxillic
acid is expected. Table VIII shows indeed a considerable
stereochemical direction of the photochemical reaction
by the imprinting procedure.
    Sarhan (112) used a copolymer (with 4-vinylpyridine
as one comonomer) imprinted with the D-mandelic ester
monomer 6 for deprotonation and selective protonation of
the racemate of mandelic acid. An enantiomeric excess of
several per cent for one enantiomer was observed.
    From the previously mentioned monomer 7 with L-DOPA
as template (60) polymers were prepared which after the
removal of the template contained chiral cavities each
having an aldehyde and a boronic acid group. This polymer
was then used for an enantioselective reaction. Glycine,
the simplest and achiral amino acid, was bound in the form
of a copper complex to the polymer (see Figure 11). This
was then reacted with acetaldehyde in the presence of a
base, giving in good yields a mixture of threonine and
allothreonine. The diastereoselectivity was low, as
expected, since this cavity was not imprinted with this
compound. But there was an enantioselectivity of more
than 30% ee. It can be assumed that the incoming aldehyde
is bound to the boronic acid and then attacks the enolate
of glycine preferably from one side (see Figure 11) (60).
    If one bears in mind that there is no asymmetric
induction center in this case but just the chiral cavity

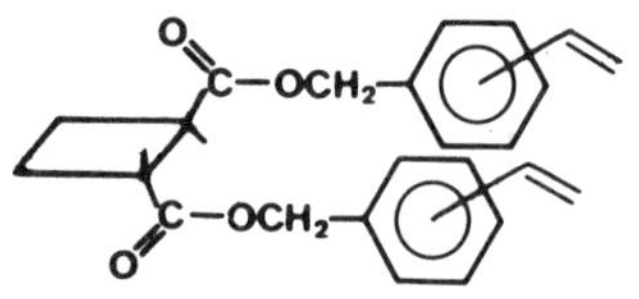

15

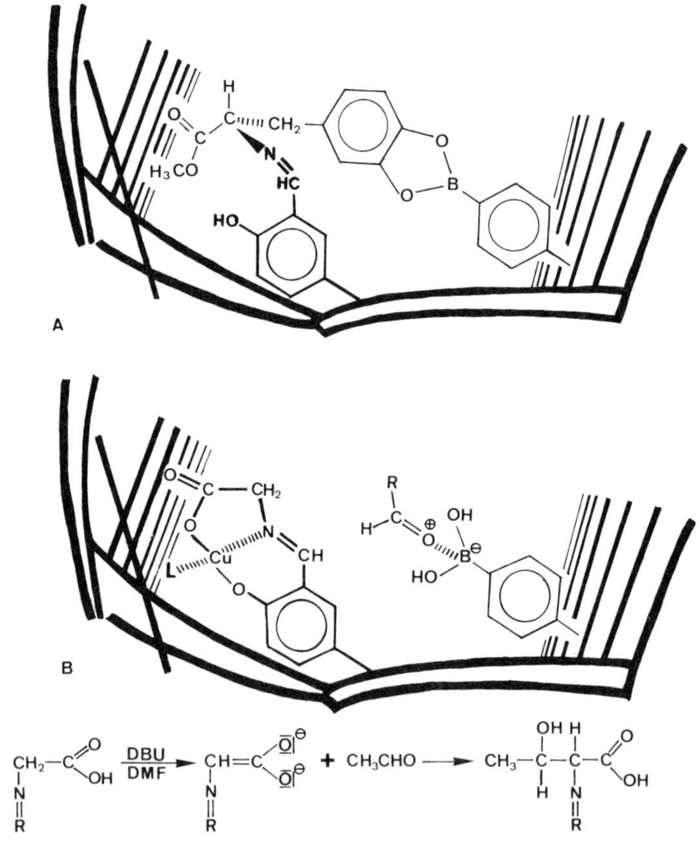

Yield 30–90%; threo/allo 0.9; ee_{threo} = 32.5%
ee_{allo} = 10.4%

Figure 11. Polymer imprinted by 8 (A) and a reaction in cavities imprinted by 8 to prepare L-threonine (B).

present, the enantioselectivity is surprisingly high. It
is planned to introduce chiral side chains into the cavity, as in conventional enantioselective synthesis, in
order to improve these results and make them comparable
or even superior to the present state of art.

Conclusion

The possibility of preparation of crosslinked polymers
with molecular recognition for a certain substance by
polymerization in the presence of templates has been
verified in many examples. Cavities in the polymer containing functional groups are thus obtained whose shape
and whose arrangement of the functional groups corres-
ponded to that of the template. The selectivity for the
optical resolution of racemates and the separation of
different diastereoisomers, configurational isomers, or
metal ions was quite high. Separations have been performed in analytical and preparative scale. It was also
possible to perform stereoselective reactions with the
substrates bound inside of the cavities.
Problems to be solved are:
i.   The selectivity in many cases has to be further
      improved.
ii.   The exchange equilibria between polymer and
      substrate should be faster. In chromatography the
      mass transfer is relatively slow and reactions
      inside the cavity in some cases show only moderate
      yields.
iii. Only a few suitable binding reactions are known up
      till now.

Further improvements are expected with investigations
along the following lines:
i.   The structure of the polymers has to be further
      optimized. Rigidity and flexibility of the polymers
      have to be adjusted in order to get high selectivity
      and fast mass-transfer. Recent results with the
      application of modified silicas show great promise
      and could be an alternative method.
ii.   New and better binding sites have to be developed.
      Possibly for each functional group in organic
      chemistry a corresponding binding site should be
      available. These template-bound binding sites
      should be able to fix the growing polymer chains in
      a defined topology. After splitting off the
      templates the binding sites should be able to under-
      go an easily reversible binding interaction with
      substrates carrying the functional group of the
      template.
iii. Strong improvements in selectivity are to be expec-
      ted by the combination of the chiral-cavity-effect
      with chiral side chains inside the cavity as has
      been shown by Fujii and co-workers (70).

It is to be expected that polymers with molecular recognition obtained by imprinting with templates can find broad application in highly selective separations both in the batch procedure and in chromatography. Furthermore within the cavities stereoselective reactions can be performed or reactive groups inside the cavity can be used as selective polymeric reagents.

The final aim in the application of these polymers is the use as catalysts working in a fashion similar to enzymes. The first steps in this direction are reached. It is now possible to prepare cavities of a specific shape Furthermore there is a method to introduce into the cavity functional groups in a defined topology and some binding sites are available which meet the requirements for substrate binding.

Difficult steps have still to be reached. Next there has to be the development of a suitable catalytic active environment. It is important to have a cooperative action of several catalytically active groups. Another part of the preparation of an enzyme model is the choice of suitable reagent groups as analogs of the coenzyme part of an enzyme. To obtain a fast turn-over in catalysis the binding of the substrate should be much tighter compared to the product. To speed up the turn-over it could be advantageous to use as the template a substrate resembling the stereochemistry of the transition state of the reaction instead of that of the substrate, or the product.

On the basis of the results obtained until now it seems that the way to polymeric enzyme models is more difficult than to those with low molecular weight models but the application of polymeric supports shows several advantages. They can mimic the natural enzymes more perfectly and they can be handled and processed more easily. So increased efforts in this direction might be worthwhile.

Acknowledgments

Thanks are due to "Fonds der Chemischen Industrie", the "Minister für Wissenschaft und Forschung des Landes Nordrhein-Westfalen" and to "Deutsche Forschungsgemeinschaft" for financial support.

Literature Cited

1. Bender, M.L.; Komiyama, M. "Cyclodextrin Chemistry"; Springer Verlag: Berlin, 1978.
2. Tabushi, I. Acc. Chem. Res. 1982, 15, 66-72.
3. Breslow, R. Science 1982, 218, 532-537.
4. Cram, D.J.; Trueblood, K.N. Topics Curr. Chem. 1981, 43-106.
5. Lehn, J.-M. Acc. Chem. Res. 1978, 11, 49-57.
6. Lehn, J.-M. Stud. Org. Chem. 13 (Biomimetic Chem.) 1983, 163-87.

7. Tabushi, I.; Yamamura, K. Top. Curr. Chem. 1983, 113, 145-182.
8. Murakami, Y. Top. Curr. Chem. 1983, 115, 107-155.
9. Keehn, P.M.; Rosenfeld, S.M., Eds.; "Cyclophanes"; Vol. II, Acad. Press: London, 1983.
10. Giglio, E. In "Inclusion Compounds"; Atwood, J.L.; Davies, J.E.D.; Mac Nicol, D.D., Eds.; Academic: London, 1984; Vol. II, pp. 207-229.
11. Barrer, R.M. In "Inclusion Compounds"; Atwood, J.L.; Davies, J.E.D.; Mac Nicol, D.D., Eds.; Academic: London, 1984; Vol. I, pp. 191-248.
12. Kunitake, T.; Shinkai, S. Adv. Phys. Org. Chem. 1980, 17, 435-487.
13. Royer, G.P. Adv. Catalysis 1980, 29, 197-227.
14. Tsuchida, E.; Nishide, H. Adv. Polymer Sci. 1977, 24, 1-87.
15. Kunitake, T.; Okahata, Y. Adv. Polymer Sci. 1976, 20, 159-221.
16. Guthrie, J.B. In "Applications of Biochemical Systems in Organic Chemistry"; Jones, J.B.; Sih, C.J. Perlman, D., Eds.; J. Wiley, New York, 1976, pp. 627-730.
17. Klotz, I.M. In "Enzyme Engineering 7"; Laskin, A.I.; Tsao, G.T.; Wingard, L.B., Eds.; Ann. N.Y. Acad. Sc. 1984, 434, pp. 302-320.
18. Schwyzer, R. Proc. Fourth Int. Congr. on Pharmacology 1970, 5, 196-209.
19. Wulff, G.; Sarhan, A. Angew. Chem. 1972, 84, 364; Angew. Chem. Int. Ed. 1972, 11, 341.
20. Wulff, G.; Sarhan, A. Patent Ger. Offen 2 242 796, 1974, Appl. P. 31.8.1972; Chem. Abstr. 1975, 83, 60300 w.
21. Wulff, G.; Sarhan, A.; Zabrocki, K. Tetrahedron Lett. 1973, 4329-32.
22. Wulff, G.; Vesper, W.; Grobe-Einsler, R.; Sarhan, A. Makromol. Chem. 1977, 178, 2799-2816.
23. Wulff, G.; Kemmerer, R.; Vietmeier, J.; Poll, H.-G. Nouv. J. Chim. 1982, 6, 681-87.
24. Dickey, F.H. Proc. Natl. Acad. Sic. U.S. 1949, 35, 227-29.
25. Dickey, F.H. J. Phys. Chem. 1955, 59, 695-707.
26. Davies, J.E.D.; Kemula, W.; Powell, H.M.; Smith, N.O. J. Inclusion Phenom. 1983, 1, 3-44.
27. Curti, R.; Colombo, U. J. Am. Chem. Soc. 1952, 74, 3961.
28. Beckett, A.H.; Anderson, P. Nature 1957, 179, 1074-75.
29. Beckett, A.H.; Anderson, P. J. Pharm. and Pharmacol. 1960, 12, 228T-236T.
30. Wulff, G. In "Festschrift 25 Jahre Fonds der Chemischen Industrie"; Fonds der Chemischen Industrie, Frankfurt 1975; pp. 135-142.
31. Wulff, G. Nachr. Chem. Techn. Lab. 1977, 25, 239-43.
32. Wulff, G.; Sarhan A. In "Chemical Approaches to Understanding Enzyme Catalysis"; Green, B.S.; Ashani, Y.; Chipman, D., Eds.; Elsevier, 1982, pp. 106-118.

33. Wulff, G. Pure Appl. Chem. 1982, 54, 2093-2102.
34. Wulff, G. In "Enzyme Engineering 7"; Laskin, A.J.;
     Tsao, G.T.; Wingard, L.B., Eds.; Ann. N.Y. Acad. Sc.
     1984, 434, pp. 327-333.
35. Akelah, A. Synthesis 1981, 413-438.
36. Shinkai, S.; Kunitake, T. Kagaku (Kyoto) 1981, 36,
     76-79; Chem. Abstr. 1981, 94, 209248x.
37. Kunieda, T. J. Synth. Org. Chem. Jap. 1982, 40,
     686.
38. Shinkai, S. Progr. Polym. Sci. 1982, 8, 1-59.
39. Halgas, J.; Toma, S. Chem. Listy 1983, 77, 949-970;
     Chem. Abstr. 1984, 100, 5341e.
40. Wulff, G.; Vesper, W. J. Chromatogr. 1978, 167,
     171-186.
41. Wulff, G.; Vietmeier, J.; Poll, H.-G., in prepara-
     tion.
42. Sarhan, A.; Wulff, G. Makromol. Chem. 1982, 183,
     1603-1614.
43. Millar, J.R.; Smith, D.G.; Kressman, T.R.E. J. Chem.
     Soc. 1965, 304-310.
44. Funke, W. Chimia 1968, 22, 111-122.
45. Andersson, L.; Sellergren, B.; Mosbach, K. Tetra-
     hedron Lett. 1984, 25, 5211-5214.
46. Wulff, G.; Lohmar, E. Isr. J. Chem. 1979, 18, 279-
     284.
47. Sarhan, A.; Wulff, G. Makromol. Chem. 1982, 183,
     85-92.
48. Wulff, G.; Zabrocki, K.; Hohn, J. Angew. Chem.
     1978, 90, 567-568; Angew. Chem. Int. Ed. Engl. 1978,
     17, 535-537.
49. Wulff, G.; Hohn, J. Macromolecules 1982, 15, 1255-
     1261.
50. Belokon, Y.N.; Tararov, V.I.; Savel'eva, T.F.; Vitt,
     S.V.; Bakhmutov, V.I.; Belikov, V.M. Makromol.
     Chem. 1980, 181, 89-104.
51. Belokon, Y.N.; Tararov, V.I.; Savel'eva, T.F.;
     Lependina, O.L.; Timofeyeva, G.I.; Belikov, V.M.
     Makromol. Chem. 1982, 183, 1921-1934.
52. Wulff, G.; Kemmerer, R. unpublished.
53. Wulff, G.; Schulze, I.; Zabrocki, K.; Vesper, W.
     Makromol. Chem. 1980, 181, 531-544.
54. Wulff, G.; Gimpel, J. Makromol. Chem. 1982, 183,
     2469-2477.
55. Wulff, G.; Best, W.; Akelah, A. Reactive Polymers
     1984, 2, 167-174.
56. Wulff, G.; Oberkobusch, D.; Minarik, M.; XVIIIth.
     Int. Solvay Conference on Chemistry, Brussels, 1983
     in "Proceedings of the XVIIIth Solvay Conference";
     van Binst, G., Ed.; Springer Verlag: Berlin, in
     press.
57. Wulff, G.; Oberkobusch, D.; Minarik, M. Reactive
     Polymers, in press.
58. Wulff, G.; Poll, H.-G., in preparation.
59. Wulff, G.; Grobe-Einsler, R.; Vesper, W.; Sarhan, A.
     Makromol. Chem. 1977, 178, 2817-2825.
60. Wulff, G.; Vietmeier, J., in preparation.

61. Wulff, G.; Dederichs, W.; Grotstollen, R.; Jupe, C.
     In "Affinity Chromatography and Related Techniques";
     Gribnau, T.C.J.; Visser, J.; Nivard, R.J.F.; Eds.;
     Elsevier: Amsterdam 1982, pp. 207-216.
62. Wulff, G.; Lauer, M.; Böhnke, H. Angew. Chem. 1984,
     96, 714-715; Angew. Chem. Int. Ed. Engl. 1984, 23,
     741-742.
63. Lauer, M.; Böhnke, H.; Grotstollen, R.; Salehnia, M.;
     Wulff, G. Chem. Ber. 1985, 118, 246-260.
64. Ferrier, R.J. Adv. Carb. Chem. Biochem. 1978, 35,
     31-80.
65. Amicon Corp. "Boronate Ligands in Biochemical
     Separations"; Amicon Corp., Danvers 1981.
66. Carlsohn, H.; Hartmann, M. Acta Polym. 1979, 30,
     420-425.
67. Sarhan, A. Makromol. Chem. Rapid. Commun. 1982, 3,
     489-493.
68. Wulff, G.; Sarhan, A.; Gimpel, J.; Lohmar, E.
     Chem. Ber. 1974, 107, 3364-3376.
69. Wulff, G.; Akelah, A. Makromol. Chem. 1978, 179,
     2647-2651.
70. Fujii, Y.; Matsutani, K.; Kikuchi, K. J. Chem. Soc.
     Chem. Commun. 1985, 415-417.
71. Damen, J.; Neckers, D.C. Tetrahedron Lett. 1980, 21,
     1913-1916.
72. Hopkins, A.; Williams, A. J. Chem. Soc., Perkin
     Trans. II, 1983, 891-896.
73. Andersson, L.; Sellergren, B.; Mosbach, K. Tetra-
     hedron Lett. 1984, 25, 5211-5214.
74. Takagishi, T.; Klotz, I.M. Biopolymers 1972, 11,
     483-491.
75. Takagishi, T.; Hayashi, A.; Kuroki, N. J. Polym.
     Sci. Polym. Chem. Ed. 1982, 20, 1533-1547.
76. Takagishi, T.; Sugimoto, T.; Hamano, H.; Lim, Y.-J.;
     Kuroki, N.; Kozuka, H. J. Polym. Sc. Polym. Lett.
     Ed. 1984, 22, 283-289.
77. Arshady, R.; Mosbach, K. Makromol. Chem. 1981, 182,
     687-692.
78. Norrlöw, O.; Glad, M.; Mosbach, K. J. Chromatogr.
     1984, 299, 29-41.
79. Shinkai, S.; Yamada, M.; Sone, T.; Manabe, O.
     Tetrahedron Lett. 1983, 24, 3501-3504.
80. Keyes, M.H. Patent Ger. Offen. 3 147 947, 8.7. 1982.
81. Saraswathi, S.; Keyes, M.H. Polym. Mater. Sci. Engl.
     1984, 51, 198-203; Chem. Abstr.1984, 101, 166015 n.
82. See also: Chem. Eng. News 1984, 62, 33.
83. Bünemann, H.; Dattagupta, N.; Schuetz, H.J.; Müller,
     W. Biochemistry 1981, 20, 2864-2874.
84. Kosturko, L.D.; Dattagupta, N.; Crothers, D.M. Bio-
     chemistry 1979, 18, 5751-5756.
85. Crowley, J.I.; Rapoport, H. Acc. Chem. Res. 1976, 9,
     135-144.
86. Wulff, G.; Schulze, I. Angew. Chem. 1978, 90, 568-
     570.
87. Wulff, G.; Schulze, I. Isr. J. Chem. 1978, 17, 291-
     297.

88. Wulff, G.; Helfmeier, G., unpublished.
89. Shea, K.J.; Dougherty, T.K. 189th ACS National
     Meeting, Miami Beach, Florida, April 28-May 3, 1985.
     Div. Org. Chem. Abstract Nr. 120.
90. Wulff, G.; Heide, B., unpublished.
91. Kabanov, V.A.; Efendiev, A.A.; Orudzhev, D.D.
     U.S.S.R. Patent 502.907, 1976, Appl. 24.4.1974;
     Chem. Abstr. 1976, 85, 6685 d.
92. Nishide, H.; Deguchi, J.; Tsuchida, E. Chem. Lett.
     1976, 169-174.
93. Nishide, H.; Tsuchida, E. Makromol. Chem. 1976,
     177, 2295-2310.
94. Efendiev, A.A.; Orudzhev, D.D.; Kabanov, V.A.
     Vysokomol. Soedin. 1977, Ser. B 19, 91-92.
95. Kabanov, V.A.; Efendiev, A.A.; Orudzhev, D.D.
     J. Appl. Polym. Sci. 1979, 24, 259-267.
96. Efendiev, A.A.; Kabanov, V.A. Pure Appl. Chem. 1982,
     54, 2077-2092.
97. Nishide, H.; Deguchi, J.; Tsuchida, E. J. Polym.
     Sci., Polym. Chem. Ed. 1977, 15, 3023-3029.
98. Kato, M.; Nishide, H.; Tsuchida, E. J. Polym. Sci.,
     Polym. Chem. Ed. 1981, 19, 1803-1809.
99. Gupta, S.N.; Neckers, D.C. J. Polym. Sci., Polym.
     Chem. Ed. 1982, 20, 1609-1622.
100. Braun, U.; Kuchen, W. Chem. Ztg. 1984, 108, 255-257.
101. Wulff, G.; Poll, H.-G.; Minarik, M. J. Liquid
     Chromatogr., in press.
102. Sagiv, J. Isr. J. Chem. 1979, 18, 346-353.
103. Davankov, V.A.; Zolotarev, Y.A. J. Chromatogr. 1978,
     155, 285-293.
104. Yamskov, I.A.; Berezin, B.B.; Davankov, V.A.; Zolo-
     tarev, Y.A.; Dostavalov, I.N.; Myasoedov, N.F.
     J. Chromatogr. 1981, 217, 539-543.
105. Koller, H.; Rimböck, K.-H.; Mannschreck, A. J.
     Chromatogr. 1983, 282, 89-94.
106. Erlenmeyer, H.; Bartels, H. Helv. Chim. Acta 1964,
     47, 46-51; 1285-1288.
107. Patrikeev, V.V.; Sholin, A.F. Molekul. Khromatogr.
     Akad. Nauk. SSSR 1964, 66-72; Chem. Abstr. 1965,
     62, 11127d.
108. Bartels, H.; Prijs, B.; Erlenmeyer, H. Helv. Chim.
     Acta 1966, 49, 1621-1625.
109. Shea, K.J.; Thompson, E.A. J. Org. Chem. 1978, 43,
     4253-4255.
110. Shea, K.J.; Thompson, E.A.; Pandey, S.D.; Beauchamp,
     P.S. J. Am. Chem. Soc. 1980, 102, 3149-3155.
111. Damen, J.; Neckers, D.C. J. Am. Chem. Soc. 1980,
     102, 3265-3267.
112. Sarhan, A. Communication at the Freiburger Makro-
     molekulares Kolloquium 1985, Abstracts p. 23.

RECEIVED October 15, 1985

Polymeric Transfer Reagents for Organic Synthesis with Self-Control

Toward Automation in Organic Synthesis

A. Patchornik, E. Nov, K. A. Jacobson, and Y. Shai

Department of Organic Chemistry, The Weizmann Institute of Science, Rehovot, Israel 76100

The use of transfer polymeric reagents (PRs) as excellent acylating agents for high yield and high purity peptide synthesis is described. Three methodologies are compared: the classical solution method, the Merrifield approach and an automated (the "mediator"-shadchan) method with continuous monitoring. The utilization of PRs as general acyl transfer reagents is also elaborated. The described approaches are not limited to peptide synthesis, but may be applicable to a wide range of organic reaction types.

In the past few years, the motivation for peptide synthesis has grown noticeably as a consequence of the increasingly rapid discoveries of new biologically active peptides. The present favorable climate for peptide developments is due to the rapid progress in synthetic, analytical and purification methods. However, the existing repetitive synthetic methods suffer from low overall yields, which is one of the causes for their high price. In the following section, a short outline of the conventional peptide synthetic methods is given, followed by a sketch of our approaches for improved peptide synthesis from the first example of a polymeric active ester of an N-protected amino acid to the novel mediator ("Shadchan") method using multiple polymers.

The Conventional Methods of Peptide Synthesis Methods

I. Classical solution method - The earliest chemical method of peptide synthesis (and one which is still being widely used today) is that in which the growing peptide chain (free amino) and the active acyl species are added as soluble components to the coupling reaction and thus the name, "solution method" (1). The major advantage of the solution method is the ability to isolate the intermediate peptides in large quantities and purify them to homogeneity. This enables one to proceed towards the final peptide with purer products and thus the target peptide is much easier to isolate in a homogeneous form. The solution method, however, suffers from some serious drawbacks: The method is very time consuming and laborious, especially for the preparation of larger peptides where problems of solubility often become formidable as the peptide grows. Because of purification and solubility (and therefore, diminishing reactive surface) problems, a practical limit of a peptide size to be constructed by this stepwise procedure is probably

0097–6156/86/0308–0231$06.00/0

about 15 to 20 amino acid (AA) residues (2). Methods of fragment condensation have increased this practical limit as in the landmark solution synthesis of ribonuclease (3), but often a large excess of one precious oligopeptide component is required.

II. Merrifield's "Solid Phase" Method (SPPS) - The isolation and purificaton of each intermediate during the assembly of a peptide chain is of importance in order to limit the presence of by-products in the final product. However, these procedures are not ideally suited to the synthesis of long chain polypeptides in solution, because of the technical difficulties mentioned above. In 1963, Merrifield reported his method of "solid phase peptide synthesis"(SPPS) (4), which simplified the preparation of long peptides and was amenable to further acceleration by automation.

Both of these goals were realized in a few short years and descriptions of the syntheses of a wide variety of peptides of chain lengths up to 124 amino acid residues (5) by SPPS have appeared in the literature. Merrifield's method possesses several clear advantages: First, the insoluble polymer-bound polypeptide is separable from the soluble reagents and by-products by simply filtering and washing the polymer, which conserves the synthetic intermediates normally lost by repeated isolation and purification steps. Second, since the soluble reagents can be removed by washing, they can be used in excess to drive the reaction to completion. Third, the synthesis is simplified and accelerated because all reactions can be conducted in a single reaction vessel containing the polymeric carrier, thus avoiding repeated transfers of intermediates and corresponding losses of material. Finally, probably the major advantage is that all operations in synthesizing a polypeptide can be automated and therefore, manpower and time expended are reduced drastically.

However, SPPS also possesses some serious disadvantages: In order to obtain nearly quantitative yields at each coupling step, large excesses of costly protected amino acids have to be used. In addition, industrial scale synthesis by the SPPS method requires huge quantities of chlorinated solvents and acidic deprotection reagents. Secondly, as a trade off for increased simplicity, speed and effeciency, one usually obtains final products in which purification becomes a great challenge: Since the intermediates in solid phase synthesis are always bound to the polymer, one cannot isolate and purify them in the conventional way. Whereas soluble impurities are easily removable, side products which remain attached to the solid support cannot be separated from the desired product at the intermediate stages of the synthesis but instead have to be removed at the end of the synthesis. Therefore, the purity of the final product depends largely upon the degree of side reactions that occur. In other words, long peptides synthesized by SPPS usually require extensive purification by conventional column chromatography or more efficient HPLC. So far SPPS is not used routinely for large scale peptide synthesis (kg quantities). The crucial drawbacks are the purification steps and the costs of reagents and solvents.

As a result, in most cases this method does not lend itself for industrial bulk production of biologically active peptides for commercial use. To date, more commercial peptide products are still synthesized by the classical solution method.

Use of Polymeric Transfer Reagents In Peptide Synthesis

In order to overcome disadvantages of the above described methods, we developed in parallel alternative approaches based on the use of functionalized polymers as acylating agents. In both of these approaches an acyl group is transferred to a nucleophile such as an amine in solution, and the leaving group remains irreversibly bound to the polymeric matrix. In this section the early examples of peptide synthesis using polymeric acyl transfer reagents are described:

Synthetic reagents bound to a polymeric backbone (polymeric reagents = PRs) have been widely used in organic chemistry during the past two decades (6). The most significant advantages of the PRs are the ease of separation from the reaction mixture and the possibility of recycling. Polymeric catalysts, polymers for specific separations, carriers for sequential synthesis, polymeric blocking groups and polymeric transfer reagents in general organic synthesis, all have utilized these advantages.

Use has also been made of the fact that chain fragments within a crosslinked polymer have restricted mobility. Under certain conditions active species attached to the polymer can thus be effectively isolated from each other at relatively high concentrations, providing the advantages of high dilution and specificity along with rapid kinetics. In other cases, properties of the backbone itself such as polarity, pore size and chirality were utilized to achieve unique reactions, the polymer providing a specific microenvironment for the reaction. These aspects of PRs have been extensively reviewed (6,7).

Many chemical reactions can be generalized as follows (~ stands for an energy-rich bond):

$$B + A{\sim}\alpha \longrightarrow BA + \alpha$$

In such a reaction, which may be termed a "transfer reaction", an active species A, is transferred to the acceptor B. In practice, an excess amount of one of the components in such a reaction often is used, in order to effect complete conversion of reagent B to the required product BA, and to shorten the reaction time. At the end of the reaction, the resulting product BA, must be separated from the excess reagent. By using polymeric transfer reagents the separation is facilitated by a simple filtration step:

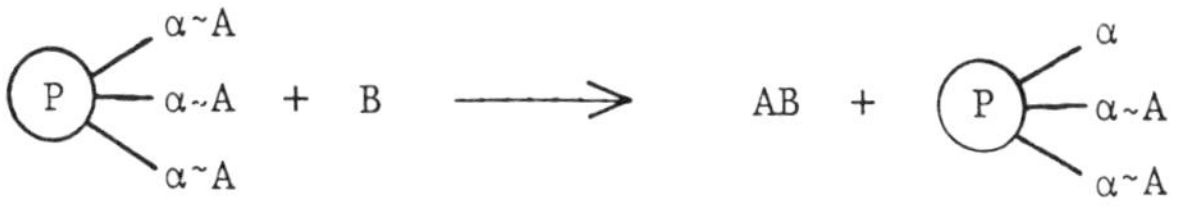

Insoluble polymeric transfer reagents (PRs) are prepared by anchoring $\alpha{\sim}A$ groups to insoluble polymers, e.g. crosslinked polystyrene (α - being a good leaving group). The common characteristics of polymeric transfer reagents are, that despite their complete insolubility in any solvent, they allow solvents and soluble reagents (B) to penetrate them, thereby enabling chemical reaction between the penetrating soluble B molecule and the insoluble ℗–$\alpha{\sim}A$. Peptide synthesis by using polymeric transfer reagents was first introduced by Fridkin, Patchornik and Katchalski in 1966 (8). The approach is shown in the following general scheme (AA_i = amino acid residue or protected derivative):

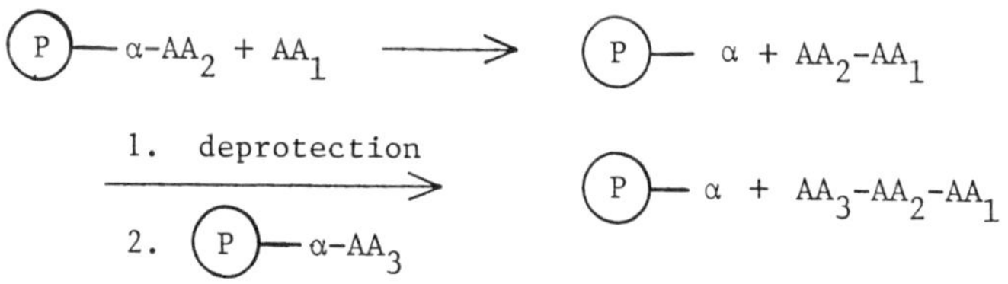

According to this approach, the unprotected amino terminus of a soluble amino acid (AA_1) ester or short peptide is allowed to react with a polymeric active ester of another protected amino acid (AA_2). As a result, a longer peptide is freed into solution, the insoluble excess reagent together with spent polymer being removed by filtration. This new peptide, can be further elongated by selective cleavage of the amine protecting group followed by additional reaction with a polymeric active ester of another amino acid (AA_3 - amine protected) and so on.

Opposite to Merrifield's approach (in which the peptide is built while anchored to the insoluble matrix and the building blocks, i.e. the protected AAs, are in solution and allowed to react with the polymer-bound growing peptide), in our method the growing peptide is in solution and its unprotected amino terminus is allowed to cleave a polymer-bound AA active ester.

All the advantages of the classical solution method described above are also inherent in our polymeric reagent approach. The PR approach has the following additional advantages: 1) Most peptide couplings proceed to completion in a short period and in good to excellent yields (typically at least 95% after purification). 2) The peptides are usually pure enough to proceed to the next step with little or no purification (8). 3) Excess polymeric acylating reagent can be used without contamination of the product since it is removed simply by filtration. 4) For peptides which contain the same amino acid residue in more than one position, batch amounts of this polymeric activated amino acid can be prepared and stored "on the shelf" (or in columns - for circulation) to be used whenever this particular residue appears in the sequence. 5) The PRs are recyclable, that is, the polymers can be regenerated and reused. The disadvantages inherent in the stepwise solution method are also seen with this PR approach, except that the excess reagent (now insoluble due to the attachment to the polymeric backbone) which is used to partially overcome the "diminishing reactive surface" problem, can be easily removed by filtration. Other limitations of the PR method are the difficulty in automation of the synthesis and the additional cost in the preparation of the PRs.

We have shown that upon applying our approach to the synthesis of soluble protected peptide intermediates, very high yields per coupling step were obtained as compared to the classical solution methods. While it is commonly accepted that on using the standard solution synthetic method the average yield per coupling step is typically approximately 80%, yields with our polymeric transfer reagent method in excess of 95% were commonly obtained. (Although by the SPPS method coupling steps are reported to reach yields of >99%, our 95% yields refer to the isolated purified product). For example, Fridkin synthesized the well known decapeptide LH/RH by the classical and PR methods in parallel. Results show a remarkable improvement in overall yield (7% vs. 40% respectively) (8).

The difference in coupling efficiency is of major importance to repetitive methods. Simple calculations show that improving each coupling yield from 80% to 95% in a 30-step synthesis will result in the saving of 97% (!) of otherwise wasted expensive protected amino acid derivatives. Specifically, 6300 equivalents of amino acid derivatives are needed for the classical solution stepwise synthesis of a 30-mer peptide compared to 155 equivalents by our method. Classical peptide solution synthesis, in spite of its disadvantages such as low yields and tedious manipulations in isolating each intermediate product, is still the preferred method for industrial scale peptide syntheses and will presumably remain so in the near future (this opinion is shared by many leaders of this field). Therefore, such an improved solution method might be of wide interest in research as well as for industrial applications.

An obvious question is the diminished efficiency of acylating a sterically hindered, high molecular weight polypeptide within the PR cavities. This phenomenon was indeed observed. However, one can overcome this problem by using either very large excess of PR (thus the acylation reaction takes place with the most available groups bound to the outer surface of the beads) or by adding a small chemical agent which penetrates the PR, releasing a soluble active ester which acylates efficiently the growing peptide. (See below the "Mediator" Approach).

Polymeric Acylating Agents

Up to the present, a variety of polymeric acyl transfer reagents have been prepared. Some of these are discussed below in terms of their activity and utilization potentials:

I. Polymeric o-nitrophenols - These polymers carry o-nitrophenolic groups to which N-protected AAs are attached as polymeric active esters. PR-1 is a polystyrene derivative in which some of the phenyl rings are substituted (8). Although the original polymeric o-nitrophenol PR-1 worked well in early trials of peptide synthesis, its disadvantage was the long contact times needed to complete the coupling reactions; that is its rate of acylation was slow. Recently, two newer PRs, PR-2 and PR-3, were developed for which the rates of acylation are about 40 times that of PR-1. Some of the results obtained with PR-2 have recently been published (9).

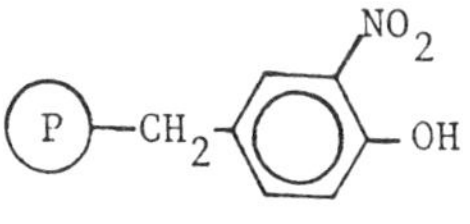

PR-1 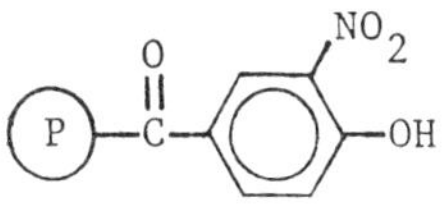   PR-2

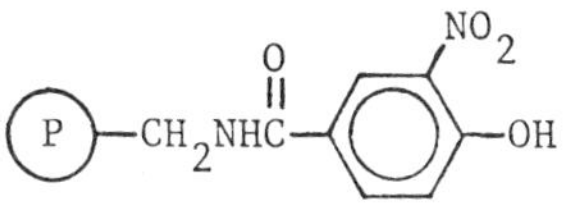

PR-3

Both PR-3 and PR-2 carry higher amounts of available OH groups for esterification (compared to PR-1) in meq/g, (2.3 vs. 1.5 meq/g of dry PR). However, PR-3 also has the advantage that it can be obtained in a purer form, and the changes between the loaded and ionized polymers are easily visualized. It is more costly to prepare because of a multistep synthesis which first requires the preparation of the aminomethyl polymer (10):

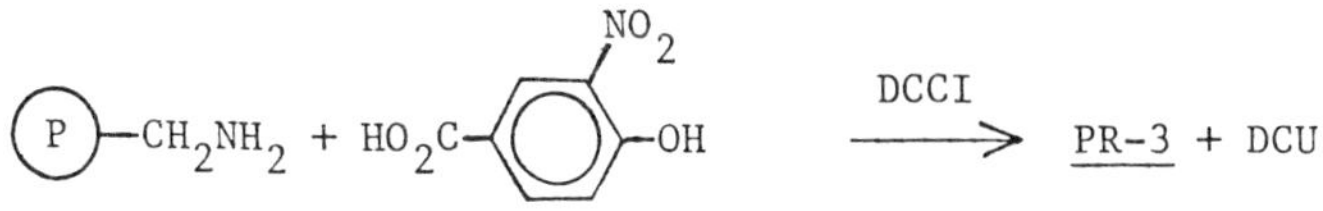

Loading (esterification) of <u>PR-2</u> and <u>PR-3</u> with protected amino acids was carried out according to the following equation:

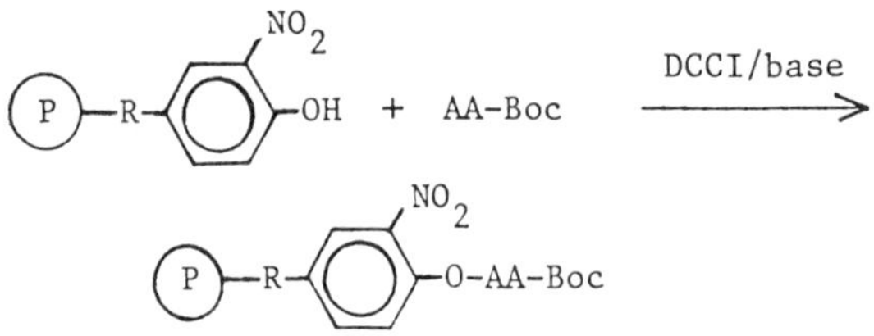

Generally, the desired protected AA (1 equiv) is shaken with a solvent swollen PR and then a mixture of DCCI (1.2 equiv) and organic base such as Et_3N (1.5 equiv) is added. The whole mixture is allowed to shake at -10 °C for 2 hours and for an additional 10 hours at 0 °C. Table I summarizes some results of loading efficiency on <u>PR-2</u> and <u>PR-3</u>.

The resulting carbonyl group serves as both an anchor to the polymer and an electron withdrawing activator for the leaving group. Synthesis of protected Leus-enkephalin, (Boc-Tyr(OBzl)-Gly-Gly-Phe-Leu-OBzl) in 92% overall yield was accomplished, with each coupling step requiring one hour (<u>9</u>). The active esters were found to be less sensitive to moisture and alcohols, making them convenient to handle.

II. Polyhydroxybenzotriazole (<u>PR-4</u>)

HOBT   PR-4

1-Hydroxybenzotriazole (HOBT) is presently used widely in solution and SPPS as an additive to both enhance rates of coupling and to reduce the danger of racemization. In some laboratories, HOBT is added routinely to all DCC-mediated couplings. It is generally accepted that the HOBT ester is the active intermediate and that the absence of racemization in stepwise couplings may be attributed to the acidity (pK_a ~6.0) of the HOBT by-product, as well as the short coupling times.

<u>Table I</u> - Loading Efficiency of <u>PR-2</u> and <u>PR-3</u> with AAs

Available OH groups[a] (meq/g)	Amino Acid	Base	Time	% Loading
1.5	BocPhe	TEA[b]	3 hrs	95-100
1.5	BocPhe	DIPEA[c]	12 hrs	95-100
1.5	BocGly	TEA	3 hrs	95-100
1.5	BocGly	DIPEA	12 hrs	95-100
1.5	BocTrp	DIPEA	12 hrs	90-95
2.8	BocGly	TEA	12 hrs	85-90
2.8	BocPhe	DIPEA	12 hrs	90-95
2.8	BocTyr(OBz)	DIPEA	12 hrs	90-95
1.5	BocMet	DIPEA	12 hrs	85-90
1.5	BocAsp	DIPEA	12 hrs	85-90
1.5	BocLys(ε-CBz)	DIPEA	12 hrs	90-95
1.5	BocGln	DIPEA	12 hrs	30

[a] <u>PR-2</u> contained 1.5 meq/g, while <u>PR-3</u> contained 2.8 meq/g of available OH groups.

[b] TEA = triethylamine.

[c] DIPEA = diisopropylethylamine.

Since the introduction by Patchornik and co-workers in 1975 (<u>11</u>), <u>PR-4</u> has been used most succesfully in peptide synthesis, as it is a good acylating agent and minimizes racemization problems. The loading of <u>PR-4</u> with Boc-AAOH was carried out similarly to <u>PR-2</u> and <u>PR-3</u> using DCCI (see above). When compared to <u>PR-1</u>, <u>PR-4</u> afforded a 100-fold improvement in the rate of acylation. For example, in the synthesis of the tetrapeptide Boc-Leu-Leu-Val-Tyr(Bzl)-OBzl (which contains bulky, slow reacting amino acids), coupling steps were completed within 20 minutes. However, because of the high reactivity of the polymeric active esters, they must be protected from moisture, as some hydrolysis occurs with time. <u>PR-4</u> has been used successfully in the synthesis of such diverse biologically active peptides as thyrotropin-releasing hormone, tuftsin and analogs and the C-terminal half of thymosin α_1 (<u>12</u>).

III. Polymeric sulfonylhydroxylamine <u>PR-5</u>.

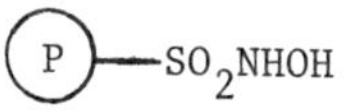

$$\text{P}\!\!-\!\!SO_2NHOH \qquad\qquad \underline{PR-5}$$

This inexpensive and easily prepared PR (from P -SO$_2$Cl treated with hydroxylamine) is another polymeric acyl support (<u>13</u>). Both nitrogen and sulfur elemental analysis showed that <u>PR-5</u> contained 3.36 meq of -NHOH groups per gram of polymer (S/N ratio = 1.0). To examine its potential for reversible active ester binding, benzoyl chloride was allowed to react with <u>PR-5</u>. Preliminary results have revealed the following information: 1) A series of ten repeated acylations and cleavages (using diethylamine) indicated that the same amount of acyl group was released after each cycle (as determined by the quantity of N,N-diethylbenzamide obtained), and thus the polymer is recyclable; 2) The average amount of cleavable groups was 1.6 meq/g polymer, whereas the amount of acyl group initially esterified to the polymer was about twice the amount released (probably due to the irreversible formation of an amide bond).

The "Mediator" Methodology for Fully Automated Peptide Synthesis (The Two Polymeric System) (14).

This section deals with the utilization of PRs as components in a fully automated system for peptide synthesis. This novel approach was applied by Shai, et.al. to the synthesis of Leu^5-enkephalin, as a model peptide (14).

Two reagents bound covalently to two different insoluble polymers will not interact, although reaction between the same reagents may occur readily in solution; such polymer-bound reagents can be made to interact, however, by the action of a soluble mediator molecule. In the past, such multiphase systems have mostly been used for mechanistic studies (15).

A review of concepts of using multipolymeric methods, including the Mediator ("Shadchan") methodology, in organic synthesis is found in ref. 16.

The actual application of such a multiphase system to peptide synthesis was carried out by constructing a system consisting of two containers in which the insoluble polymers, donor (I) and acceptor (II), are placed. A solution of a soluble "mediator" molecule (e.g. imidazole) circulates between the containers (Figure 1). Polymer I carries the bound N-protected amino acid as an active ester (9). Polymer II is a Merrifield-type polymer carrying an amino acid or peptide with a free amino group.

The procedure involves the circulation of acyl carrier or "shadchan" (17) in solution between the two polymers - polymer I (carrying 0.5 - 0.8 meq/g Boc amino acid as a nitrophenyl ester) and polymer II (carrying 0.2 - 0.5 meq/g of free amino group of the attached peptide or amino acid). The "shadchan" (e.g. imidazole), interacts with polymer I to release the soluble acyl imidazole. The latter reacts with polymer II to elongate the peptide by an additional amino acid. The concurrently released imidazole is now free to repeat the cycle until polymer II is saturated. On-line monitoring of the reaction progress is possible by comparing the UV absorptions of the circulating solution on entering and leaving polymer II compartment; zero difference indicates the end of the reaction. (Although an on-line continuous monitoring of couplings during Merrifield's SPPS is usually not possible, a continuous-flow variation of Merrifield's method, that recently appeared on the market, permits spectrophotometric monitoring of the coupling of Fmoc-protected residues (18). However, such a system used very high concentrations (~4 M). Monitoring the decrease of such high concentration can never be as accurate and sensitive as compared to 0.3 M (and less) solutions used in our reaction systems). Polymer II is then washed thoroughly, the Boc protecting group is removed, the polymer is neutralized and is ready for the next coupling step. Some typical results are shown in Table II. Thus on using polymer I, the methyl ester of the protected Leu$_5$-enkephalin was obtained in 85% overall yield and 99% purity directly from the polymer without further purification. This procedure involved four cycles of transacylation, followed by alcoholysis in methanol and triethylamine (removal of the Boc group after each cycle was performed at -10 °C). On the other hand, the deprotected free enkephalin was obtained in 75% yield and 80% purity by cleavage with hydrogen bromide in trifluoroacetic acid (the Boc group was removed after each step with TFA at 25 °C).

Pratically quantitative yields of couplings were obtained within 24 hours as indicated by ninhydrin detection (19). A more critical test was used to prove that the coupling was quantitative indeed: After each coupling step some of the polymer carrying growing peptide was cleaved and subjected to TLC test to observe contamination that could have arisen from previous incomplete couplings. No such contaminations were observed. However, by increasing the concentration of the "shadchan" the reaction time could be shortened.

The high yield synthesis of peptides, such as the enkaphalin obtained by this system, encouraged us to expand the "mediator" methodology to other than acylation reactions, such as sulfonation and phosphorylation. Towards this end it was necessary to develop a "bank" polymer to store these groups in an active form.

The "shadchan" methodology may be viewed in general terms as shown in Figure 2. Thus, to expand the scope of the "shadchan" reactions we examined new possibilities for immobilized electrophilic banks.

1-Acyl-4-dimethylaminopyridinium salts have been reported to be stable in monomeric form, and we demonstrated that they are also stable when bound to a polymer (14,20), for example, the polymeric 4-dimethylaminopyridine (DMAP), PR-6. This PR proved a useful polymer of choice, forming N-substituted pyridinium salts, PR-7, for subsequent transfer to a mediator ("shadchan") molecule. Conveniently, a number of polymer-bound forms of the acylation catalyst of the 4-dialkylaminopyridine type have been reported (21-23).

PR-6 was prepared according to Shinkai (21) from N-methylaminomethyl-polystyrene and 4-chloropyridine. When acylated with an acyl chloride and exhaustively washed with anhydrous methylene chloride, a stable 1-acylpyridinium adduct remained on polymer PR-7 with negligible leakage. With benzoyl chloride, derivatization was in the range of 0.5 - 0.8 meq/g. Acetyl chloride, tosyl chloride, isobutyl chloroformate, 2-chlorophenyl isopropyl phosphorochloridate and other electrophilic species were similarly retained by the polymer.

The applicability of PR-6 as a storable acylium "bank" and its versatility were demonstrated by the following experiments.

When exposed to an excess of imidazole in chloroform, the benzoylated polymer was discharged to give soluble N-benzoylimidazole (Figure 2, E = PhCO, H-α = imidazole), which reacted quantitatively with benzylamine to form N-benzyl-benzamide. Moreover, since the acyl-DMAP support is more reactive than the corresponding polymeric o-nitrophenyl ester PR-2, the time required for each transfer step was reduced.

PR-7 was allowed to react also with various nucleophiles in solution, such as amines, alcohols, and thiols (Figure 3), without the addition of imidazole or tertiary amine base. Since no added base is needed, purification of products often consists of simple filtration and evaporation. The reactions were carried out in methylene chloride using two equivalents of acylating polymer, and the nucleophile concentration was approximately 0.1 M. The acyl polymers were found to be sufficiently active to form symmetric and mixed anhydrides with carboxylic acids. Some typical results are given in Table III. PR-6 and isobutyl chloroformate yielded PR-8 (typically 0.72 meq/g), which acts as an activating agent for carboxylic acids, to form the mixed carbonic-carboxylic anhydride, which may then acylate various nucleophiles. (See Figure 4). Benzoic acid, on activation as the mixed anhydride (by using PR-8), separation (by filtration) and addition of excess benzylamine to the filtrate, gave N-benzylbenzamide which was isolated in 97% yield. A series of dipeptides was also synthesized in solution by this activation technique.

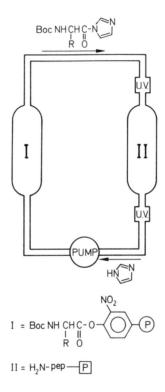

Figure 1. Schematic "shadchan" synthesis system. Reproduced with permission from reference 14. Copyright 1985 American Chemical Society.

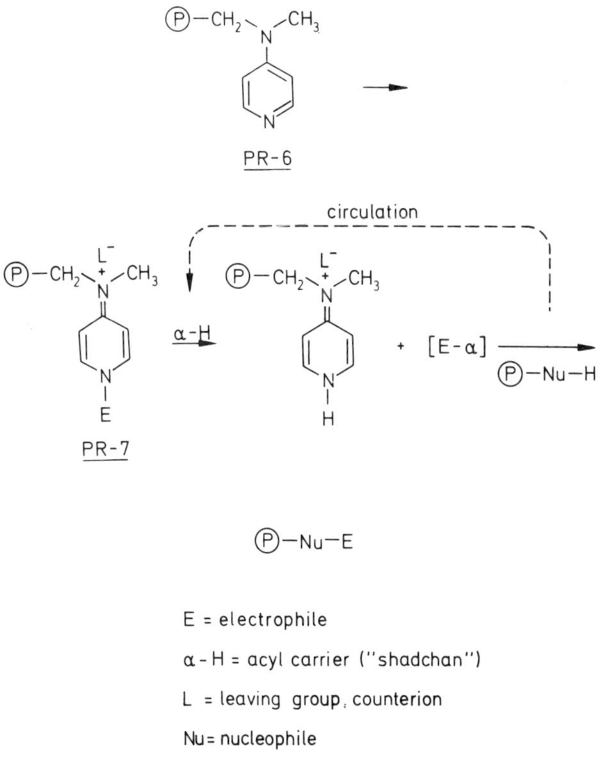

Figure 2.   A schematic acylation reaction carried out by the DMAP poplymer as an acyl-transfer support in the "Shadchan" system.   Reproduced with permission from reference 14. Copyright 1985 American Chemical Society.

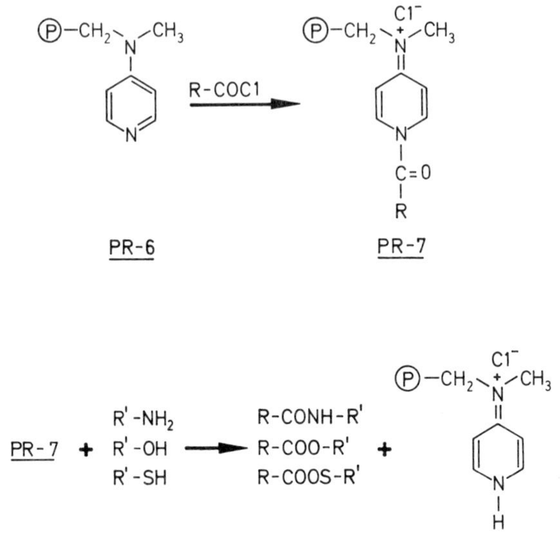

Figure 3. Acylation of nucleophiles by polymer-bound 1-acyl-
DMAP chloride.

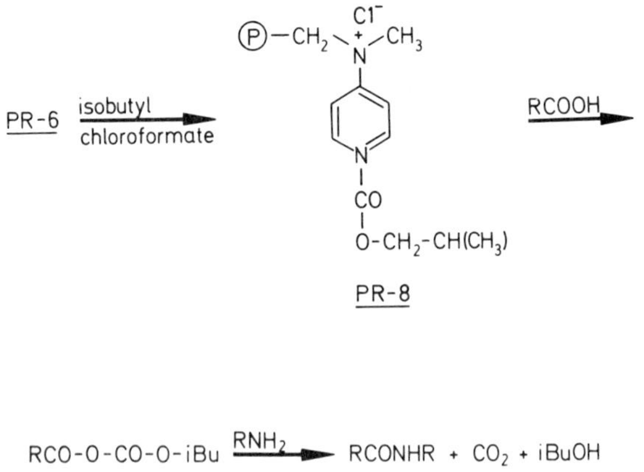

RCO-O-CO-O-iBu $\xrightarrow{RNH_2}$ RCONHR + CO_2 + iBuOH

Figure 4. Direct amide formation by DMAP polymer via mixed
anhydride formation.

Table II. Peptides Synthesized Using the Shadchan Method

I[a]	α[b]	temp.	II[c]	products after cleavage[d]	coupling yield[e]
Boc-Phe-O-(P)	A	0°C	NH₂Leu-OCH₂-(P)	H₂NPhe-Leu-OH	>99.8%
Boc-Phe-O-(P)	A	25°C	NH₂Leu-OCH₂-(P)	H₂NPhe-Leu-OH	low yield[f]
Boc-Phe-O-(P)	B	0°C	NH₂Leu-OCH₂-(P)	H₂NPhe-Leu-OH	>99.8%
Boc-Phe-O-(P)	B	25°C	NH₂Leu-OCH₂-(P)	H₂NPhe-Leu-OH	>99.8%
Boc-Ala-O-(P)	A	0°C	NH₂Leu-OCH₂-(P)	H₂NAla-Leu-OH	>99.8%
Boc-Ala-O-(P)	A	25°C	NH₂Leu-OCH₂-(P)	H₂NAla-Leu-OH	low yield[f]
Boc-Gly-O-(P)	A	0°C	NH₂Phe-Leu-OCH₂-(P)	H₂NGly-Phe-Leu-OH	>99.8%
Boc-Gly-O-(P)	A	25°C	NH₂Phe-Leu-OCH₂-(P)	H₂NGly-Phe-Leu-OH	>99.8%
Boc-Gly-O-(P)	B	0°C	NH₂Gly-Phe-Leu-OCH₂-(P)	H₂NGly-Gly-Phe-Leu-OH	>99.8%
Boc-Gly-O-(P)	B	25°C	NH₂Gly-Phe-Leu-OCH₂-(P)	H₂NGly-Gly-Phe-Leu-OH	>99.8%
Boc-Tyr(OBz)-O-(P)	A	0°C	NH₂Gly-Gly-Phe-Leu-OCH₂-(P)	Boc-Tyr(OBz)-Gly-Gly-Phe-Leu-OCH₃	>99.8%
Boc-Tyr(OBz)-O-(P)	A	25°C	NH₂Gly-Gly-Phe-Leu-OCH₂-(P)	Boc-Tyr(OBz)-Gly-Gly-Phe-Leu-OCH₃	low yield[f]
Boc-Tyr(OBz-2,6-Cl₂)-O-(P)	A	0°C	NH₂Gly-Gly-Phe-Leu-OCH₂-(P)	H₂NTyr-Gly-Gly-Phe-Leu-OH	>99.8%

Footnotes on next page

Table II, footnotes.

a) $-O-\boxed{P}$ = $-O-\langle O \rangle-\overset{\overset{\textstyle O}{\|}}{C}-\boxed{P}$, where $\boxed{P}$ is macroporous cross-linked polystyrene (Xe-305 from Rohm & Haas Company)

b) α = the "shadchan" molecule. A = imidazole; B = N-hydroxybenzo-triazole.

c) $-O-CH_2-\boxed{P}$ = 1% cross-linked polystyrene carrying 0.3 meq/g oxymethyl sites or pam-oxymethyl sites.(8)

d) Overall yields of peptides cleaved from polymer II were 70%-90% depending upon the type of polymer II used. Higher yields were obtained when using the pam resin.(8)

e) Yield was estimated on T. L. C. plates (about 1000 μmol peptide was cleaved from the polymer, which allowed detection of 0.2% impurities on T. L. C. plate after spraying with ninhydrin).

f) The acylimidazole molecule is unstable at room temperature.

Table III. Reactions of PR-7 with nucleophiles in solution.

E=	Nu-H=	product	m/e	% yield[b]
CH_3CO	$HOCH_2Ph$	$CH_3CO_2CH_2Ph$	150	82
PhCO	$HOC_6H_4-p-NO_2$	$PhCO_2C_6H_4-p-NO_2$	243	100
PhCO	$HSC_6H_4-p-NO_2$	$PhCOSC_6H_4-p-NO_2$	259	100
PhCO	menthol	menthol benzoate	260	30
PhCO	PhCOOH	benzoic anhydride	226	92[c]
PhCO	CH_3COOH	CH_3CO_2COPh[a]	164	54[c]
PhCO	HF (48%)	PhCOF[a]	124[c]	
i-BuOCO	H_2NCH_2Ph	i-BuOCONHCH_2Ph	207	
p-tosyl	H_2NCH_2Ph	p-tosyl-NHCH_2Ph	261	100
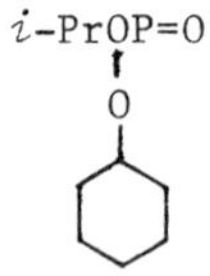 i-PrOP=O (O cyclohexyl)	$HNMe_2$	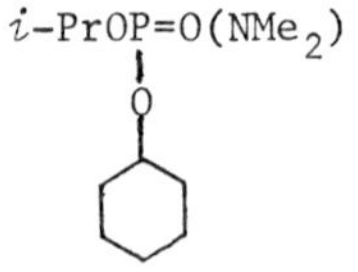 i-PrOP=O(NMe_2) (O cyclohexyl)	277	
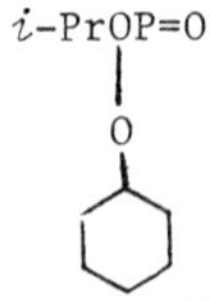 i-PrOP=O (O cyclohexyl)	MeOH	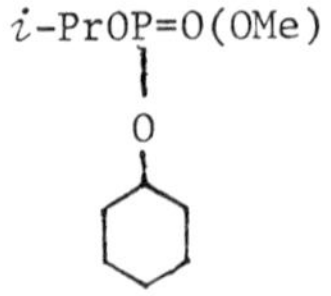 i-PrOP=O(OMe) (O cyclohexyl)	248	

[a]Benzoic anhydride present as impurity. [b]After 16 h, product is isolated unless noted. [c]Product not isolated, measured by GC.

Reproduced from reference 14. Copyright 1985 American Chemical Society.

In considerably lower yields, Boc-protected amino acids could be esterified to o-nitrophenyl resin activated as the mixed anhydrides by circulation over <u>PR-8</u> (Figure 2).

The potential applicability of the mediator methodology to the synthesis of oligonucleotides was demonstrated by reacting polymer <u>PR-6</u> with the 5'-(dimethoxytrityl)deoxythymidine-3'-(2-chlorophenyl) phosphorochloridate. The incorporation on the polymer, according to dimethoxytrityl assay, was 3.7 mmol/g. Upon treatment with tetrazole carrier in the presence of a polymer bound 5'-free-hydroxyl thymidine, transfer of the dimethoxytrityl-containing species was demonstrated qualitatively.

Conclusions

1) Convenient and highly effecent condensation reactions were shown to take place by transferring polymer bound electrophiles (i.e. active esters) via a mediator (shadchan) to polymer bound nucleophiles (i.e. amines).

2) It was also shown that the possibility of on-line monitoring can be achieved - a most essential feature in planning a fully automatic, self-controlled machine for multistep syntheses.

In order to widen the application scope of PRs we have also developed more simple, convenient methods for preparation of polymers. Now we are able to maximize the loading of the PRs with protected amino acid and other reactive species, which should greatly expand the use of these methods.

The mediator methodology developed here is believed not to be limited to acylation and related processes only, but also to be applicable to other chemical processes that involve the formation of activated intermediates. These possibilities are currently under investigation.

Acknowledgment

The authors wish to thank the Etta P. Schiff Trust and the Bantrell Fund for financial support.

Literature Cited

1. Bergmann, M.; Zervas, L. <u>Chem. Ber.</u> 1932, *65*, 1192.
2. Bodanzsky, M. "Principles of Peptide Synthesis"; Springer-Verlag: Berlin, 1984.
3. Yajima, H.; Fujii, N. <u>J. Chem. Soc., Chem. Commun.</u> 1980, 115.
4. Merrifield, R. B. <u>J. Am. Chem. Soc.</u> 1963, *85*, 2149.
5. Gutte, B.; Merrifield, R. B. <u>J. Am. Chem. Soc.</u> 1969, *91*, 501.
6. Akelah, A.; Sherrington, D. C. <u>Chem. Rev.</u>, 1981, *81*, 557.
7. Kraus, M. A.; Patchornik, A. <u>Isr. J. Chem.</u> 1978, *17*, 298.
8. Fridkin, M.; Patchornik, A.; Katchalski, E. <u>J. Am. Chem. Soc.</u> 1966, *88*, 3164; Fridkin, M.; et. al. <u>J. Solid - Phase Biochem.</u> 1977, *2*, 175.
9. Cohen, B. J.; Karoly-Hafeli, H.; Patchornik, A. <u>J. Org. Chem.</u> 1984, *49*, 922.
10. Warshawsky, A. ; Deshe, A. ; Rossey, G.; Patchornik, A. <u>Reactive Polymers,</u> 1984, *2*, 301.
11. Kalir, R.; Warshawsky, A.; Fridkin, M.; Patchornik, A. <u>Eur. J. Biochem.</u> 1975, *59*, 55.
12. Mokotoff, M.; Patchornik, A. <u>Int. J. Peptide Protein Res.</u> 1983, *21*, 145.
13. Nov, E.; Patchornik, A., Unpublished results.

14. Shai, Y.; Jacobson, K. A.; Patchornik, A. J. Am. Chem. Soc. 1985, *107*, 4249.
15. Rebek, J., Jr. Tetrahedron Lett. 1979, *35*, 723.
16. Patchornik A.; Nouv. J. Chim., 1982, *6*, 639.
17. "Shadchan" ("ch" is pronounced as in German) is the Hebrew term for a mediator, matchmaker, go-between or agent.
18. Atherton, E.; Dryland, A.; Sheppard, R. C.; Wade, J. D. in "Peptides: Structure and Function", Proceedings of the 8th American Peptide Symposium, V. Hruby and D. Rich, eds. p. 45, Pierce, Rockford, Illinois (1983).
19. Sarin, N. K.; Kent, S.B.H.; Tan, J. P.; Merrifield, R.B. Analyt. Biochem. 1981, *117*, 147.
20. Jacobson, K. A.; Patchornik, A. Abstracts of the 50th Anniversary Meeting of the Israel Chemical Society, Jerusalem, Israel, April 1984; Jacobson, K. A.; Shai, Y.; Patchornik, A. Abstracts of the 189th Meeting of The American Chemical Society, Symposium on Polymeric Reagents and Catalysts, Miami Beach, Florida, May 1985.
21. Shinkai, S.; Tsuji, Y.; Hara, Y.; Manabe, O. Bull. Chem. Soc. Japan 1981, *54*, 631.
22. Tomoi, M.; Akada, Y.; Kakiuchi, H. Makromol. Chem., Rapid Commun. 1982, *3*, 537.
23. Guendouz, F.; Jacquier; Verducci, H. Tetrahedron Lett. 1984, *25*, 4521.

RECEIVED February 7, 1986

Site Isolation Organic Synthesis in Polystyrene Networks

Warren T. Ford

Department of Chemistry, Oklahoma State University, Stillwater, OK 74078

Reagents supported in solvent-swollen, cross-linked polysty-renes are less mobile than reagents in solution. As a result, the rate of reaction between two polymer-supported species can be retarded relative to the rate of reaction of a polymer-supported species with a soluble reagent or with itself. With 2-10% cross-linked polystyrenes in swelling solvents, reduced mobility provides 1-10 minute lifetimes of polymer-supported reactive intermediates such as an ester enolate, a benzyne, and a glycine active ester. Trapping of the reactive intermediate in less than the lifetime gives kinetically effective site isolation synthesis on a polymer support. Polymer chain mobility decreases with decreased swelling, increased cross-linking, and decreased temperature. Organometallic catalysts can be effectively site isolated by low coverage on the surface of a highly cross-linked macroporous polymer. Within a polymer gel, amine and phosphine ligands provide equilibria between supported organometallic complexes, catalytic activities, and selective reactions different from those in solution.

Polymers often have been called immobilizing media for chemical reactions, yet reagents and catalysts bound to solvent-swollen polymers have substantial motional freedom. This is not a contradiction. There would be far fewer misunderstandings about the nature of polymer supports used for reagents and catalysts in organic synthesis if the supports were always called gels, not solids, and if the functional groups were described as being "in", not "on", the polymer. Motion of reagents and catalysts bound into polymer gels is slower than the motion of the analogous reagents and catalysts in solution. Reduced mobility can have dramatic effects on the courses of chemical reactions. This chapter reviews the evidence for, and the consequences of, restricted motion of polymeric reagents.

The idea that a polymer support could help to isolate polymer-bound reactive species from one another was suggested and illustrated shortly after the first solid phase peptide syntheses. Cyclic tetrapeptides were obtained in higher yields from polymer-bound 2-nitrophenyl esters than from analogous micromolecular active esters (Scheme 1) (1). Polymer-bound ester enolates were formed at 0 °C and trapped with alkyl bromides and carboxylic acid chlorides with no competing self-condensation (Scheme 2) (2). Soluble analogs gave primarily self-condensation.

0097-6156/86/0308-0247$10.75/0

Scheme 1

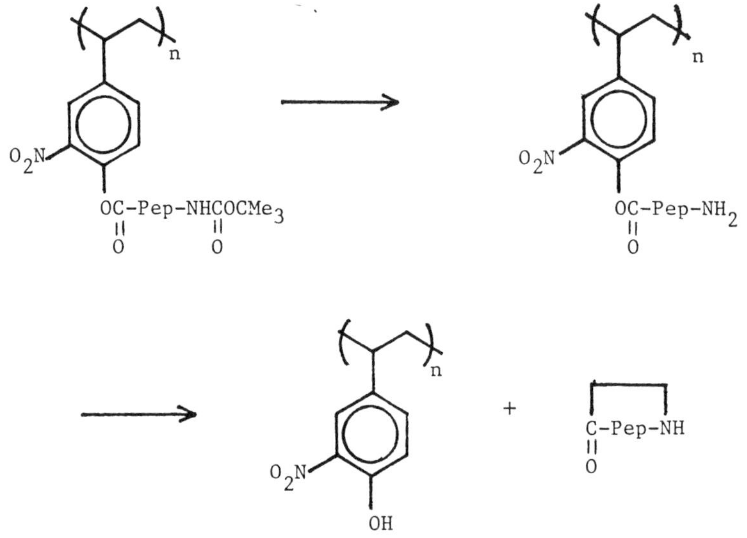

The success of both peptide cyclizations and ester enolate trapping is due to the lesser mobility of polymer-bound species, which reduces the rates of the bimolecular reactions that lead to higher oligomeric peptides or to ester self-condensation.

Early successes in use of polymer supports as immobilizing media led to optimistic predictions about the future of the synthetic technique of "site isolation on" a polymer support. Failures of the site isolation method soon were discovered. Dieckmann cyclization of polymer-bound diesters gave acceptable yields of six-membered ß-keto esters, but valiant attempts to synthesize the nine-membered ß-keto ester failed (Equation 1) (3,4). (Zero yield was found also in attempted high dilution Dieckmann cyclization to the nine membered ring in solution (5).) The Dieckmann cyclization results led to a pessimistic review (6) of the potential of site isolation synthesis in polymer gels, which has discouraged further research.

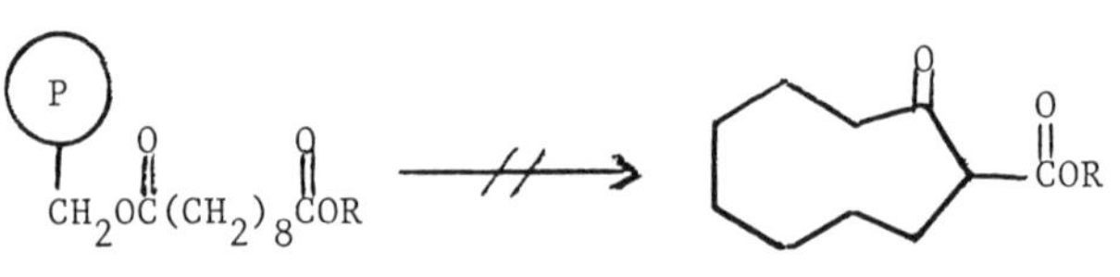

$$ (1) $$

Kinetic investigations of the lifetimes of polymer-bound benzyne (7,8) and of an N-deprotected amino acid active ester (9) put the concept of site isolation into proper perspective. Polymer-bound reactive intermediates have substantially longer lifetimes than the analogous micromolecular species, but they are not completely isolated, and in time react with other polymer-bound species. Several reviews summarize the field as of 1978-82 (10-17).

Scheme 2

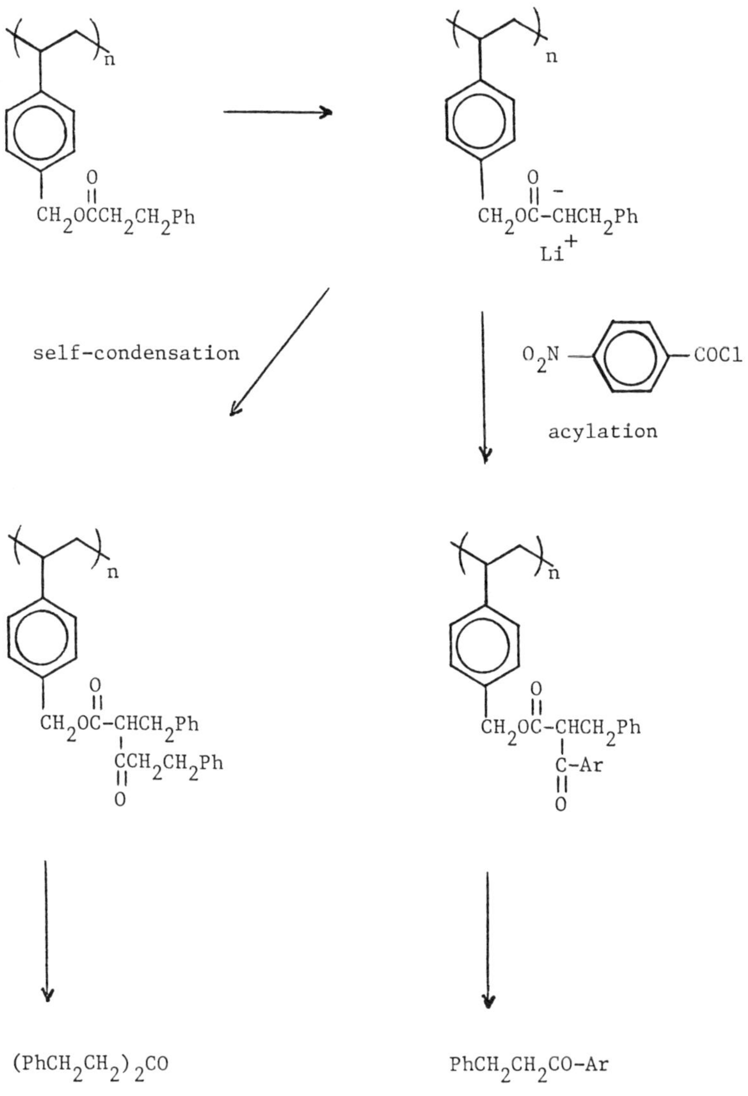

$CH_2OCCH_2CH_2Ph$
$CH_2OC-CHCH_2Ph$
Li^+
self-condensation
O_2N
COCl
acylation
$CH_2OC-CHCH_2Ph$
CCH_2CH_2Ph
C-Ar
$(PhCH_2CH_2)_2CO$
$PhCH_2CH_2CO-Ar$

This review emphasizes more recent results and selects earlier work to illustrate specific points. Most research on polymeric reagents has used 1-2% cross-linked polystyrenes that are quite mobile when solvent-swollen. The mobility of polymer-bound reactive groups depends critically on the distribution of those groups within the network and on the surface, the chemical nature of the network, the degree of cross-linking, and the swelling solvent. Often those factors have not been investigated, even in attempts at site isolation syntheses, yet they can be manipulated to provide either enhanced site isolation or enhanced cooperation between polymer-bound functional groups.

General Considerations of Reactivity within Polymer Networks

Notation. Throughout this chapter the degree of functionalization of polystyrene is reported as DF, the fraction of rings substituted. The % yield of a transformation on a polymer is 100 x DF(product)/DF(reactant). The % cross-linking of a polystyrene is reported as wt % divinylbenzene (DVB) in the monomer mix at the start of copolymerization. Technical DVB typically contains 55% active DVB (*meta* and *para*) and 45% ethylvinylbenzenes. Thus a 2% cross-linked polystyrene also contains 1.6% ethylvinylbenzene. Circled P is used for polystyrene, either all *para* or mixed *meta* and *para* isomers.

DF = degree of functionalization

% cross-linking = wt % divinylbenzene

Polymer Heterogeneity. The microstructures of all common polymer networks are heterogeneous. Copolymerization of styrene (M_1) and divinylbenzene (M_2) incorporates the first double bond of p-divinylbenzene ($r_1 = 0.30$, $r_2 = 1.02$) into the copolymer much faster than styrene, and the first double bond of m-divinylbenzene ($r_1 = 0.62$, $r_2 = 0.54$) at about the same rate as styrene ([18]). The polymer formed early in a high conversion copolymerization is more highly cross-linked with p-divinylbenzene than the polymer formed late. If a polymer-supported reagent is synthesized via copolymerization of a functional monomer with styrene and divinylbenzene, copolymer reactivity may concentrate the functional groups in the more highly cross-linked regions (p-styryldiphenylphosphine, for example), or in the less cross-linked regions, or they may be nearly randomly distributed throughout the network (m- and p-chloromethylstyrene, for example). Even functional groups incorporated randomly by copolymerization are not in identical environments ([19]). Within a single main polymer chain, some are closer to cross-links than others, and some are closer to like functional groups than others. In an atactic vinyl polymer there are many different stereochemical environments.

Microstructural heterogeneity of functional groups can have profound effects on their reactivity (20). Reactions of functional groups in polymer networks with large excesses of external reagents commonly follow pseudo first order kinetics to partial conversion, but slow markedly at higher conversion, and often fail to reach completion. Even the most precise kinetics available for a reaction of a cross-linked polymer, in which acetylation by acetic anhydride of 1 mol % of aminostyrene repeat units in a poly(methyl methacrylate) was determined by fluorescence spectroscopy over seven half-lives, gave deviations from first order kinetics with cross-linked polymers but not with soluble polymer (21). The kinetics of reaction of a polymer with a difunctional reagent are still more complex because the degree of cross-linking increases as the reaction proceeds.

Macroporous and Microporous Polymers. The common 1-2% cross-linked polystyrenes are microporous, meaning they have no porosity in the dry state. Pores are formed as solvent swells the polymer. They are commonly called gels. "Gel" will be used in this chapter even though the term is slightly misleading, because the network of a macroporous polymer containing solvent is also a gel.

During copolymerization of styrene with divinylbenzene in the presence of a solvent, the polymer precipitates as it forms. At high conversion the polymer consists of submicroscopic fused polymer particles and solvent filled pores (22,23). Removal of the solvent leads either to collapse of the network or to permanent pores. Polymers with permanent porosity are called macroporous or macroreticular. The more highly cross-linked the network, and the poorer the solvent used as diluent during polymerization, the more likely the product is to be macroporous.

Macroporous polymers appear heterogeneous when viewed with an electron microscope. Their surface areas and average pore sizes can range from <5 m^2/g and >1 μm to >800 m^2/g and <5 nm. Only in the highest surface area macroporous polymers do as much as 10% of the repeat units lie on the internal surface. The highest surface area polymers are synthesized with the highest concentrations of divinylbenzene. Consequently, their networks are the least penetrable by external reagents, and a significant fraction of their reactive sites are on the surface.

Macroporous supports are potentially advantageous for site isolation syntheses because of high cross-link density and porosity. In principle, the diffusivities of small molecules within the polymer network are the same for microporous and macroporous polymers. Diffusion to the active sites of a macroporous polymer proceeds mainly through the solvent-filled macropores, leaving only a short path through the network, whereas the entire diffusion path lies in the network of a microporous polymer. Slow intraparticle diffusion may limit the functionalization of highly cross-linked supports to the polymer surface. [Since the surface areas of 200/400 mesh and larger microporous particles are <1 m^2/g, the capacities of surface-functionalized microporous polymers are <0.01 mmol/g, too low for synthetic purposes.] If site isolation synthesis is the goal, even a high surface area macroporous polymer may require a low degree of functionalization to limit reactions between the functional groups that are concentrated on the surface.

Distribution of Active Sites. If the chemical reaction that functionalizes the polymer proceeds slower than diffusion of the reagent into the polymer, functional groups will be distributed uniformly throughout the particle. If chemical reaction proceeds faster than diffusion, the polymer bead is functionalized first at its surface, and the reaction proceeds into the bead by a shell diffusive mechanism (24). Partial conversion, shell diffusive reactions give particles functionalized predominantly near the surface. The progress of shell diffusive reactions in microporous polymers can be observed with beads under a microscope.

Distribution of functional groups in polymer-supported reagents and catalysts has been studied with a scanning electron microprobe. Chloromethylation of 2% cross-linked 300-600 μm polystyrene beads with chloromethyl methyl ether and stannic chloride to 0.67 mequiv Cl/g (25) followed by phosphination with lithium diphenylphosphide gave uniform distribution of phosphorus throughout the bead (26). Treatment with a deficient quantity of chlorobis(cyclooctene)rhodium(I) dimer (Scheme 3) after 12 h gave a shell of rhodium near the surface. With excess reagent after 4 days the rhodium was incorporated uniformly. Cyclopentadienyltitanium catalysts bound to 2% and 20% cross-linked macroporous polystyrenes had uniform titanium distribution when initially functionalized by chloromethylation (27). Surface binding of Pd from $(PhCN)_2PdCl_2$ on 2% cross-linked, tertiary amine-substituted polystyrene was found by X-ray photoelectron spectroscopy (28).

Functionalized XE-305 (a 3% cross-linked macroporous polystyrene with most probable pore size 130 nm), XAD-4 (a poly(divinylbenzene) with surface area >700 m^2/g and most probable pore size <5 nm), and PSP-12 (a macroporous poly(divinylbenzene) with most probable pore size <5 nm) also have been studied by scanning electron microprobe (29). Chloromethylation proceeded uniformly throughout all three polymers. Reaction with lithium diphenylphosphide for 18 h in THF at room temperature, and photochemical metalation with phenanthrenechromium tricarbonyl, proceeded uniformly in XE-305, but primarily near the surfaces of XAD-4 and PSP-12.

The reagent diffusion processes in microporous and macroporous polymers differ. Resolution of the scanning electron microprobe is a few μm, insufficient to distinguish pores from polymer in a macroporous bead. Uniform functionalization of microporous beads means that the reagents have diffused from the surface to the center of the bead entirely through the solvent-swollen polymer network. Uniform functionalization of XE-305 means that reagents have diffused throughout the macropores. Surface functionalization of XAD-4 and PSP-12 means than the reagents have failed to diffuse throughout the <5 nm pores during the time of the experiment. Electron microprobe analyses reveal nothing about penetration of reagents into the networks of macroporous polymers.

Scheme 3

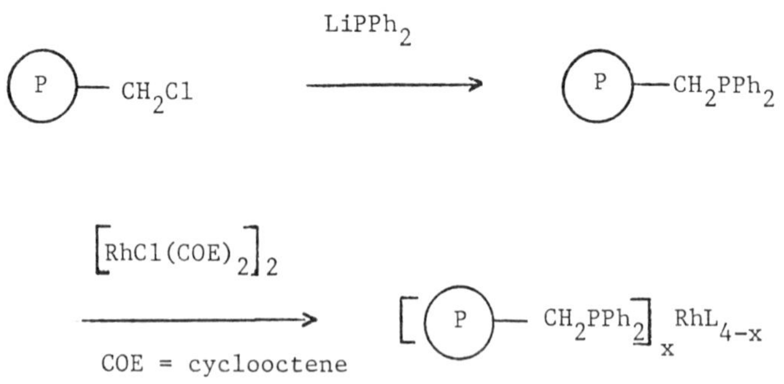

Spectroscopic Determination of Mobility in Polystyrene Networks

The courses of bimolecular reactions in a polymer gel depend upon the mobilities of the polymer-supported functional group and the micromolecular reactants. Functional group mobility is necessary for reaction between two polymer-bound species. In principle, only solute mobility is necessary for reactions between a polymer-bound species and a soluble reagent. Motions of both micromolecular solutes and macromolecular chains have been studied by spectroscopic techniques that are sensitive to rates of translational and rotational diffusion.

The self-diffusion coefficients of toluene in polystyrene gels are approximately the same as in solutions of the same volume fraction polymer, according to pulsed field gradient NMR experiments (30). Toluene in a 10% cross-linked polystyrene swollen to 0.55 volume fraction polymer has a self-diffusion coefficient about 0.08 times that of bulk liquid toluene. Rates of rotational diffusion (molecular Brownian motion) determined from ^{13}C NMR spin-lattice relaxation times of toluene in 2% cross-linked ((polystyryl)methyl)tri-*n*-butylphosphonium ion phase transfer catalysts are reduced by factors of 3 to 20 compared with bulk liquid toluene (31). Rates of rotational diffusion of a soluble nitroxide in polystyrene gels, determined from ESR linewidths, decrease as the degree of swelling of the polymer decreases (32).

Rotational diffusion rates of 2,2,6,6-tetramethyl-4-piperidinol-1-oxyl, bound to 2% cross-linked polystyrene with DF 0.02, are slower than those of the soluble nitroxyl (33). The rates (as the inverse rotational correlation time τ_c^{-1}) increase with increased swelling of the polymer, from 3×10^7 s^{-1} with no solvent or with the nonsolvents ethanol and 2-propanol, to 3×10^9 s^{-1} with benzene, to $>10^{10}$ s^{-1} for a benzene solution of the corresponding soluble polystyrene. Increased cross-linking (4% and 12% DVB) gives decreased swelling and decreased rotational diffusion rates.

The linewidths in single pulse ^{13}C NMR spectra of polystyrene gels decrease as the degree of swelling of the polymer increases (34). ^{13}C NMR spin-lattice relaxation times (T_1) of functional groups increase with distance from the polymer backbone. Increased T_1 and decreased linewidth are due to faster rotational diffusion. The terminus of a long chain (>10 carbon atoms) bound to a $CDCl_3$-swollen, 2% cross-linked polystyrene has a T_1 reduced by less than a factor of two from that of the corresponding carbon of a soluble micromolecule (35). This indicates only a small difference in their rates of rotational diffusion. Increased cross-linking of polystyrene, either by DVB or by intrapolymer Friedel-Crafts alkylation with chloromethyl groups, increases the ^{13}C NMR spectral linewidths (34,36).

These dynamic NMR and ESR results have important implications for the design of polymer-supported site isolation syntheses. The rates of diffusion of polymer-bound species vary more with the degree of swelling and the degree of cross-linking of the polymer than do the rates of diffusion of the micromolecular species in polystyrene gels. More cross-linking and less swelling should lead to a greater decrease in the frequency of encounters between two polymer-bound species than between one polymer-bound species and a soluble reagent, thus favoring reactions of the polymer-bound species with a soluble reagent or with itself, as in a cyclization.

Reactions of Difunctional Reagents with Polymer Networks

Reactions of diamines, diols, dihaloalkanes, and other polyfunctional reagents with polymer networks have been carried out for many purposes. If both groups of the reagent react with the polymer, a new cross-link is formed (Equation 2). If only one group reacts, the difunctional molecule is monosubstituted (Equation 3). Whether one or both groups react depends upon the DF of the polymer, shell diffusive or intrinsic reactivity control of the reaction, the degree of swelling of the polymer, and possible enhanced or reduced reactivity of the second functional group after reaction of the first functional group. Effective single binding of a difunctional reagent (Equation 3) can enable use of the polymer as a protecting group (15,37).

$$\text{(P)}-X \;+\; Y-R-Y \;\longrightarrow\; \text{(P)}-X-Y-R-Y-X-\text{(P)} \qquad (2)$$

$$2\,\text{(P)}-X \;+\; 2\,Y-R-Y \;\longrightarrow\; 2\,\text{(P)}-X-Y-R-Y \qquad (3)$$

Cross-linking of Polymers with Difunctional Reagents. Difunctional reagents have been used to produce cross-linked polymers that are more swellable than the standard networks produced by copolymerization with a difunctional monomer, such as divinylbenzene. Alkylation of soluble polystyrene with α,α'-dichloro-p-xylene gives a cross-linked network (Equation 4) that swells even in nonsolvents and has been called "isoporous" (38). The ease of swelling in nonsolvents is due to the formation of the network in a swollen state (39). The polymer chains in the isoporous network are conformationally relaxed in the swollen state, whereas the polymer chains in a microporous styrene/divinylbenzene network produced by polymerization without solvent are conformationally relaxed in the contracted state. Chloromethylation of polystyrene also produces new cross-links, identified by the conversion of soluble polystyrene to an insoluble polymer and by decreased swelling of chloromethylated polymers compared with the starting polymers (25). Reactions of poly(4-vinyl-pyridine) (40) and of polystyrene (41) with α,ω-dibromoalkanes provide isoporous networks for the preparation of polymer-supported phase transfer catalysts.

Reactions of Chloromethylated Polystyrenes with Difunctional Reagents. Tertiary amine anion exchange resins are prepared by the reaction of highly chloromethylated (DF >0.8) cross-linked polystyrenes with dimethylamine (Equation 5). Small amounts (<10% of functional sites) of quaternary ammonium groups usually are formed in this process. This amount of difunctionalization is lower than in most of the examples in the following paragraphs for three reasons: 1) Dimethylamine diffuses into the polymer more rapidly than larger reagents. 2) The polymers used for ion exchange resins are more highly cross-linked than the 1-2% cross-linked polystyrenes usually employed as supports for organic synthesis. 3) The dimethylamino-methyl group formed from the first alkylation is less mobile than functional groups located farther from the polymer backbone.

Treatment of 2% cross-linked, DF 0.12 and 0.06 chloromethylated polystyrenes with N,N,N',N'-tetramethylhexane-1,6-diamine (1.0 mequiv N/mequiv Cl) in benzene at 70 °C for up to 240 h replaced 75% and 66% of the chlorine in the two polymers (Equation 6) (42). N,N,N',N'-tetramethylethylenediamine gave 66% and 47% replacement of chlorine. Dimethyl-n-butylamine reacted completely under the same conditions. Thus when N/Cl = 1/1, no more than half of the singly-bound amines were able to react further. The extent of the second, cross-linking reaction decreased as the DF of the polymer decreased and as the length of the diamine chain decreased. Similar results have been reported with ethylenediamine, diethylene-triamine, and triethylenetetramine (43). Deliberate syntheses of unsymmetrical diamides from symmetrical 1,ω-diaminoalkanes (C_4 to C_{12}) by monoprotection with 1-2% cross-linked, chloromethylated (DF 0.13) polystyrene gave 80% single binding and 20% double binding of the diamines to the polymer (44,45).

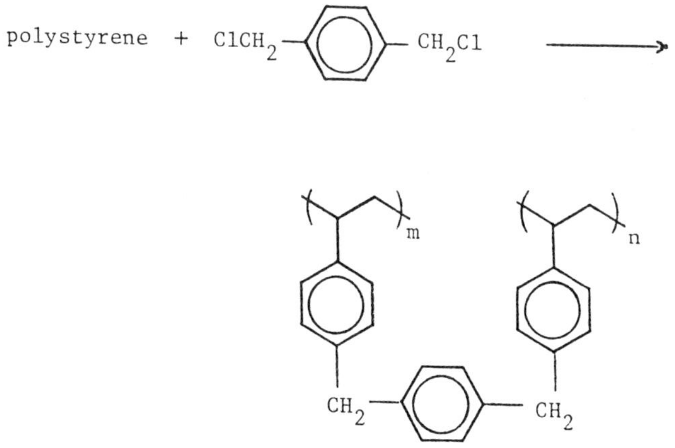
polystyrene + ClCH2 CH2Cl
m
n
CH2
CH2

(4)

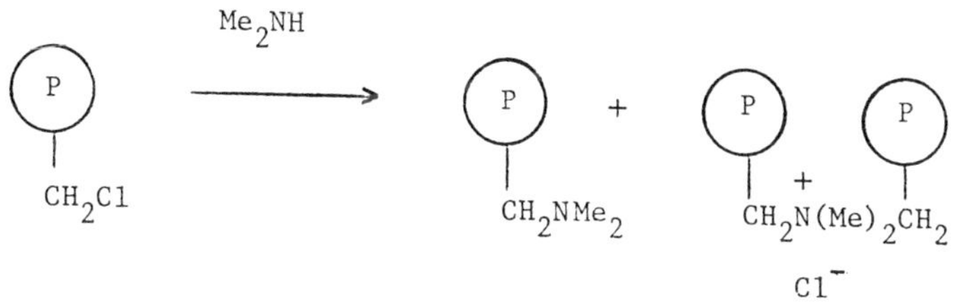
Me2NH
P
CH2Cl
CH2NMe2
+
CH2N(Me)2CH2
Cl−

(5)

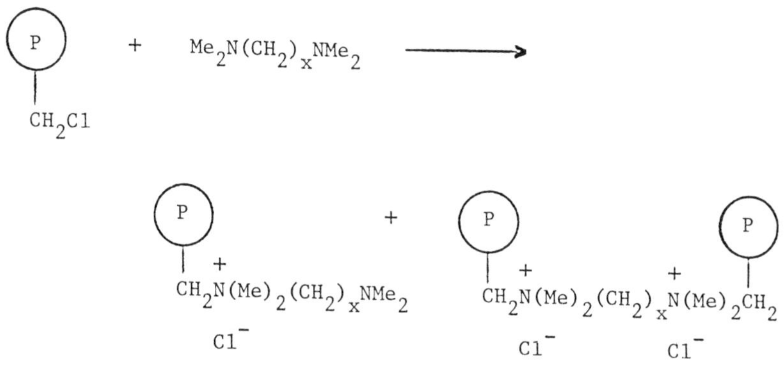
P
CH2Cl
+
Me2N(CH2)xNMe2
CH2N(Me)2(CH2)xNMe2
Cl−
CH2N(Me)2(CH2)xN(Me)2CH2
Cl−
Cl−

(6)

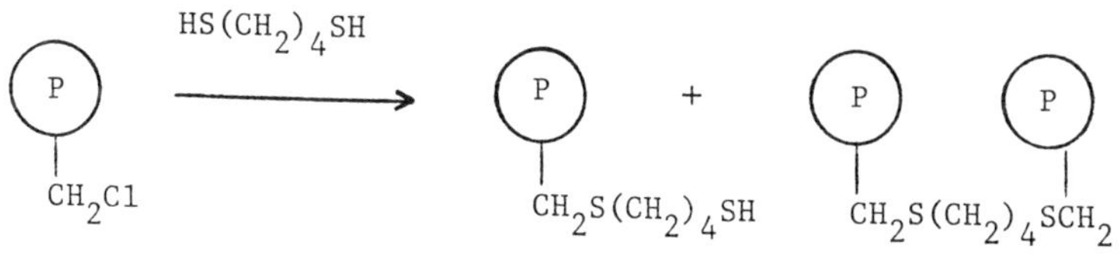
HS(CH2)4SH
P
CH2Cl
CH2S(CH2)4SH
+
CH2S(CH2)4SCH2

(7)

As much as 94%, or as little as 5%, cross-linking (double binding) resulted from treatment of a 1% cross-linked, chloromethylated (DF 0.24) polystyrene with 1,4-butanethiol (Equation 7) (46). Maximum single binding was attained under phase transfer catalysis conditions, during which only the monoanion of the dithiol was present in the aqueous phase, and the dithiol was used in large excess over the chloromethylated polymer. Maximum double binding was attained with excess base and a 3.5/1 molar ratio of dithiol to chloromethyl groups.

Attempts at single binding of diols to polystyrene networks have given widely varied results. With triarylmethyl chlorides (DF <0.14) in 2% cross-linked polystyrene, 1,4-, 1,7-, and 1,10-alkanediols gave about 70% single binding (47). Acylation and acidic cleavage from the polymer gave the diol monoacetates in 60% recovered yield (Scheme 4). Recovered diol from double binding accounted for most of the remaining polymer-bound diol. This monoprotection method has been extended to the synthesis of insect pheromones (48,49). Similar experiments with both a 1% cross-linked polystyrene (DF 0.17-0.45) and a 20% cross-linked macroporous polystyrene (DF 0.07-0.12) gave monobenzoates of symmetrical diols in yields up to 78% based on resin chloride (50). No recovered diol was reported, but cleavages of the monobenzoates from the polymers with HBr or trifluoroacetic acid produced alkyl bromides and alkyl trifluoroacetates as well as alcohols. 1,2,4-Butanetriol gave rearranged benzoates. Polystyrylboronic acid is a more selective reagent for protection of cis-diols and glycosides (51,52). A 2% cross-linked polystyrene containing benzoyl chloride groups (DF 0.12-0.17) gave mainly single binding with dihydroxyaromatic compounds (53). Alkylation and deprotection gave alkoxyphenols.

Reactions of sodium salts of poly(ethylene glycol) with chloromethylated polystyrenes often give substantial double binding. Macroporous 3% cross-linked polystyrene XE-305 (DF 0.67-0.92) gave >90% double binding with the disodium salts of diols from ranging in size from diethylene glycol to poly(ethylene glycol)-600 (average 13-mer) in dioxane, even when a large excess of the dialkoxide was used (54). Greater than 90% of the chloromethyl groups in XE-305 reacted in most cases. More unreacted sites remained when the much more highly cross-linked macroporous resins XAD-1 and XAD-4 were used. Use of the disodium salt of the glycol was designed to produce double binding. Single binding was achieved with a chloromethylated (DF 0.14), 2% cross-linked polystyrene and the monosodium salt of triethylene glycol (55). A DF 0.40 chloromethylated polystyrene with the monosodium salt of tetraethylene glycol gave 28% single binding, and loss of all of the chlorine indicated double binding at the rest of the sites (55). Thus only the low DF chloromethylated polystyrenes give mainly single binding. Presumably the macroporous polystyrenes could also give singly bound poly(ethylene glycol)s if low DF resins and monosodium salts were used in the preparations.

Scheme 4

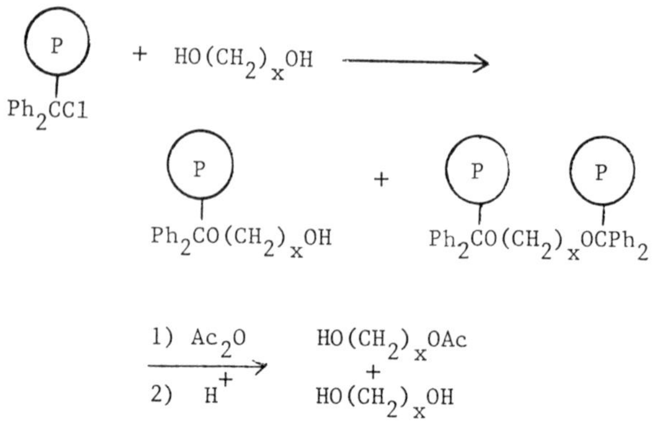

Single binding of other symmetrical difunctional compounds has succeeded when the degree of functionalization of the polymer was low. Examples are mono-protection of symmetric dialdehydes with polymer-bound diols (37) and monoprotection of symmetric dicarboxylic acid chlorides with benzyl alcohol resins (56,57,58). Diol-functionalized resins from a 60/40 glycidyl methacrylate / ethylene dimethacrylate copolymer gave 2/1 double to single binding of terephthalaldehyde (59). Use of a high capacity (DF 0.53) 1% cross-linked polystyrene diol resin to protect terephthalaldehyde gave only 52% single binding, but a diketosteroid was more selectively monoprotected (64-77%) because of unequal reactivity of the two carbonyl groups (Equation 8) (60). The same results were obtained by binding the diketosteroid to the diol (DF 0.09) in a 20% cross-linked macroporous polystyrene. Monomethyl esters of symmetrical aromatic and aliphatic dicarboxylic acids are formed with >97% selectivity by treatment of 50-70 mg of the diacid adsorbed onto 1 g of alumina with diazomethane (61). Larger, less reactive esterification reagents than diazomethane gave much lower selectivity.

Oxidation of 1,7-heptanediol with chlorine and 3% cross-linked polystyrene-bound aryl methyl sulfide (DF 0.07-0.13) gave selective formation of the mono-aldehyde, but only 50% yield (Equation 9) (62).

A 3% cross-linked polystyryl methyl sulfide (DF 0.21) was converted to its lithium derivative. Reaction with 1,4-diiodobutane followed by methyl iodide and sodium iodide gave single and double homologation (Equation 10) (63). No conditions were found for selective single homologation. The experiment most selective for double homologation gave 21% 1,4-, 8% 1,5-, and 71% 1,6-diiodoalkanes.

A 3% cross-linked polystyryl-n-butyltin dihydride (DF 0.17) reduced terephthalaldehyde to an 86:14 mixture of monoalcohol and diol in 91% yield (Equation 11) (64).

In summary, monoprotection of symmetrical difunctional reagents by polymer-bound protecting groups has been accomplished with high selectivity only by use of polystyrenes with low degrees of functionalization. More highly cross-linked macroporous polymers have shown no advantage for single binding at loading levels high enough for synthetic utility. Maximum single binding should be obtained by use of a low DF resin, high swelling with the difunctional reagent, and low temperature. The binding must proceed with intrinsic reactivity control, not by a shell diffusive mechanism.

<u>Binding of Two Polymeric Sites</u>

The objective of the experiments of the preceeding section was monoprotection of a difunctional reagent. Usually simultaneous single and double binding of the difunctional reagent could not be avoided. A few experiments have been performed to test how many doubly bound species can be formed deliberately, and to create shape-selective cavities in the polymer, as described in detail in the chapter by Wulff.

<u>Disulfide Formation in Polystyrene Networks</u>. Polymer-bound thiols were prepared by copolymerizations of bis(p-vinylbenzyl)disulfide with other divinyl monomers followed by diborane reduction (Scheme 5) (65). The initially formed thiols were juxtaposed for reoxidation to disulfides. Polymer-bound thiols were prepared also by copolymerization of p-vinylbenzyl thiolacetate with divinyl monomers followed by hydrolysis (Scheme 6). The latter thiols were distributed randomly throughout the polymer network. The copolymer reactivity ratios for p-vinylbenzyl thiolacetate and styrene are unknown, but should be similar to those of styrene (M_1) and p-vinyl-benzyl chloride (M_2): $r_1 = 0.6$, $r_2 = 1.1$ (66). Copolymerizations with equal volumes of monomers and 1/1 acetonitrile/toluene produced macroporous 40-48% DVB-cross-linked networks (65).

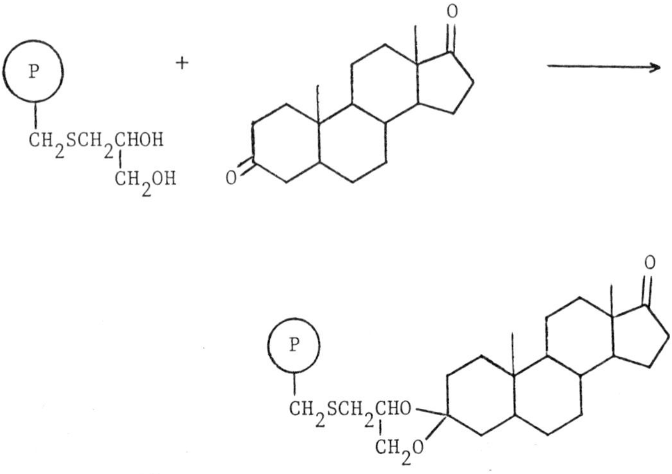

(P)   +   [steroid diketone structure]   ⟶

CH_2SCH_2CHOH
          |
          CH_2OH

(P)   [steroid ketone structure]   (8)

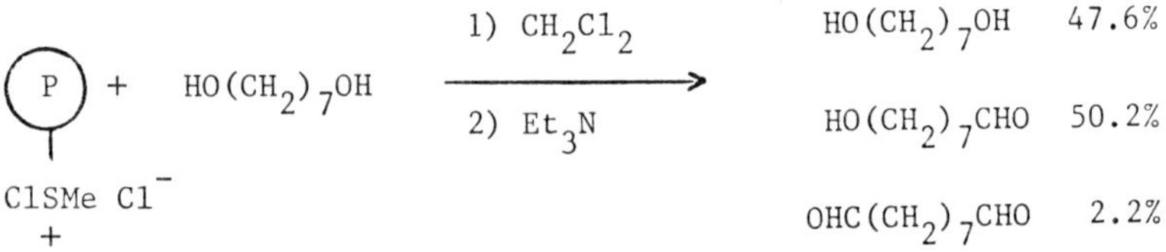

CH_2SCH_2CHO
          |
          CH_2O

(P)   + $HO(CH_2)_7OH$ $\xrightarrow[\text{2) Et}_3\text{N}]{\text{1) CH}_2\text{Cl}_2}$ $HO(CH_2)_7OH$ 47.6%

                                                                              $HO(CH_2)_7CHO$ 50.2%   (9)

$ClSMe\ Cl^-$
   + $OHC(CH_2)_7CHO$ 2.2%

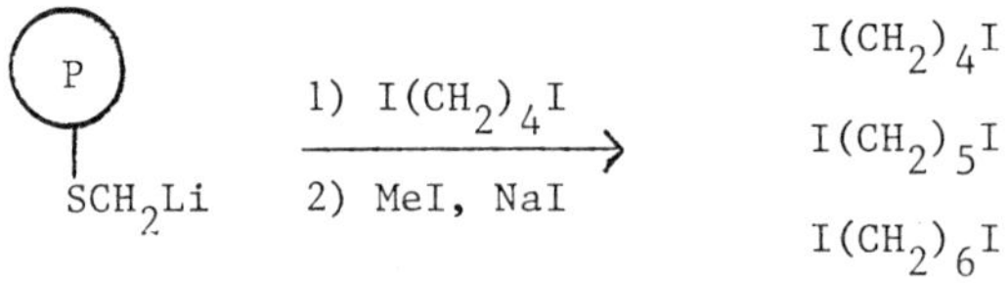

(P) $\xrightarrow[\text{2) MeI, NaI}]{\text{1) I(CH}_2)_4\text{I}}$ $I(CH_2)_4I$

SCH_2Li $I(CH_2)_5I$ (10)

                                                        $I(CH_2)_6I$

(P)   +   [benzene with CHO, CHO]   ⟶   [benzene CHO, CH_2OH]   +   [benzene CH_2OH, CH_2OH]   (11)

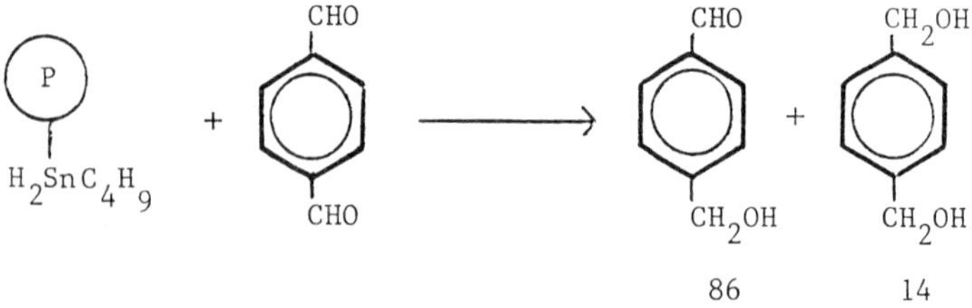

$H_2SnC_4H_9$

                                                        86   14

Table I. Oxidation of Polymer-Bound Thiols to Disulfides [*]

	Polymer[a]					
	A1	A2	B1	B2	C1	C2
wt % disulfide	24.2	--	20.2	--	2.3	--
wt % thiol acetate	--	24.7	--	26.7	--	2.9
wt % DVB	0.0	1.5	43.0	39.5	48.8	48.5
DF[b]	0.18	0.15	0.18	0.20	0.02	0.02
% thiol groups[c]	97	97	68	96	95	96
% oxidation in methanol	99	69	98	33	95	<5
% oxidation in toluene	98	96	36	62	--	--

[a] Remaining monomers were styrene and ethylvinylbenzene.
[b] Fraction of rings substituted with disulfide or thiol acetate groups in first-formed polymer.
[c] % Conversion to thiol by diborane reduction of disulfide or hydrolysis of thiol acetate.
[*] Abridged from ref. 65 with permission from the Israel Journal of Chemistry.

Oxidations of the polymer-bound thiols to disulfides were determined as functions of the degree of cross-linking, the swelling solvent, and the method of incorporation of the thiols into the polymer. Copolymer compositions and the percentages of thiols oxidized to disulfides with I_2 in methanol during 27 h at 20 °C are shown in Table I. The thiols synthesized by copolymerization of the disulfide monomer were reoxidized to disulfides in at least 95% yield, whereas the thiols incorporated by random copolymerization gave low conversion to disulfide, even in the case of a 1.5% DVB-cross-linked polymer. The highly cross-linked polymers maintained the proximity of the thiol groups formed by reduction of disulfides, and prevented disulfide formation from most of the randomly incorporated thiols. Use of better swelling solvents than methanol, such as toluene (Table I), for the oxidation decreased the conversion to disulfide when the thiols were incorporated via the disulfide monomer. Apparently greater swelling allowed polymer chains to reach new conformations in which the thiols were separated from one another. Greater swelling with toluene increased conversion to disulfide when the thiols were incorporated randomly, because greater mobility of polymer chains enabled more of the thiols to "find" each other. These solvent effects on oxidation yields show that a poor solvent for polystyrene, methanol, permits little movement of thiols from their original sites, whereas a good swelling solvent, toluene, promotes mobility of the polymer-bound thiols.

Scheme 5

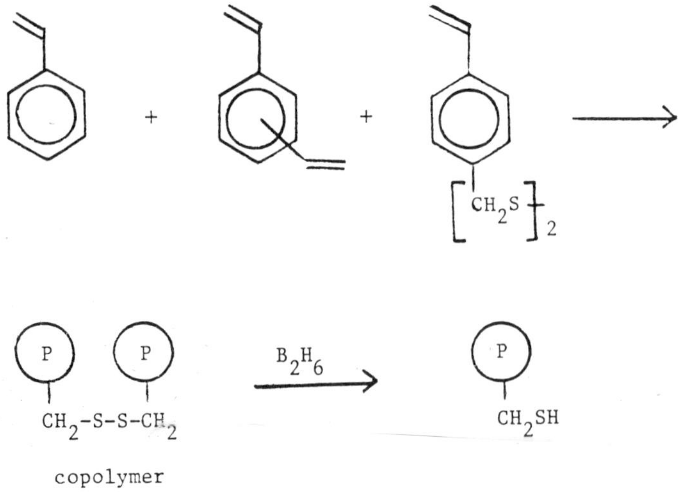

Scheme 6

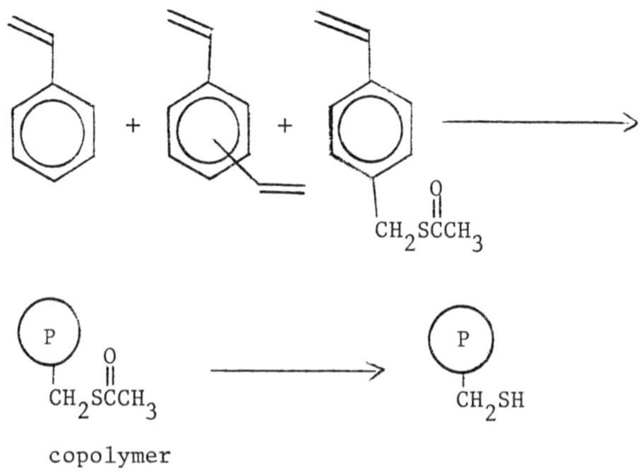

<u>Intrapolymeric Anhydrides</u>. Anhydrides were formed from polystyrene-bound carboxylic acids (DF 0.10) by treatment with dicyclohexylcarbodiimide (DCC) in dichloromethane for 50 h at room temperature (<u>67</u>). Carboxylic acid groups that failed to produce an anhydride gave an N-acylurea (Equation 12), and unreacted carboxylic acid groups were found in some of the more highly cross-linked polymers, indicating that DCC failed to penetrate to some of the sites originally functionalized by bromination, lithiation, and carboxylation. Relative amounts of the various functional groups were analyzed semi-quantitatively by intensities of carbonyl stretching bands in infrared spectra. The intensity of the anhydride band was strongest for 1-2% cross-linked polymers, but N-acylurea and carboxylic acid bands were stronger in the 4% cross-linked polymer. A 20% cross-linked macroporous polymer gave band intensities almost the same as those of the 4% cross-linked polymer.

$$\underset{P}{\bigcirc}\!-\!CO_2H \quad \xrightarrow{DCC} \quad \underset{P}{\bigcirc}\!-\!\overset{O}{\overset{\|}{C}}\!-\!O\!-\!\overset{O}{\overset{\|}{C}}\!-\!\underset{P}{\bigcirc} \;+\; \underset{P}{\bigcirc}\!-\!\overset{O}{\overset{\|}{C}}\!-\!\underset{\underset{C_6H_{11}}{|}}{N}\!-\!\overset{O}{\overset{\|}{C}}NHC_6H_{11} \qquad (12)$$

Isolation and Aggregation of Polymer-Bound Organometallic Groups

Polymer and silica gel supports have been used to try to prevent formation of dimeric organometallic complexes. The catalytic activities of many bound complexes have been tested, and in some cases the major species in the support have been identified spectroscopically. However, the structure observed spectroscopically is not necessarily that of the active catalyst.

<u>Spectroscopic Studies</u>. Iron porphyrins in solution add oxygen and form a μ-oxo dimer (Equation 13). Attempts to prevent formation of the oxygen-bridged dimer failed using a cross-linked polystyrene-supported imidazole (<u>68</u>) but succeeded with silica gel-supported imidazole, 0.65-0.73 mmol/g (<u>69</u>).

$$Fe(II) + O_2 \rightleftharpoons Fe\text{-}O_2 \xrightarrow{Fe(II)} Fe(III)\text{-}O\text{-}Fe(III) \qquad (13)$$

The effects of polymer swelling on coordination of one vs. two polystyryldiphenylphosphines (prepared by lithium diphenylphosphide treatment of brominated polystyrene) to cobalt from $Co(NO)(CO)_3$ (Equation 14) were studied by the intensities of the NO stretching vibrations of the monophosphine (1:1) and bisphosphine (2:1) complexes (<u>70,71</u>). With 2% cross-linked, DF 0.14, phosphine resin and P/Co = 2/1 in benzene at room temperature, the 1:1 complex is formed kinetically (<u>70</u>). Heating to 70 °C in *m*-xylene, a good swelling solvent, for 240 h produced mainly the 2:1 complex. In a poor solvent, hexadecane, only the 1:1 complex was observed. As little as 0.4 g of *m*-xylene per g of polymer led to 70% of the 2:1 complex. Swelling of the corresponding 20% cross-linked macroporous resin (DF 0.08) in excess *m*-xylene gave 65% of the 2:1 complex. In a macroporous copolymer of 4-styryldiphenylphosphine and 58% active DVB, only the 1:1 P:Co complex was observed even in a good solvent. Thus under forcing conditions in a good swelling solvent, even a majority of the phosphine sites in a 20% cross-linked, DF 0.08, macroporous resin are not isolated from one another. On a molecular level, however, the sites in the 20% cross-linked resin may have been clustered in the most accessible regions of the polymer matrix, with large regions of the matrix completely unfunctionalized. The site isolation with >50% DVB could be due to either the incorporation of the phosphine throughout the polymer or the higher degree of cross-linking.

$$\boxed{P}-PPh_2 \;+\; Co(NO)(CO)_3 \longrightarrow$$

$$\boxed{P}-P(Ph)_2Co(NO)(CO)_2 \;+\; \left[\boxed{P}-P(Ph)_2\right]_2 Co(NO)(CO) \qquad (14)$$

ESR spectral analysis of bis(diphenylglyoximato)cobalt(II) coordinated by imidazole showed that a 1:1 complex was formed in a THF-swollen, 10% cross-linked, DF 0.01 polystyrene, and a 2:1 complex was formed in a 15% cross-linked, DF 0.06 polystyrene and in a macroporous, 20% cross-linked, DF 0.20 polystyrene (72). The gel polymers were functionalized randomly by copolymerizations with chloromethylstyrenes, whereas the macroporous polymer was functionalized by chloromethylation, a process that may concentrate functional groups near the internal surface at low DF.

Site-site interactions have been studied with rhodium complexes because of their importance as hydrogenation and hydroformylation catalysts. cis-Complexes were observed in IR spectra of macroporous resins (73). Equation 15 shows one example. Only a dimeric complex was observed in solution. Extended X-ray absorption fine structure (EXAFS) spectra indicate that (polystyrylmethyl)diphenylphosphine forms two different complexes with RhBr(PPh₃)₃, depending on the degree of cross-linking (74). At high P/Rh ratios a dimeric structure was observed with a 2% cross-linked gel resin and 0.22 mmol of Rh/g, and a monomeric structure was observed with a macroporous 20% cross-linked resin and 0.14 mmol of Rh/g.

$$\boxed{P}-CH_2NMe_2 \;+\; Rh_2(CO)_4Cl_2 \longrightarrow \boxed{P}-CH_2NMe_2\underset{Rh}{\overset{Cl\diagdown\;\;\diagup CO}{\diagup\;\;\diagdown CO}} \qquad (15)$$

Polystyrene-supported rhodium catalysts have been studied by IR spectroscopy under hydroformylation conditions (75). Spectra of RhH(CO)(PPh₃)₃, bound to 1.8% cross-linked, DF 0.07 (polystyrylmethyl)diphenylphosphine with a 6/1 P/Rh ratio, were obtained at ambient temperature and 18 atm total pressure of hydrogen and carbon monoxide. At 0.67 CO/H₂ the monomer and dimer in Equation 16 were identified. The relative amounts of the catalytically active monomer and the inactive dimer varied with the CO/H₂ ratio.

$$2\left[\boxed{P}-P(Ph)_2\right]_2 RhH(CO)_2 \underset{H_2}{\overset{CO}{\rightleftharpoons}} \left[\left(\boxed{P}-P(Ph)_2\right)_2 Rh(CO)_2\right]_2 \qquad (16)$$

The rhodium complex $[RhCl(CO)_2]_2$ was bound to polystyrene and to silica gel with phosphine-amine ligands such as that in Equation 17 (76). The macroporous polystyrenes were prepared by copolymerizations of 0.02-0.27 mol fraction chloro-methylstyrenes, styrene, and 5-20% divinylbenzene with heptane diluent, and con-verted to a variety of phosphine-amine resins. At high ligand/metal ratios and DF <0.8, phosphorus, but not nitrogen, bound to rhodium according to IR and X-ray photoelectron spectra, but exact structural assignments could not be made. The least cross-linked (5%), most highly functionalized (DF 0.17) resin gave a *trans*-bis(phos-phine)carbonyl rhodium complex, but the presence of a small amount of a dimeric rhodium complex could not be excluded.

$$
\begin{array}{c}
\text{1) } \underline{n}\text{-BuLi, TMEDA} \\
\text{2) } \boxed{P}\text{—}CH_2Cl
\end{array}
$$

(17)

IR and photoelectron spectroscopy have been used also to study equilibria of complexes prepared from soluble $PdCl_2(pyridine)_2$ or $PdCl_2(benzonitrile)_2$ and either (polystyrylmethyl)diphenylphosphine or polystyryldiphenylphosphine (77). Comp-lexes prepared with 2/1 P/Pd ratios from $PdCl_2(pyridine)_2$ in dichloromethane, using three different 2% cross-linked polystyrenes of DF 0.07, 0.10, and 0.25, still con-tained 0.5-1.0 pyridine per Pd, indicating a large number of inaccessible phosphines. It is not clear why the three polymers differed so much in the fractions of inaccessible phosphine sites. IR spectra of the complexes formed from $PdCl_2(benzonitrile)_2$ indicated the presence of structures of the type $Ar_3P\text{-}(PdCl_2)_n\text{-}PAr_3$ with n > 2. Thus in 2% cross-linked resins in swelling solvents most, but not all, of the phosphine sites were capable of coordination with two phosphines per Pd atom.

In summary, polymer supports have succeeded in breaking dimeric metal complexes into monomeric species in several cases where the polymeric amine or phosphine ligand is present in large excess. Organometallic complexes at high concentrations in solvent-swollen polymers still tend to exist in dimeric form. However, many of the spectral studies have failed to mention that polymer-bound phosphines can be oxidized to phosphine oxides, a facile reaction with adventitious air in a solvent-swollen phosphine polymer (78). Many of the samples may have contained phosphine oxide as well as phosphine.

Organometallic Complexes in Macroporous Polymers. Frequently the catalytic ac-tivity of two or more organometallic complexes in equilibrium is due only to one species. The composition, as we have seen from spectroscopic studies, depends on the degree of functionalization of the polymer, the degree of cross-linking, and the swelling solvent. Reviews of catalysis by polymer-supported organometallic complexes are available (79-84). The focus here is on evidence for site isolation or site cooperation of metal species.

The activities of polystyrene-bound titanocene catalysts, prepared with a macroporous 20% cross-linked support as in Equation 18, were 25-120 times higher for hydrogenation of 1-hexene than those of the corresponding soluble catalysts obtained by reduction of titanocene dichloride or benzyltitanocene dichloride (85). Higher activity of the polymer-bound catalyst was attributed to the presence of a significant concentration of an active monomeric titanium species. Only a dimeric titanium species was found in solution. The catalytic activity as a function of the DF of the polymer reached a maximum at 0.14 mmol of Ti/g (86). In a model of the polymer internal surface the titanocenes were assumed to be randomly distributed over square or circular sites. Titanocenes located in adjacent sites were assumed to be inactive: all activity was attributed to isolated sites. From the fraction of sites occupied at maximum activity, and the 90 m^2/g surface area of the polymer, the area occupied by one titanocene was calculated to be 22 $\overset{\circ}{A}{}^2$. Hydrogenation rate constants calculated from the model as a function of mmol Ti/g fit the experimental data extraordinarily well, probably because both the functionalization of the polymer with cyclopentadiene rings and the hydrogenations were carried out in non-swelling solvents. Even 20% cross-linked polystyrenes swell in good solvents. Use of a swelling solvent during the functionalization would have resulted in some titanocene sites in the interior of the polymer, and the surface model would not have been appropriate. The surface area of the polymer was determined by nitrogen intrusion under completely non-swelling conditions. This model should be applied to the design of future surface-active, site-isolated catalysts. It should be tested also under conditions where the assumption of active sites only on the surface is not likely valid, such as with a catalyst bound to a macroporous 20% cross-linked polymer under swelling conditions, or with a reaction carried out in a swelling solvent.

$$
\begin{array}{ccc}
 & \text{1) MeLi} & \\
 & \xrightarrow{\hspace{2cm}} & \\
 & \text{2) TiCpCl}_3 &
\end{array}
$$

$$
\xrightarrow{\underline{n}\text{-BuLi}} \quad \text{catalyst} \tag{18}
$$

A similar analysis of the surface area available and the area occupied by the polymer bound species was used to design the synthesis of bis(di-n-butylchlorotin)-tetracarbonylosmium shown in Scheme 7 (87). The analogous reaction sequence in solution gives bis(μ-dibutyltin-tetracarbonylosmium), a cyclic dimer. A calculation based on a Poisson distribution of sites on a 100 m^2/g, 20% cross-linked macroporous polystyrene led to the choice of a substitution level of 0.01-0.02 mmol/g, practical only for small scale syntheses.

Scheme 7

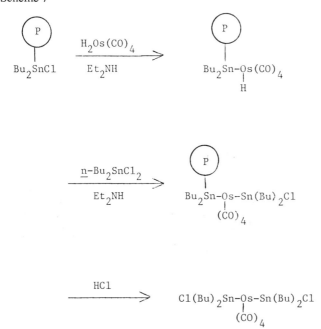

Kinetics of Hydrogenation and Hydroformylation Reactions. Much of the evidence for equilibria between polymer-supported organometallic complexes comes from kinetic experiments. Many studies have employed rhodium catalysts prepared from polymer-bound phosphines and a soluble rhodium complex such as $RhCl(PPh_3)_3$, $RhCl(CO)(PPh_3)_2$, or $[Rh(COD)Cl]_2$ (COD = 1,4-cyclooctadiene). With 2% cross-linked, DF 0.10 polystyryldiphenylphosphine (prepared by bromination, lithiation, and phosphination) both phosphines of $RhCl(CO)(PPh_3)_2$ and of the corresponding iridium complex were replaced by polymer-bound phosphines (70). Treatment of $[Rh(COD)Cl]_2$ with the same polystyryldiphenylphosphine formed a singly bound complex by splitting the chloro bridge (70). Under an atmosphere of CO, Rh[(COD)-Cl]$_2$ split to give the singly bound rhodium phosphine complex, $Rh(CO)_2Cl(ArPPh_2)$ (Ar = polystyryl), on both DF 0.17 and DF 0.83 polystyryldiphenylphosphine resins, and a diphosphine complex, $Rh(CO)Cl(ArPPh_2)_2$, on (polystyrylmethyl)di-phenylphosphine resins (88). $(ArPPh_2)_3RhCl$ and $(ArPPh_2)_xRhH(CO)(PPh_3)_y$ are the most common polymer-supported hydrogenation and hydroformylation catalysts.

The goal of most hydroformylations of 1-alkenes is a large ratio of normal to branched (n:b) aldehydes, because of the greater commercial value of the normal isomers. Polymer-bound complexes prepared by ligand exchange from $(PPh_3)_3$-RhH(CO) gave n:b >10 in hydroformylation of 1-pentene when DF >0.40, 1% cross-linked phosphine resins were employed at high P/Rh ratios (89). The complex with the more phosphine ligands on the left side of Equation 19 is associated with the high n:b ratios. It is favored in a phosphine polymer with a high DF more than in solution because of the higher effective concentration of phosphines in the polymer network ($\underline{80}$). Thus hydroformylation catalyzed by polymer-bound rhodium complexes involves favorable site-cooperation. However, the bis-(phosphine)dicarbonyl complex observed directly by IR spectroscopy under hydroformylation conditions ($\underline{75}$) may not be the active catalyst. Hydroformylations of allyl alcohol and of methyl methacrylate with the same type of catalyst do not give high n:b ratios under the conditions used for 1-pentene. With allyl alcohol only H_2/CO ratios of >10 gave high n:b aldehydes ($\underline{90}$). With methyl methacrylate the opposite selectivity was observed: The n:b ratio decreased as the P/Rh ratio increased ($\underline{91}$).

$$\left[\text{P}—PPh_2 \right]_2 RhH(CO)_2 \quad \rightleftharpoons$$

$$\text{P}—PPh_2RhH(CO)_2 \quad + \quad \text{P}—PPh_2 \qquad (19)$$

Ruthenium complexes analogous to the preceeding rhodium complexes also gave higher n:b aldehyde ratios from 1-pentene when the catalyst had a high P/Rh ratio, in support of the bisphosphine complex in Equation 20 as the species that favors the normal aldehyde ($\underline{92}$).

$$\text{P}—PPh_2RuH_2(\text{alkene})(CO)_2 \quad + \quad \text{P}—PPh_2 \quad \rightleftharpoons$$

$$\left[\text{P}—PPh_2 \right]_2 RuH_2(\text{alkene})(CO) \quad + \quad CO \qquad (20)$$

Phosphine rhodium complexes bound to 10-20% of the sulfonate sites in macroporous ion exchange resins (Equation 21) were more active than the corresponding 2% cross-linked polystyryldiphenylphosphine complex for 1-octene hydrogenation ($\underline{93}$). Activity increased with the surface area of the ion exchange polymers. The n:b aldehyde ratios were 2.0-2.3. Even at P/Rh ratios of >3, the IR spectra of the resins suggested that only the *trans*-chlorodicarbonylrhodium phosphine complex was present. Considering that the ligands lie primarily near the surfaces of highly cross-linked macroporous resins, and that ligand motion is not restricted except by electrostatic attraction to its counterion, it is surprising that diphosphine complexes were not observed. Apparently the macroporous polymer enabled site isolation of the phosphines, and either the monophosphine complex or the ionic environment increased catalytic activity.

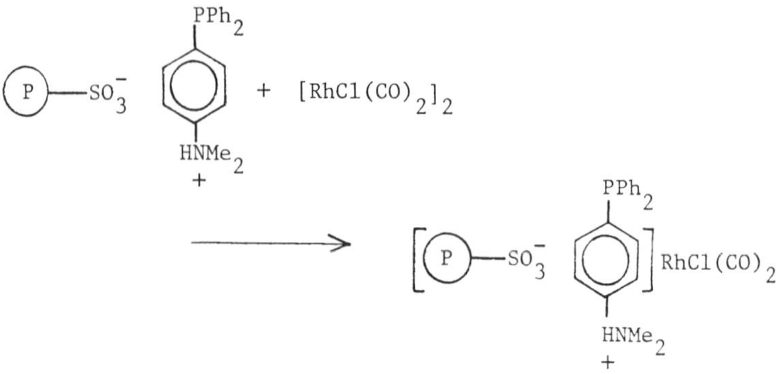

$$\text{P}—SO_3^- \quad \overset{PPh_2}{\bigcirc} \quad + \quad [RhCl(CO)_2]_2$$
$$\underset{\overset{+}{HNMe_2}}{}$$

$$\longrightarrow \quad \left[\text{P}—SO_3^- \; \overset{PPh_2}{\bigcirc} \right] RhCl(CO)_2 \qquad (21)$$
$$\underset{\overset{+}{HNMe_2}}{}$$

Strong evidence for a dinuclear rhodium species as the active catalyst in hydroformylation of styrene was reported for the silica gel-supported species in Equation 22 (94). Relative amounts of diphosphine and tetraphosphine rhodium species were determined from UV/visible spectra. The rates of hydroformylation decreased markedly as % site isolation on the silica surface increased. All of the supported catalysts were less active than the corresponding soluble catalysts, which were not site isolated. The results were interpreted by a dinuclear elimination step in the mechanism of hydroformylation (Equation 23).

$$\boxed{SiO_2} \!\!-\!\! OSi(Me)_2 \!-\!\! \bigcirc \!\!-\!\! OCH_2 \underset{CH_2P(Ph)_2}{\overset{}{CHP(Ph)_2}} \quad \overset{+}{Rh}(norbornadiene) \quad \overset{-}{B}F_4$$

$$\xrightarrow[CO]{H_2} \quad L_n(CO)_m \overset{+}{Rh}(H_2)\overset{-}{B}F_4 \qquad (22)$$

$$L_n M\!-\!H \;+\; L_m \overset{O}{\overset{\|}{M}}CR \quad \longrightarrow \quad L_n M\!-\!ML_m \;+\; H\overset{O}{\overset{\|}{C}}R \qquad (23)$$

The initial goals of binding homogeneous transition metal complexes to insoluble polymers were to facilitate recovery and reuse of the catalyst, and to improve catalytic activity for hydrogenation by forming more coordinatively unsaturated metal centers in the polymer-bound catalyst than in the soluble catalyst. Less complete binding of polymeric phosphines than of soluble phosphines was expected on the basis of polymer rigidity. Some of the early polymer-bound catalysts were discovered to have considerable mobility (70). Nevertheless, higher activities of polymer-bound catalysts than of homogeneous catalysts have been observed. Examples include polystyrene-bound iridium complexes (95,96), polystyrene-bound ruthenium complexes (97), and poly(phenylsiloxane) rhodium phosphine complexes on silica gel (98).

The catalytic activities often are determined at constant phosphine to metal ratio, deliberately kept high on the polymer to minimize leaching of the metal from the support. (The problem of physical and chemical loss of metal complexes from polymer supports is reviewed in the chapter by Garrou.) High phosphine to metal ratios reduce the fraction of coordinatively unsaturated sites and reduce activity. In some cases higher activity of polymer-bound, more coordinatively unsaturated complexes may have been masked by diffusional limitations, discussed in the chapter by Ekerdt.

The dimerization-alkoxylation of butadiene (Equation 24) is catalyzed by soluble $Pd(PPh_3)_4$ and by the analogous 1% cross-linked polystyryldiphenylphosphine catalyst (99). At similar P/Pd ratios the rates per Pd are higher with the polymeric catalysts, and the polymer retards the agglomeration and precipitation of the catalyst observed in solution.

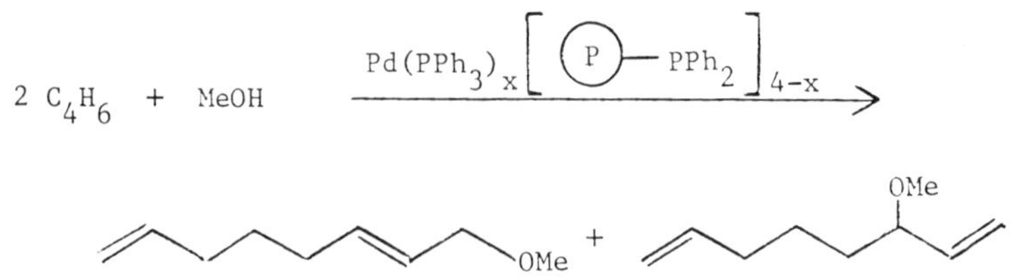

$$2 \ C_4H_6 \ + \ \text{MeOH} \ \xrightarrow{\text{Pd(PPh}_3)_x \left[\textcircled{P} - \text{PPh}_2 \right]_{4-x}}$$

(24)

Supports also have been used to prevent a high degree of aggregation of metal clusters under hydrogenation conditions (100,101). Polymer-bound analogs of $H_4Ru_4(CO)_{12-x}(PPh_3)_x$ (x = 1, 3, or 4) were prepared by ligand exchange in cross-linked polystyrene membranes functionalized with phosphines (102). The membranes were active catalysts for hydrogenation of ethylene with no decrease in activity for thousands of turnovers.

Kinetics of Oxidation Reactions. Numerous transition metal oxidation catalysts have been prepared with the metal bound into a polymer as a porphyrin or a phthalo-cyanine. In solution, metalloporphyrins and metallophthalocyanines commonly aggregate, but aggregation often is prevented in polystyrene or silica gel. The ESR spectra of copper derivatives of polymer-bound tetraamino and tetracarboxytetra-phenylporphyrin show an absence of aggregated structures (103). The corres-ponding cobalt tetraphenylporphyrins (CoTPP) are active catalysts for the basic autoxidation of 1-butanethiol (103). Cobalt phthalocyanine (CoPc) tetrasulfonates and tetracarboxylates bound to anion exchange resins (103,104), to polyacrylamide (105), and to poly(vinylamine) (106) have high activities for autoxidation of thiols to disulfides, apparently because they are not aggregated.

In solution Fe(II)TPP reacts rapidly and irreversibly with oxygen to form a μ-oxo dimer (Equation 13). Fe(II) supported on silica gel does not dimerize (69). On lightly cross-linked polystyrenes dimerization occurs (68), but with 20-30% cross-linked, DF 0.01, macroporous copolymers of styrene and 4-aminostyrene, dioxygen binds reversibly (107,108). Similarly, Fe(III)Pc(COOH)$_4$ bound covalently to linear polystyrene (109) and to linear poly(styrene-co-2-vinylpyridine) (110) is active for catalase-like decomposition of hydrogen peroxide.

Another method of binding metals to polymers is via Schiff base complexes, such as in Scheme 8 (111). On a microporous resin (DF 0.10, % cross-linking not repor-ted) ESR spectra indicated that the Fe(II) complex autoxidized irreversibly and the Co(II) complex formed a dioxygen adduct, but not a dimer. The monomeric, poly-mer-bound Co complex served as a catalyst for autoxidation of 2,6-dimethylphenol to 2,6-benzoquinone and 2,2',6,6'-tetramethyldiphenoquinone (Equation 25).

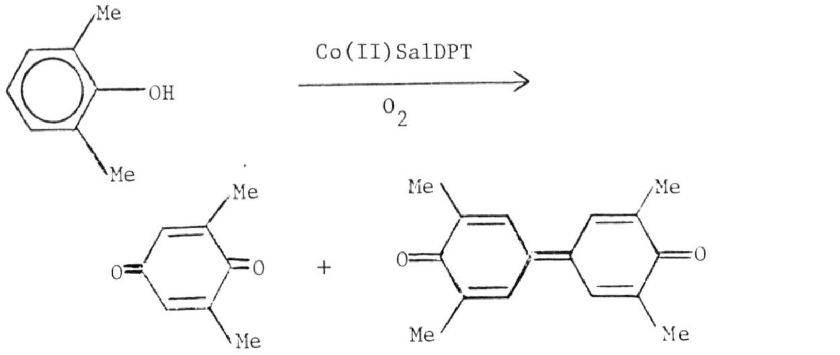

(25)

Scheme 8

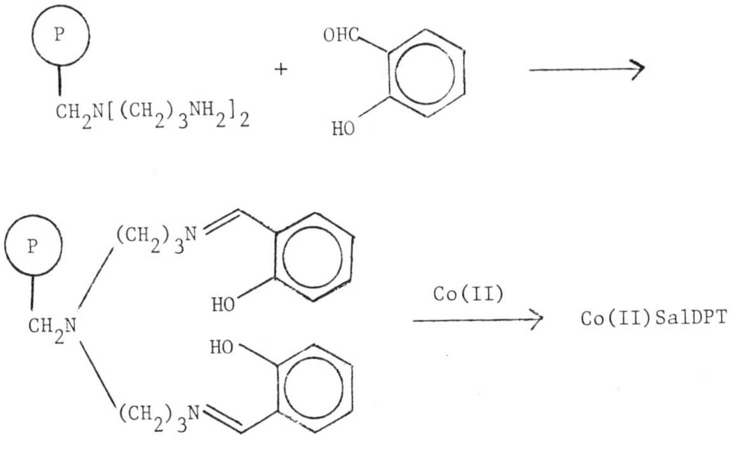

A rhodium sulfide oxidation catalyst on silica gel has highest activity when the rhodium sites are isolated (<u>112</u>). Mercaptopropyl-functionalized silica gel gives varied amounts of the monomeric and dimeric species in Equation 26. The lower the DF, the higher the activity per rhodium atom for the autoxidation of 1-hexene to 2-hexanone with Cu(II) as a cocatalyst.

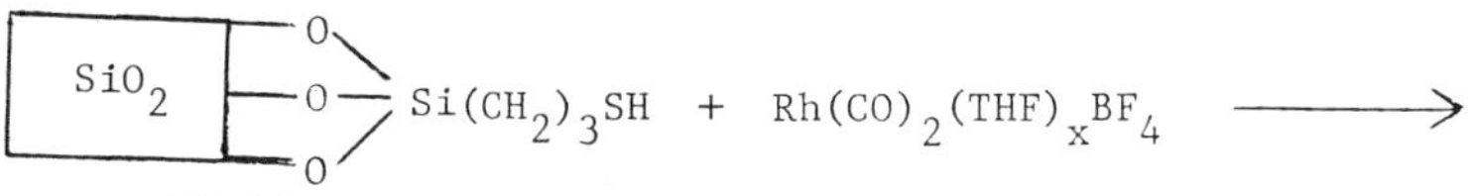

$$(SG-S)_2Rh_2(CO)_4 \quad \text{or} \quad (SG-S)Rh(CO)_2(THF)_{x-1} \qquad (26)$$

Reactions between Two Polymer-bound Species

The rates of reactions between polymer-bound species decrease with increased cross-linking, decreased DF, decreased solvent swelling, and decreased temperature. Seldom have all of these variables been explored for a single reaction.

Amination of a DF 0.13, 1% cross-linked poly(chloromethylstyrene-*co*-styrene) with 0.5 molar equivalent of *n*-butyldimethylamine, followed by exchange of the chloride ion for acetate, gave a resin containing approximately equal amounts of quaternary ammonium acetate groups and chloromethyl groups (<u>113</u>). The kinetics of quaternization with excess amine in toluene-swollen resin were pseudo-first-order and were slow enough to expect that the amine reacted uniformly throughout the bead. Heating caused intraresin displacement of chloride by acetate (Equation 27) with second order kinetics to partial conversion, and then no further reaction was observed. Part of the sites were kinetically isolated. Control experiments demonstrated that the reaction was not due to acetate ion in solution, that the unreacted acetate sites were accessible to benzyl chloride, and that the unreacted chloromethyl sites were accessible to soluble *n*-butyldimethylbenzylammonium acetate. In good swelling solvents (dioxane and toluene), in a nonsolvent (hexane), and in a toluene/water mixture, the % kinetically isolated sites decreased as temperature increased. The % kinetically isolated sites at 60 °C increased in the order dioxane < toluene < hexane < toluene/water, from a minimum of 10% to a maximum of 75%. Apparently the chloromethyl sites are solvated by toluene, the ammonium acetate sites are solvated by water, and the toluene and water in the resin form separate phases. When cross-linking of the resin was increased from 1% to 4% to 10%, the rate of the intraresin reaction in dioxane decreased and the % kinetic isolation increased. The effect of DF was not studied in this intraresin reaction.

$$ \text{(27)} $$

Alcohols yield alkyl chlorides with polystyryldiphenylphosphine in carbon tetrachloride. The reaction occurs by competing mechanisms that are monomolecular (A) and bimolecular (B) in phosphine (Scheme 9). The relative contributions of the two mechanisms have been studied as a function of % cross-linking of the polystyrene (114,115). Gel copolymers of *p*-styryldiphenylphosphine (DF 0.12), styrene, and divinylbenzene with 15, 25, and 37% cross-linking were used for chlorination of 1-octanol (114). With two molar equivalents of the phosphine resins, the yield of 1-chlorooctane decreased from 100% to 0% as the cross-linking increased from 15% to 37%. With a series of 37% cross-linked resins, yields increased from 0% with a gel resin to 47-72% with macroporous resins of increasing porosity. The rates of conversion of alcohols to alkyl chlorides in carbon tetrachloride were up to 20 times faster with 1-2% cross-linked polystyryldiphenylphosphine than with triphenylphosphine in solution (115). The rates could be explained by a mechanism involving site-cooperation of phosphines, or an acceleration of the reaction by the polar phosphonium ion sites in the resin. By the bimolecular mechanism (B) in Scheme 9, a polymer-bound dichloromethylidenetriarylphosphorane should be formed. Reaction of the phosphine resins with benzaldehyde in carbon tetrachloride produced the expected dichloroalkene (Equation 28) in yields that decreased from 63% to 0% as the cross-linking of the resin increased from 1% to 37% (116,117). The phosphines were site isolated in the highly cross-linked polymer.

$$ \text{(28)} $$

Scheme 9

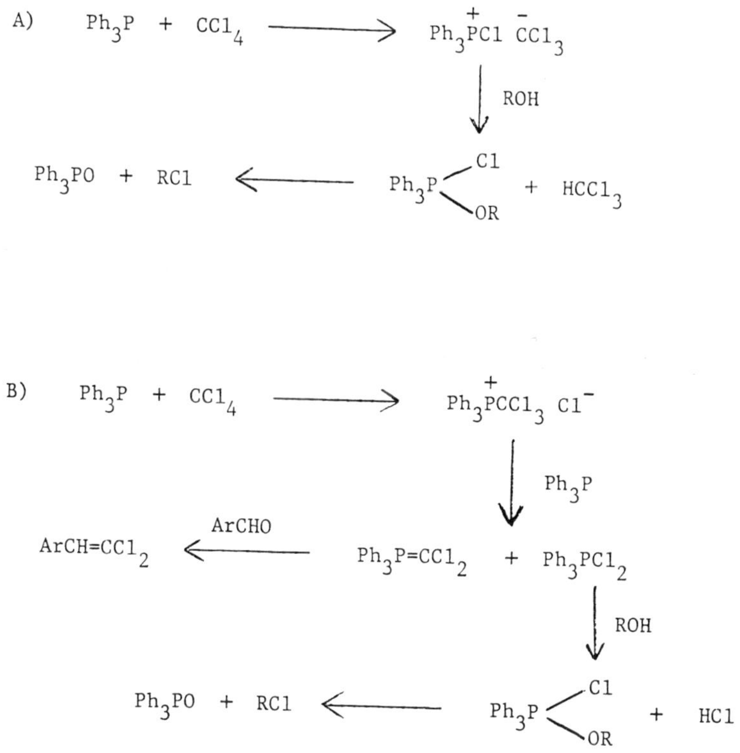

A) Ph_3P + CCl_4 $\longrightarrow$ $Ph_3\overset{+}{P}Cl$ $\overset{-}{C}Cl_3$

$\downarrow$ ROH

Ph_3PO + RCl $\longleftarrow$ $Ph_3P\overset{Cl}{\underset{OR}{<}}$ + $HCCl_3$

B) Ph_3P + CCl_4 $\longrightarrow$ $Ph_3\overset{+}{P}CCl_3$ Cl^-

$\downarrow$ Ph_3P

$ArCH=CCl_2$ $\overset{ArCHO}{\longleftarrow}$ $Ph_3P=CCl_2$ + Ph_3PCl_2

$\downarrow$ ROH

Ph_3PO + RCl $\longleftarrow$ $Ph_3P\overset{Cl}{\underset{OR}{<}}$ + HCl

Scheme 10

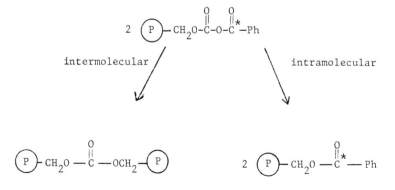

Decomposition of a ^{14}C-labeled mixed anhydride (DF 0.13, 1.0 mmol/g) in a popcorn polystyrene produced polymer-bound carbonate and benzoate esters (Scheme 10) (118). At 25 °C for one year or at 100 °C for 10 h the dry resin gave 20% or 25% intermolecular decomposition. The resin in dioxane at 90 °C for 10 h or at 37 °C for 15 h with added triethylamine gave 84% or 82% intermolecular decomposition. The swelling solvent provided the polymer chain mobility needed for reaction between two polymer-bound functional groups.

Lifetimes of Reactive Intermediates

Reactions in cross-linked polymers are only part of a large family of chemical reactions that have been studied in solid and in solid/fluid heterogeneous environments. Diffusion of small molecules in high viscosity fluids and in crystals is much slower than in low viscosity fluids, making intramolecular reactions more probable. Photolysis of azobisisobutyronitrile (AIBN) in fluid solution, in glassy benzyl benzoate, in frozen mixtures and in pure crystals gives increasing amounts of cage recombination products, up to 91% in crystalline AIBN (119). Adsorption of aryl esters onto silica gel also increases the yields of radical cage products from the photochemical Fries rearrangement (120). Some of the effects of micelles, membranes, and microemulsions (121) on chemical reactivity, on stereochemistry of reactions, and on isotopic enrichment of the products of photolysis of benzyl ketones (122) can be attributed to slower diffusion of reactive intermediates in heterogeneous environments.

Ester Enolates. Acylations and alkylations of polymer-bound ester enolates at ambient temperature were among the earliest examples of altered chemical reactions due to lower mobility of polymer-bound species (2). In the highest yield of the early acylation experiments, a 2% cross-linked, polystyrene-bound, DF 0.025 (0.23 mmol/g) ester was converted to the enolate with triphenylmethyllithium in THF at 0 °C. Immediate addition of p-nitrobenzoyl chloride gave 43% yield of the desired ketone and no detected self-condensation of the polymer-bound ester, as shown in Scheme 2 (123). If no acylating agent was added, 15% of the self-condensation product was isolated from the polymer. Under the same conditions the corresponding soluble benzyl ester produced 67% self-condensation product. It is possible to generate unhindered ester enolates with organolithium bases and trap them in solution, but only at dry ice temperature. These first polymer-bound ester enolate reactions demonstrated the principle of use of reduced mobility of polymer-bound species for synthetic objectives, but the yields were too low to be practical.

Reinvestigation of the ester enolate acylations and alkylations in more highly cross-linked polystyrene supports has provided high yield syntheses (124,125). With 2%, 6%, and 10% cross-linked gel and 4%, 6%, and 20% cross-linked macroporous polystyrene supports, yields of acylation product increased, and the amount of self-condensation product decreased with increased cross-linking of the support. However, 15-44% of unreacted ester remained on the polymer when the reagent was prepared by complete esterification of the chloromethyl groups of the original poly(chloromethylstyrene-co-styrene). After partial esterification of only the more kinetically accessible sites in the polymer, higher yields were obtained in the subsequent acylations, as shown in Table II for a 10% cross-linked gel polystyrene support. Yields of acylation decreased, self-condensation increased, and unreacted ester increased when the polymer was esterified to higher conversion. The residual chloromethyl groups had no effect on the enolate generation and trapping. Using a

DF 0.09, 10% cross-linked gel polystyrene, the enolate generated with triphenyl-methyllithium at room temperature and trapped as soon as the red color of the base faded, gave 94-97% GC yields from acylation with *p*-nitrobenzoyl chloride or acetyl chloride and from alkylation with 1-bromobutane or benzyl bromide. Isolated yields were 73-87% on a 16-34 mmol scale, and 77% in one example on a 100 mmol scale (175 g of dry polymeric reagent). The polymer was recycled with no decrease in acylation yield. The analogous benzyl ester in solution gave 59% self-condensation and 17% acylation.

When the enolate was generated (Scheme 2) in a 20% cross-linked macroporous polystyrene with lithium diisopropylamide at room temperature and allowed to stand for ten minutes before addition of the *p*-nitrobenzoyl chloride, 1% acylation and 79% self-condensation were found. That establishes the lifetime, τ, of the polymer-bound enolate as $1 < \tau \ll 10$ min.

These experiments show that a polymeric reagent can provide <u>kinetically effective site isolation</u> for a high yield synthetic transformation. Moreover, the reaction can be scaled up. Polystyrene, even at a low DF, can be used in large quantities. It is cheaper than many solvents, and if it is highly cross-linked to limit swelling, not too much solvent is required. Because of decreased swelling and increased yield, the reinvestigation (<u>124,125</u>) gave ester acylation with no competing self-condensation at an overall concentration twenty times higher than in the original investigation, in which a 2% cross-linked polystyrene was used (<u>2,123</u>).

<u>Benzyne</u>. The lifetime of a polymer-bound benzyne has been estimated from delayed trapping experiments (<u>8</u>). Oxidation of an aminobenztriazole bound to soluble, 2% cross-linked microporous, and 20% cross-linked macroporous polystyrenes with lead tetraacetate in 1,2-dichloroethane in the presence of tetraphenylcyclopentadienone gave the Diels-Alder adduct in >90% yield (Scheme 11). In the absence of the diene the polymer-bound benzyne is converted to a mixture of aryl acetates, a reaction not observed for soluble monomeric aminobenztriazoles. A soluble analog gave 68% of isomeric benzyne dimers. Use of iodobenzene diacetate as the oxidant gave no aryl acetates, and in time led to polymer-bound benzyne dimers. Delayed trapping experiments, in which the diene was added to the polymer at various times after benzyne generation with iodobenzene diacetate, established the lifetime of benzyne in the 2% cross-linked, DF 0.11 polymer as $0.6 < \tau < 15$ min.

<u>Glycine Active Ester</u>. Delayed trapping experiments with a radio-labeled 2-nitrophenyl ester of glycine, DF 0.12 in a 4% cross-linked macroporous polystyrene, gave a lifetime of a polymer-bound reactive intermediate similar to that of the polymer-bound benzyne (<u>9</u>). Simultaneous addition to the polymer of triethylamine to form the free amine, and acetic anhydride to acylate the amine, followed by benzylamine cleavage of the 2-nitrophenyl ester, gave 80% acetylation (Equation 29). The cyclodimer, diketopiperazine, formed by reaction of the free amine with a second polymer-bound active ester, accounted for 20% of the radioactivity released. Oligomers accounted for the rest of the radioactivity. No appreciable amount of free glycine was released. In separate experiments, delayed addition of acetic anhydride 5 min and 30 min after formation of the free base trapped 20% and <5% of the monomeric active ester left in the polymer. Thus an empirical lifetime of the free amine of 1-2 min was established.

Table II

Effect of Degree of Esterification on Yields of

p-Nitrobenzoylation of a Polymer-Bound Ester

			% yield [a]	
mmol/g	DF	acylation	self-condensation	unreacted
0.41	0.05	96	0	0
0.68	0.09	90	0	6
0.97	0.13	75	10	11
1.12	0.15	64	14	18

[a] By gas chromatographic analysis of products cleaved from resins.

Scheme 11

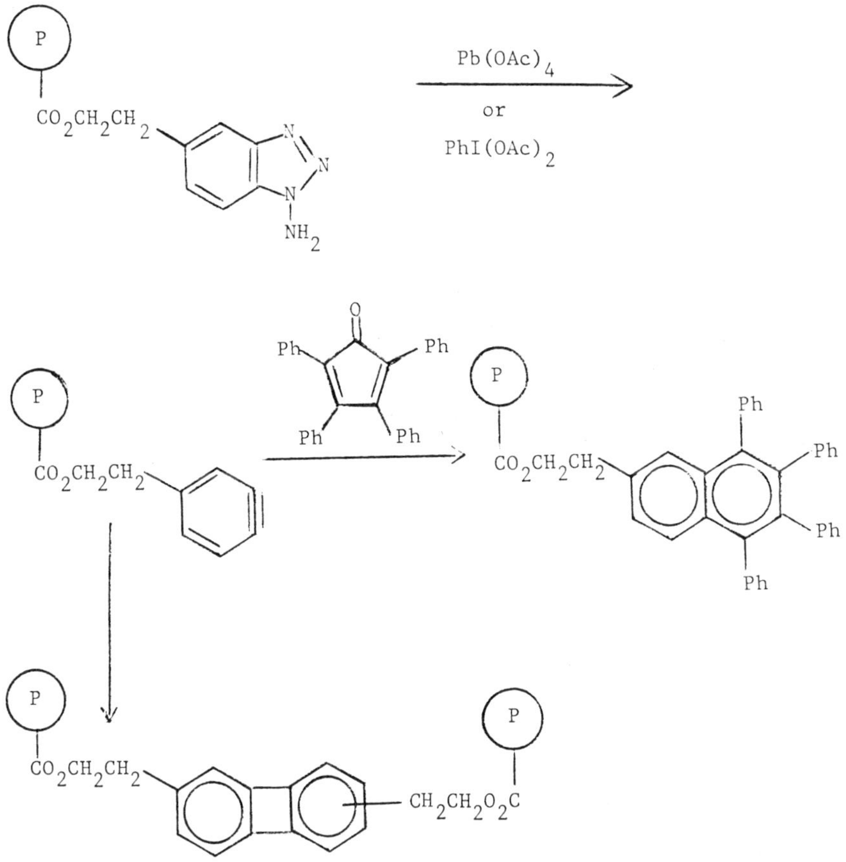

The delayed trapping experiments of an ester enolate in 10% cross-linked polystyrene, a benzyne in 2% cross-linked polystyrene, and a glycine active ester in 4% cross-linked polystyrene all gave lifetimes of about 1-10 minutes. Site isolation syntheses in the most common solvent-swollen polystyrene supports can be achieved if the reactive species is trapped quickly. More highly cross-linked supports and poorly swelling solvents may give longer lifetimes of reactive intermediates, but no data are available.

Synthesis of Medium and Large Ring Compounds

One of the major challenges in synthetic chemistry is the preparation of large ring compounds. A large decrease in entropy is required to join the ends of a long chain molecule. Important large ring compounds include polyethers and amines as complexing agents for metal ions, simple ketones and esters used for their fragrances, and polyfunctional esters as macrolide antibiotics. Cyclic constituents of high polymers have marked effects on their properties. For a thorough analysis of the probability of ring-forming reactions, see (126).

Lactones. Polymer supports have been tested as media for cyclizations of ω-hydroxy-alkanoic acids to lactones (Equation 30). Lactones larger than 7-membered are difficult to prepare because of a large decrease in entropy. There is also an unfavorable enthalpy of cyclization to the strained 8-12 membered rings. The side reactions that complete with unimolecular cyclization are bimolecular formation of cyclic and acyclic dimers and higher oligomers (Equations 30 and 31).

$$\text{HO(CH}_2)_n\text{COX} \xrightarrow{\quad k_{intra} \quad} \lfloor (\text{CH}_2)_n\text{COO} \rfloor \qquad (30)$$

$$2 \text{ HO(CH}_2)_n\text{COX} \xrightarrow{\quad k_{inter} \quad} \text{HO(CH}_2)_n\text{CO}_2(\text{CH}_2)_n\text{COX} \qquad (31)$$

The ratio of rate constants k_{intra}/k_{inter}, in units of moles per liter, has been called the cyclization constant (C) (127,128) or the effective molarity (EM) (129). C is the substrate concentration at which the rates of the intramolecular and intermolecular reactions are equal. Substrate concentrations lower than C are required to achieve high cyclization yields. For acid-catalyzed cyclizations of ω-hydroxyalkanoic acids, C ranges from 8×10^{-6} M for the 8-membered lactone to 6×10^{-3} M for the 17-membered lactone (128). Consequently, lactonizations are usually carried out by the "high dilution" method in which substrate and reagents are added to a large volume of solvent over many hours, so that the substrate concentration is kept extremely low throughout the reaction, and the concentration of product in the final solution is 1-5 mM. High dilution processes are impractical for large scale syntheses. Only an upper limit to C can be calculated from the cyclization yield and the final product concentration of a high dilution experiment. The need for high dilution in principle could be overcome by acceleration of the intramolecular cyclization or by retardation of the intermolecular reaction. Attempted site isolation cyclizations of polymer-supported ω-hydroxyalkanoic acid derivatives are in the latter category. The lower mobility of the polymer-bound species than of the analogous species in solution should retard intermolecular reactions in the polymer. However, improved medium and large ring synthesis via site isolation in polymer supports has seldom been reduced to practice. Cyclizations of ω-hydroxyalkanoic acids to 16-19 membered lactones (Equation 30, n = 14-17) proceed in 77-78% isolated yields by heating 7.8 mM starting material and an equal weight of 3% cross-linked macroporous polystyrene beads in 0.1 M boron trifluoride etherate in 1,2-dichloroethane (130). The yield was 58% in the absence of the polystyrene. Yields of 13- and 15-membered rings by the polystyrene method were 41-46%. GC yields in all cases were 2-6% higher than isolated yields. The mechanism of enhancement of cyclization by the polystyrene is unknown. The 13-membered lactone (Equation 30) has been obtained from 0.15 M 12-hydroxydodecanoic acid, 2% cross-linked polystyryldiphenylphosphine, and diethyl azodicarboxylate in 10% GC yield (131).

Cyclization of the methanesulfonate of 12-hydroxydodecanoic acid in toluene with aqueous sodium bicarbonate and a 1% cross-linked, DF 0.04, (polystyrylmethyl)tri-n-butylphosphonium ion phase transfer catalyst gave 66% yield (132). However, the concentration of substrate in the final reaction mixture was 2.3 mM, no higher than in most high dilution experiments. A more promising method is solid-liquid phase transfer catalysis. A 0.1 M suspension of the potassium salt of 12-bromododecanoic acid in toluene containing 2.5 Mm tetra-n-butylammonium bromide was heated at 90 °C to give 95% of the 13-membered lactone (Equation 32) (133). Effective concentrations within the reactive phase are unknown for these processes, but estimates of the cyclization constants C based on moles of reactant per liter of heterogeneous mixture and lactone yields are 4.5 mM for the triphase reaction (132) and 0.7 M for the biphase reaction (133).

$$K^+ \, {}^-O_2C(CH_2)_{11}Br \xrightarrow[\text{toluene}]{n-Bu_4NBr} \boxed{ O_2C(CH_2)_{11} } \qquad (32)$$

$$\text{solid}$$

A 3% cross-linked, DF 0.6 chlorouracil-substituted polystyrene was reported to give 23-24% yields of 11- and 12-membered lactones from 0.10 M solutions of the ω-hydroxyalkanoic acids in acetonitrile at reflux (134). From experience in our laboratory with a similar cyclization using cyanuric chloride as the reagent, and from inadequate characterization of the products in the original paper, we suspect the products actually were cyclic dimers, not monomeric lactones (135).

A 12-hydroxydodecanoic thiol ester in 2% cross-linked polystyrene cyclizes to the diolide (cyclic dimer) upon treatment with potassium *t*-butoxide in THF or to a mixture of monolide and diolide upon treatment with mercuric trifluoroacetate in dichloromethane (Equation 33) (136). The best yield was 13% of monolide and 19% of diolide with DF 0.03 at an overall concentration of 0.03 M of the thiol ester. For comparison, 0.0096 M of the corresponding benzyl thiol ester in dichloromethane with mercuric trifluoroacetate gave only 2% each of monolide and diolide. The benzyl thiol ester in acetonitrile with mercuric trifluoroacetate gave 26% of monolide and 35% of diolide, but acetonitrile failed to swell the polymer or to effect the cyclization of the polymer-bound thiol ester.

$$\text{(P)}-CH_2\overset{O}{\overset{\|}{S}}C(CH_2)_{11}OH \xrightarrow[\text{Hg}(O_2CCF_3)_2, \, CH_2Cl_2]{KOCMe_3, \, THF \atop \text{or}}$$

$$\boxed{\overset{O}{\overset{\|}{C}}(CH_2)_{11}O} \quad + \quad \overset{O}{\underset{O(CH_2)_{11}\overset{\displaystyle C}{}}{\overset{\displaystyle C(CH_2)_{11}O}{|}}{\diagdown_O} \qquad (33)$$

Macrolides have been synthesized by carbon-carbon bond formation using a polymer-supported analog of tetrakis(triphenylphosphine)palladium as the catalyst (137). The palladium catalyst opens a vinylepoxide, proton transfer gives a bis(phenylsulfonyl) carbanion, and the carbanion cyclizes with the π-allylpalladium complex as shown in Scheme 12. Using a 3% cross-linked macroporous polystyrene, phosphine DF 0.3, and P/Pd = 5, macrolides were formed in 70-87% yields with 0.1-0.5 M substrate and 5 mol % catalyst. This is the highest concentration macrolide synthesis of which the author is aware. It is too bad that the method is not more general. More effort should be devoted to catalytic methods of medium and large ring synthesis. Pd⁰ complexes bound to both polystyrene and silica gel also catalyze stereospecific replacement of allylic cyclohexenyl acetates by secondary amines (138).

Scheme 12

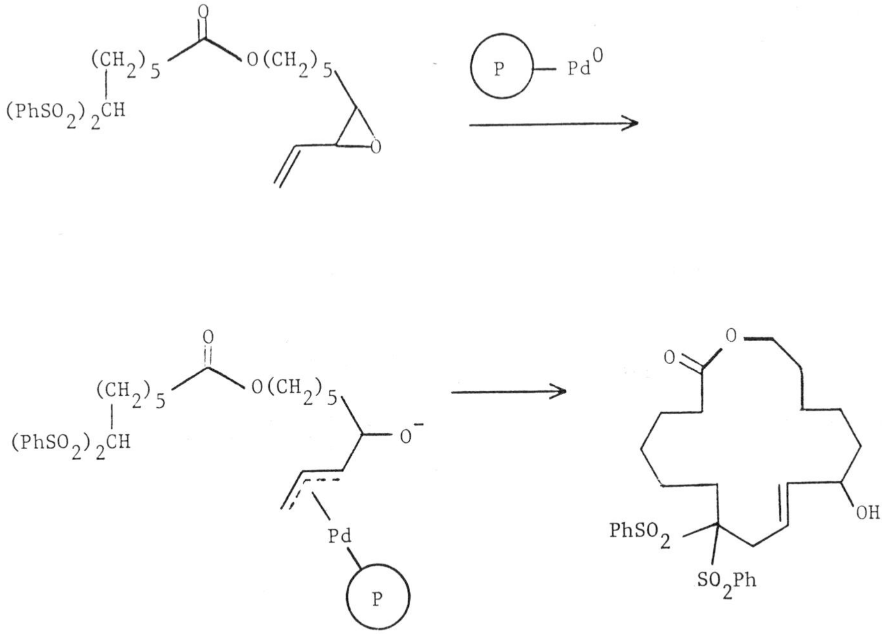

A polystyrene-bound phenylmercuric perchlorate promotes closure of the 8-membered ring of the bicyclic piperazinedione shown in Equation 34 and avoids a chromatographic separation in the workup (139).

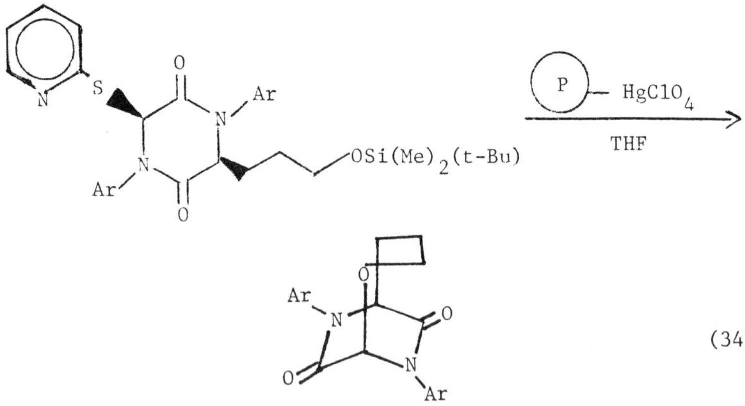

(34)

Cyclic Peptides. Polymer supports aid in the synthesis of cyclic peptides (1,10,13). A recent example is the synthesis of cyclo-(Gly-His)₃ in 42% crude chromatographic yield in a 1% cross-linked, 0.20 mmol/g aminomethyl resin by a bidirectional method (140). The actual yield of the cyclization step (Scheme 13) is unknown, because the overall yield includes the yields of initial attachment, five deprotection-coupling steps, and cleavage from the resin. Most importantly, the cyclic hexapeptide contained no detectable amounts of linear or oligomeric cyclic peptides.

Cyclic peptide oligomers have been formed by multiple intrapolymeric reactions during attempts to synthesize cyclic tripeptides or cyclid tetrapeptides (141). Polymer-bound active esters of BocProProSar, BocSarSarPro, and the trapeptide partial sequence of valinomycin upon deprotection gave oligomers with about 20 to more than 80 ring atoms as shown by gel chromatographic and mass spectrometric analysis. Specific details have not been reported, but large rings are said to be favored by swelling solvents and long reaction times, whereas high cross-linking decreases yields of large rings. Similar results were obtained with polymer-bound active esters of ε-aminocaproic acid (141).

Dieckmann Cyclization. The failure to obtain any 9-membered keto ester from treatment of polystyrene-bound mixed sebacic diesters prompted a critical review (6) in 1976 that set back by years further research on site isolation synthetic methods. However, that review concluded correctly that increased cross-linking, decreased swelling, and lower reaction temperatures should improve site separation. Dieckmann cyclizations to 6-membered keto esters are among the most thoroughly and carefully studied reactions on polymer-supports. Extensive ^{14}C isotopic labeling experiments with benzyl alkyl pimelates and the corresponding (polystyryl)methyl alkyl pimelates established that only reaction of the highly hindered t-butyl and triethylcarbinyl esters with potassium triethylcarbinolate as base avoided extensive scrambling of the labeled carbonyl carbon atoms and competing transesterification reactions during Dieckmann cyclization (142). The example that gave the least scrambling of label is shown in Equation 35. The reactions were carried out by adding the 2% cross-linked, DF 0.021-0.045 resin to a refluxing solution of 4.5 molar equivalents of base in toluene and cooling the reaction mixture after 1.5-5 min. The autocleaved keto ester showed essentially all keto ^{14}C label, while the ^{14}C in the resin-retained keto ester was almost completely scrambled by an intraresin reaction. Less hindered bases and mixed diesters led to scrambling in the autocleaved keto ester also, and to somewhat lower overall yields. The polymer-bound pimelic esters provided improved yields over solution cyclizations, which were attributed to greater ease of product purification, and enabled easy isolation of the product of unidirectional cyclization of mixed diesters.

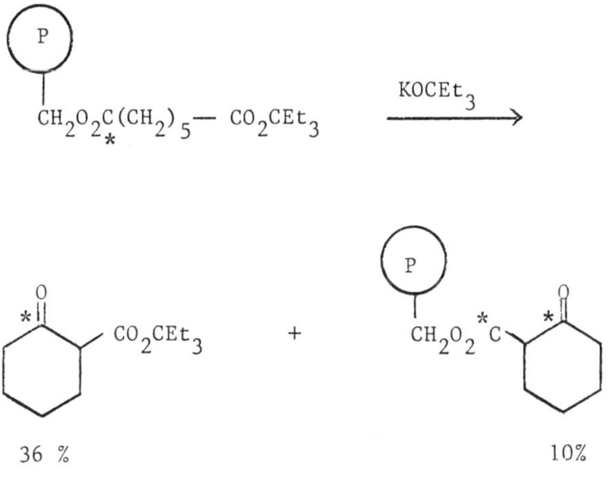

$$CH_2O_2C(CH_2)_5 — CO_2CEt_3 \xrightarrow{\text{KOCEt}_3}$$

(35)

36 %   10%

Scheme 13

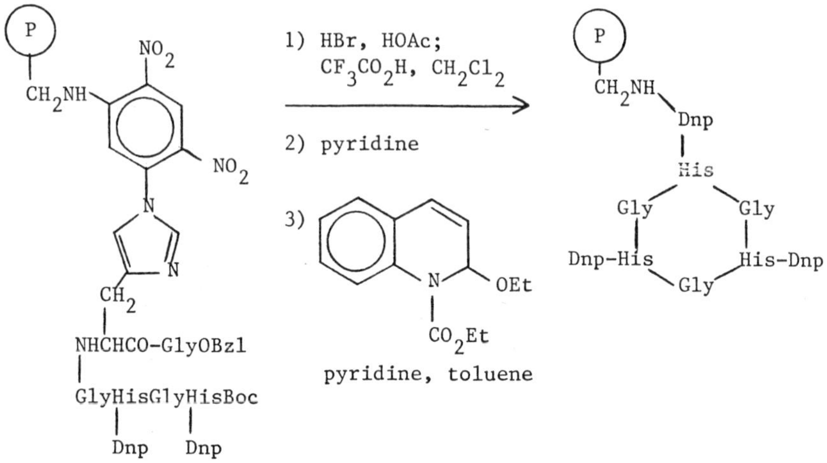

Attempted 9-membered keto ester synthesis produced small amounts (exact yields were not reported) of both autocleaved and resin retained cyclodimers (Equation 36) (4). An analogous 9-cyanononanoic thiol ester resin treated with lithium diethylamide gave 5% of 2-cyanocyclononanone, 10% of cyclodimer, and 40% of the cleaved cyanoamide (Equation 37). A more hindered base, lithium bistriethyldisilazide, gave 19% yield of cyclodimer and 75% cleavage of the uncyclized starting material from the resin. The predominant dimerizations are due to the failure of the resin to separate polymer-bound ester enolates and α-cyanocarbanions from other polymer-bound esters. One must recognize, however, that the Dieckmann cyclization of a sebacic diester to a 9-membered keto ester has never been reported under any conditions. The synthesis of medium rings on polymer supports should be reexamined with more highly cross-linked polymers and by lower temperature methods.

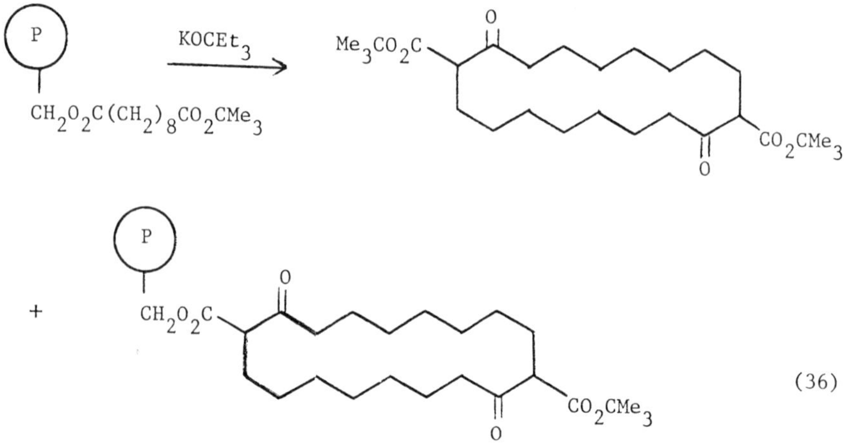

(36)

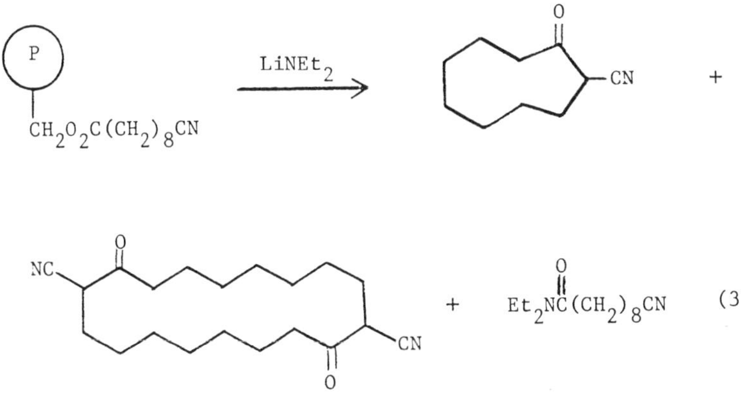

$$P - CH_2O_2C(CH_2)_8CN \xrightarrow{LiNEt_2}$$

reaction (37)

Conclusions

The number of site isolation syntheses on polymer supports so far is small, but sufficient information is available to aid in the design of new methods. Site isolation is favored by high cross-linking and low swelling of the polymer. These conditions also lead to slow transport of soluble reagents to active sites within the polymer network. Highly cross-linked, high surface area macroporous polymers should be advantageous for site isolation reactions. The fraction of isolated sites on a macroporous polymer under non-swelling conditions can be calculated from the surface area and the number of randomly distributed functional groups. Lifetimes of reactive intermediates in microporous polystyrenes indicate that rapid reactions are required for site isolation syntheses. A low degree of functionalization increases site isolation in both gel and macroporous polymers, but the DF required in some cases may be so low that only catalytic, not stoichiometric, application of polymer-bound species is practical. In the future special attention should be paid to catalytic methods and to processes in which the polymeric reagent is regenerable.

Literature Cited

1. Fridkin, M.; Patchornik, A.; Katchalski, E. J. Am. Chem. Soc. 1965, 87, 4646.
2. Patchornik, A.; Kraus, M. A. J. Am. Chem. Soc. 1970, 92, 7587.
3. Crowley, J. I.; Rapoport, H. J. Am. Chem. Soc. 1970, 92, 6363.
4. Crowley, J. I.; Harvey, T. B., III; Rapoport, H. J. Macromol. Sci. Chem. 1973, A7, 1117.
5. Leonard, N. J.; Schimelpfenig, C. W., Jr. J. Org. Chem. 1958, 23, 1708.
6. Crowley, J. I.; Rapoport, H. Acc. Chem. Res. 1976, 9, 135.
7. Jayalekshmy, P.; Mazur, S. J. Am. Chem. Soc. 1976, 98, 6710.
8. Mazur, S.; Jayalekshmy, P.; J. Am. Chem. Soc. 1979, 101, 677.
9. Rebek, J., Jr.; Trend, J. E. J. Am. Chem. Soc. 1979, 101, 737.
10. Kraus, M. A. ; Patchornik, A. Isr. J. Chem. 1978, 17, 298.
11. Warshawsky, A. Isr. J. Chem. 1979, 18, 318.
12. Daly, W. H. Makromol. Chem., Suppl. 1979, 2, 3.
13. Kraus, M. A.; Patchornik, A. J. Polym. Sci., Macromol. Rev. 1980, 15, 55.
14. Mathur, N. K.; Narang, C. K.; Williams, R. E. "Polymers as Aids in Organic Chemistry"; Academic Press: New York, 1980; pp. 138-151.

15. Fréchet, J. M. J. In: "Polymer-supported Reactions in Organic Synthesis"; Hodge, P., Sherrington, D. C., Eds.; Wiley-Interscience; Chichester, 1980; pp. 293-342.
16. Akelah, A. ; Sherrington, D. C. Chem. Rev. 1981, 81, 557.
17. Akelah, A.; Sherrington, D. C. Polymer, 1983, 24, 1369.
18. Frick, C. D.; Rudin, A.; Wiley, R. H. J. Macromol. Sci. Chem. 1981, A16, 1275.
19. Sherrington, D. C. Nouv. J. Chim. 1982, 6, 661.
20. Harwood, H. J. In: "Reactions on Polymers"; Moore, J. A., Ed.; D. Reidel: Boston, 1973; pp. 188-225.
21. Pan, S; Morawetz, H. Macromolecules 1980, 13, 1157.
22. Seidl, J.; Malinsky, J.; Dusek, K.; Heitz, W. Adv. Polym. Sci. 1967, 5, 113.
23. Guyot, A.; Bartholin, M. Progr. Polym. Sci. 1982, 8, 277.
24. Schmuckler, G.; Goldstein, S. In: "Ion Exchange and Solvent Extraction"; Marinsky, J. A.; Marcus, Y., Eds. Marcel Dekker: New York, 1977, Vol. 7, pp. 1-28.
25. Pepper, K. W.; Paisley, H. M.; Young, M. A. J. Chem. Soc. 1953, 4093.
26. Grubbs, R. H.; Sweet, E. M. Macromolecules 1975, 8, 241.
27. Chandrasekaran, E. S.; Grubbs, R. H.; Brubaker, C. H., Jr. J. Organomet. Chem. 1976, 120, 49.
28. Terasawa, M.; Sano, K.; Kaneda, K.; Imanaka, T.; Teranishi, S. J. Chem. Soc., Chem. Commun.1978, 650.
29. Tatarsky, D.; Kohn, D. H.; Cais, M. J. Polym. Sci., Polym. Chem. Ed. 1980, 18, 1387.
30. Pickup, S.; Blum, F., Polym. Mat. Sci. Eng. 1985, 53, 108.
31. Ford, W. T.; Periyasamy, M.; Spivey, H. O. Macromolecules 1984, 17, 2881.
32. Regen, S. L. Macromolecules 1975, 8, 689.
33. Regen, S. L. J. Am. Chem. Soc. 1974, 96, 5275.
34. Ford, W. T.; Balakrishnan, T. Macromolecules 1981, 14, 284.
35. Ford, W. T.; Mohanraj, S.; Periyasamy, M. Brit. Polym. J. 1984, 16, 179.
36. Mohanraj, S.; Ford, W. T. Macromolecules 1985, 18, 351.
37. Leznoff, C. C. Acc. Chem. Res. 1978, 11, 327.
38. Davankov, V. A.; Tsyurupa, M. P. Angew. Makromol. Chem.1980, 91, 127.
39. Flory, P. J. "Principles of Polymer Chemistry", Cornell Univ. Press: Ithaca, N. Y., 1953, p. 579.
40. Noguchi, H.; Sugawara, M.; Uchida, Y. Polymer 1980, 21, 861.
41. Tundo, P. Synthesis 1978, 315.
42. Regen, S. L.; Lee, D. P. J. Am. Chem. Soc. 1974, 96, 294.
43. Shambhu, M. B.; Theodorakis, M. C.; Digenis, G. A. J. Polym. Sci., Polym. Chem. Ed. 1977, 15, 525
44. Dixit, D. M.; Leznoff, C. C. J. Chem. Soc., Chem. Commun. 1977, 798.
45. Dixit, D. M.; Leznoff, C. C. Isr. J. Chem. 1978, 17, 248.
46. Farrall, M. J.; Fréchet, J. M. J. J. Am. Chem. Soc. 1978, 100, 7998.
47. Fyles, T. M.; Leznoff, C. C. Can. J. Chem. 1976, 54, 935.
48. Fyles, T. M.; Leznoff, C. C.; Weatherston, J. Can. J. Chem. 1977, 55, 4135.
49. Svirskaya, P. I.; Leznoff, C. C. J. Chem. Ecology 1984, 10, 321.
50. Fréchet, J. M. J.; Nuyens, L. J. Can. J. Chem. 1976, 54, 926.
51. Fréchet, J. M. J.; Seymour, E. Isr. J. Chem. 1978, 17, 253.
52. Fréchet, J. M. J.; Nuyens, L. J.; Seymour, E. J. Am. Chem. Soc. 1979, 101. 432.
53. Leznoff, C. C.; Dixit, D. M. Can. J. Chem. 1977, 55, 3351.

54. Warshawsky, A.; Kalir, R.; Deshe, A.; Berkovitz, H.; Patchornik, A. J. Am. Chem. Soc. 1979, *101*, 4249.
55. Heffernan, J. G.; MacKenzie, W. M.; Sherrington, D. C. J. Chem. Soc., Perkin II, 1981, 514.
56. Leznoff, C. C.; Goldwasser, J. M. Tetrahedron Lett. 1977, 1875.
57. Goldwasser, J. M.; Leznoff, C. C. Can. J. Chem. 1978, *56*, 1562.
58. Leznoff, C. C.; Yedidia, V. Can. J. Chem. 1980, *58*, 287.
59. Fréchet, J. M. J.; Bald, E.; Svec, F. Reactive Polym. 1982, *1*, 21.
60. Hodge, P.; Waterhouse, J. J. Chem. Soc., Perkin Trans. I 1983, 2319.
61. Ogawa, H.; Chihara, T.; Taya, K. J. Am. Chem. Soc.1985, *107*, 1365.
62. Crosby, G. A.; Weinshenker, N. M.; Uh, H. S. J. Am. Chem. Soc. 1975, *97*, 2232.
63. Crosby, G. A.; Kato, M. J. Am. Chem. Soc. 1977, *99*, 278.
64. Weinshenker, N. M.; Crosby, G. A.; Wong, J. Y. J. Org. Chem. 1975, *40*, 1966.
65. Wulff, G.; Schultze, I. Isr. J. Chem. 1978, *17*, 291.
66. Braun, D.; Czerwinski, W.; Disselhoff, G.; Tüdös, F.; Kelen, T.; Turcsányi, B. Angew. Makromol. Chem. 1984, *125*, 161.
67. Scott, L. T.; Rebek, J.; Ovsyanko, L.; Sims, C. L. J. Am. Chem. Soc. 1977, *99*, 625.
68. Collman, J. P.; Reed, C. A. J. Am. Chem. Soc. 1973, *95*, 2048.
69. Leal, O.; Anderson, D. L.; Bowman, R. G.; Basolo, F.; Burwell, R. L., Jr., J. Am. Chem. Soc. 1975, *97*, 5125.
70. Collman, J. P.; Hegedus, L. S.; Cooke, M. P.; Norton, J. R.; Dolcetti, G.; Marquardt, D. N. J. Am. Chem. Soc. 1972, *94*, 1789.
71. Regen, S. L.; Lee, D. P. Macromolecules 1977, *10*, 1418.
72. Drago, R. S.; Gaul, J. H. Inorg. Chem. 1979, *18*, 2019.
73. Rollman, L. D. Inorg. Chim. Acta 1972, *6*, 137.
74. Reed, J.; Eisenberger, P.; Teo, B. K.; Kincaid, B. M. J. Am. Chem. Soc. 1978, *100*, 2375.
75. Park, S. C.; Ekerdt, J. G. J. Mol. Catal. 1984, *24*, 33.
76. Michalska, Z. M. J. Mol. Catal. 1983, *19*, 345.
77. Andersson, C.; Larsson, R. J. Catal. 1983, *81*, 179.
78. Grubbs, R. H.; Su, S. C. H. In "Enzymic and Non-Enzymic Catalysis"; Dunnill, P., Wiseman, A., Blakebrough, N., Eds.; Wiley/Halstead: Chichester, 1980, p. 223.
79. Chauvin, Y.; Commereuc, D.; Dawans, F. Progr. Polym. Sci. 1977, 5, 95.
80. Pittman, C. V., Jr. In: "Polymer-supported Reactions in Organic Synthesis"; Hodge, P.; Sherrington, D. C., Eds.; Wiley-Interscience: Chichester, 1980, pp. 267-273.
81. Bailey, D. C.; Langer, S. H. Chem. Rev. 1981, *81*, 109.
82. Whitehurst, D. D. CHEMTECH 1980, *10*, 44.
83. Ciardelli, F.; Braca, G.; Carlini, C.; Sbrana, G.; Valentini, G. J. Mol. Catal. 1982, *14*, 1.
84. Holy, N. L. In: "Homogeneous Catalysis by Metal Phosphine Complexes"; Pignolet, L. H., Ed.; Plenum Pub. Corp., New York, 1983, pp. 443-484.
85. Bonds, W. D., Jr.; Brubaker, C. H., Jr.; Chandrasekaran, E. S.; Gibbons, C.; Grubbs, R. H.; Kroll, L. C. J. Am. Chem. Soc. 1975, *97*, 2128.
86. Grubbs, R.; Lau, C. P.; Cukier, R.; Brubaker, C., Jr. J. Am. Chem. Soc. 1977, *99*, 4517.
87. Burlitch, J. M.; Winterton, R. C. J. Am. Chem. Soc. 1975, *97*, 5605.
88. Sanger, A. R.; Schallig, L. R. J. Mol. Catal. 1977, *3*, 101.
89. Pittman, C. V., Jr.; Hanes, R. M. J. Am. Chem. Soc. 1976, *98*, 5402.
90. Pittman, C. U., Jr.; Honnick, W. D. J. Org. Chem. 1980, *45*, 2132.

91. Pittman, C. U., Jr.; Honnick, W. D.; Yang, J. J. J. Org. Chem. 1980, 45, 684.
92. Pittman, C. U., Jr.; Wilemon, G. M. J. Org. Chem. 1981, 46, 1901.
93. Ford, M. E.; Premecz, J. E. J. Mol. Catal. 1983, 19, 99.
94. Collman, J. P.; Bellmont, J. A.; Brauman, J. I. J. Am. Chem. Soc. 1983, 105, 7288.
95. Pittman, C. U., Jr.; Jacobson, S. E.; Hiramoto, H. J. Am. Chem. Soc. 1975, 97, 4774.
96. Jacobson, S.; Clements, W.; Hiramoto, H.; Pittman, C. U., Jr. J. Mol. Catal. 1975, 1, 73.
97. Pittman, C. U., Jr.; Wilemon, G. Ann. N. Y. Acad. Sci. 1980, 67.
98. Conan, J.; Bartholin, M.; Guyot, A. J. Mol. Catal. 1975, 1, 375.
99. Pittman, C. U., Jr.; Ng, Q. J. Organomet. Chem. 1978, 153, 85.
100. Gates, B. C.; Lieto, J. CHEMTECH 1980, 10, 195.
101. Gates, B. C.; Lieto, J. CHEMTECH 1980, 10, 248.
102. Otero-Schipper, Z.; Lieto, J.; Gates, B. C. J. Catal. 1981, 63 , 175.
103. Rollmann, L. D. J. Am. Chem. Soc. 1975, 97, 2132.
104. Skorobogaty, A.; Smith, T. D. J. Mol. Catal. 1982, 16, 131.
105. Maas, T. A. M. M.; Kuijer, M.; Zwart, J. J. Chem. Soc., Chem. Commun. 1976, 86.
106. Brouwer, W. M.; Traa, P. A. M.; de Weerd, T. J. W.; Piet, P.; German, A. L. Angew. Makromol. Chem. 1984, 128, 133.
107. Ledon, H.; Brigandat, Y. J. Organomet. Chem. 1979, 165, C25.
108. Ledon, H.; Brigandat, Y. J. Organomet. Chem. 1980, 190, C87.
109. Shirai, H.; Maruyama, A.; Kobayashi, K.; Hojo, N.; Urushido, K. J. Polym. Sci., Polym. Lett. Ed. 1979, 17, 661.
110. Shirai, H.; Higaki, S.; Hanabusa, K.; Hojo, N. J. Polym. Sci., Polym. Lett. Ed. 1983, 21, 157.
111. Drago, R. S.; Gaul, J.; Zombeck, A.; Straub, D. K. J. Am. Chem. Soc. 1980, 102, 1033.
112. Nyberg, E. D.; Drago, R. S. J. Am. Chem. Soc. 1981, 103, 4966.
113. Kim, B.; Kirszensztejn, P.; Bolikal, D.; Regen, S. L. J. Am. Chem. Soc. 1983, 105, 1567.
114. Sherrington, D. C.; Craig, D. J.; Dagleish, J.; Domin, G.; Taylor, J.; Meehan, G. V. Eur. Polym. J. 1977, 13, 73.
115. Hodge, P.; Hunt, B. J.; Khoshdel, E.; Waterhouse, J. Nouv. J. Chim. 1982, 6, 617.
116. Hodge, P.; Richardson, G. J. Chem. Soc., Chem. Commun. 1975, 622.
117. McKenzie, M. W.; Sherrington, D. C. J. Polym. Sci., Polym. Chem. Ed. 1982, 20, 431.
118. Martin, G. E.; Shambhu, M. B.; Shakhshir, S. R.; Digenis, G. A. J. Org. Chem. 1978, 43, 4571.
119. Jaffe, A. B.; Skinner, K. J.; McBride, J. M. J. Am. Chem. Soc. 1972, 94, 8510.
120. Abdel-Malik, M. M.; de Mayo, P. Can. J. Chem. 1984, 62, 1275.
121. Fendler, J. H. "Membrane Mimetic Chemistry"; Wiley-Interscience, New York, 1982; Chapter 11.
122. Turro, N. J.; Gratzel, M.; Braun, A. M. Angew. Chem., Int. Ed. Engl. 1980 19, 675.
123. Kraus, M. A.; Patchornik, A. J. Polym. Sci. Polym. Symp. 1974, 47, 11.
124. Chang, Y. H.; Ford, W. T. J. Org. Chem. 1981, 46, 3756.
125. Chang, Y. H.; Ford, W. T. J. Org. Chem. 1981, 46, 5364.

126. Winnik, M. A. Chem. Rev. 1981, *81*, 491.
127. Stoll, M.; Rouve, A.; Stoll-Comte, G. Helv. Chim. Acta 1934, *17*, 1289.
128. Stoll, M.; Rouve, A.; Helv. Chim. Acta 1935, *18*, 1087.
129. Illuminati, G.; Mandolini, L. Acc. Chem. Res. 1981, *14*, 95.
130. Scott, L. T.; Naples, J. O. Synthesis 1976, 738.
131. Amos, R. A.; Emblidge, R. W.; Havens, N. J. Org. Chem. 1983, *48*, 3598.
132. Regen, S. L.; Kimura, Y. J. Am. Chem. Soc. 1982, *104*, 2065.
133. Kimura, Y.; Regen, S. L. J. Org. Chem. 1983, *48*, 1533.
134. Kondo, K.; Murakami, M.; Takemoto, K. Makromol. Chem. 1983, *184*, 497.
135. Mohanraj, S.; Ford, W. T.; unpublished results.
136. Mohanraj, S.; Ford, W. T. J. Org. Chem. 1985, *50*, 1616.
137. Trost, B. M.; Warner, R. W. J. Am. Chem. Soc. 1983, *105*, 5940.
138. Trost, B. M.; Keinan, E. J. Am. Chem. Soc. 1978, *100*, 7779.
139. Dung, J. S.; Armstrong, R. W.; Williams, R. M. J. Org. Chem. 1984, *49*, 3416.
140. Isied, S. S.; Kuehn, G. G.; Lyon, J. M.; Merrified, R. B. J. Am. Chem. Soc. 1982, *104*, 2632.
141. Rothe, M.; Löhmuller, M.; Taiber, W.; Fischer, W.; Hornung, K., reported at the 20th IUPAC International Symposium on Macromolecules, The Hague, The Netherlands, 1985.
142. Crowley, J. I.; Rapoport, H. J. Org. Chem. 1980, *45*, 3215.

RECEIVED January 13, 1986

Author Index

Subject Index

Production and indexing by Keith B. Belton
Jacket design by Pamela Lewis

Elements typeset by Hot Type Ltd., Washington, DC
Printed and bound by Maple Press Co., York, PA

RECENT ACS BOOKS

"Organic Marine Geochemistry"
Edited by Mary L. Sohn
ACS Symposium Series 305; 440 pp; ISBN 0-8412-0965-0

"Fungicide Chemistry: Advances and Practical
Applications"
Edited by Maurice B. Green and Douglas A. Spillker
ACS Symposium Series 304; 184 pp; ISBN 0-8412-0963-4

"Petroleum-Derived Carbons"
Edited by John D. Bacha, John W. Newman and
J. L. White
ACS Symposium Series 303; 416 pp; ISBN 0-8412-0964-2

"Coulombic Interactions in Macromolecular Systems"
Edited by Adi Eisenberg and Fred E. Bailey
ACS Symposium Series 302; 282 pp; ISBN 0-8412-0960-X

"Mineral Matter and Ash in Coal"
Edited by Karl S. Vorres
ACS Symposium Series 301; 552 pp; ISBN 0-8412-0959-6

"Equations of State: Theories and Applications"
Edited by K. C. Chao and R. Robinson
ACS Symposium Series 300; 608 pp; ISBN 0-8412-0958-8

"Xenobiotic Conjugation Chemistry"
Edited by Guy Paulson, John Caldwell, and Julius J. Menn
ACS Symposium Series 299; 368 pp; ISBN 0-8412-0957-X

"Strong Metal-Support Interactions"
Edited by R. T. K. Baker, S. J. Tauster, and J. A. Dumesic
ACS Symposium Series 298; 238 pp; ISBN 0-8412-0955-3

"Chromatography and Separation Chemistry:
Advances and Developments"
Edited by Satinder Ahuja
ACS Symposium Series 297; 304 pp; ISBN 0-8412-0953-7

"Historic Textile and Paper Materials: Conservation
and Characterization"
Edited by Howard L. Needles and S. Haig Zeronian
Advances in Chemistry Series 212; 464PP; ISBN 0-8412-0900-6

"Multicomponent Polymer Materials"
Edited by D. R. Paul and L. H. Sperling
Advances in Chemistry Series 211; 354 pp; ISBN 0-8412-0899-9

For further information contact:
American Chemical Society, Sales Office
1155 16th Street NW, Washington, DC 20036
Telephone 800-424-6747